MBSE with SysML and Eclipse Papyrus

Philipp Helle

MBSE with SysML and Eclipse Papyrus

by Philipp Helle

Published by MBSE4U, www.mbse4u.com

/MBSE4U

Printed by Lulu Press, Inc.

ISBN: 978-3-911081-06-1

Für Tim und Daniel

Contents

Foreword . . . . **i**

About MBSE4U . . . . **iii**

About Philipp . . . . **v**

Acknowledgments . . . . **vii**

Chapter 1 - Introduction **1**
1.1 About this book 2
1.1.1 Target Audience 4
1.1.2 Using This Book 5
1.1.3 Downloading the Examples 5
1.1.4 Tool Version 6
1.1.5 Permitted Use of the Examples 6
1.2 Running Example - The Coffee Machine 6

Chapter 2 - Background and Foundations **9**
2.1 Systems Engineering 9
2.1.1 Origins of Systems Engineering 11
2.1.2 Systems Thinking 12
2.1.3 Key Concepts of Systems Engineering 13
2.1.4 Recurring Themes in Systems Engineering 14
2.2 Modeling 17
2.2.1 Models in Engineering 19
2.2.2 Process, Method, and Tool 20
2.3 Model-Based Systems Engineering 21
2.3.1 MBSE Methodology 24
2.3.2 Existing MBSE Methods and Methodologies 25
2.3.3 MBSE Vision and Current State of Practice 26

2.3.4 Misconceptions About MBSE and Clarifications 28

Chapter 3 - The Systems Modeling Language (SysML) **31**
3.1 A Brief History of SysML 32
3.2 The SysML Diagrams 33
3.2.1 The Model versus the Diagrams 36
3.3 Common Misconceptions about SysML 37
3.4 SysML Pitfalls 38

Chapter 4 - Eclipse Papyrus **41**
4.1 Eclipse Rich-Client Platform 42
4.2 Papyrus GUI Overview 44
4.3 Papyrus Preferences and Default Settings 46

Chapter 5 - MBSE with Papyrus **49**
5.1 Manage Requirements 53
5.1.1 Requirement Management Pitfalls 53
5.2 Define System of Interest 55
5.2.1 SysML Blocks 56
5.2.2 Model the System of Interest for the Coffee Machine 57
5.3 Identify Stakeholders 66
5.3.1 Stakeholder Analysis Pitfalls 68
5.3.2 Stakeholders for the Coffee Machine 70
5.3.3 Modeling the Coffee Machine Stakeholders 71
5.3.4 Stakeholder Requirements 89
5.3.5 Stakeholder Requirements for the Coffee Machine 90
5.3.6 Modeling the Coffee Machine Stakeholder Requirements 93
5.4 Define System Goals 101
5.4.1 System Goals Definition Pitfalls 103
5.4.2 Coffee Machine Measures of Effectiveness 104
5.4.3 Value Types, Quantity Kinds and Units 105
5.4.4 Value Types Relevant to the Coffee Machine 107
5.4.5 Modeling the Coffee Machine Measures of Effectiveness and Value Types 109
5.5 Define Context and High-Level Interfaces 127
5.5.1 Context Analysis Pitfalls 127
5.5.2 Context Elements and High-Level Interfaces of the Coffee Machine . . . 128
5.5.3 Modeling the Coffee Machine Context 129
5.6 Identify Use Cases 145
5.6.1 Use Case Identification Pitfalls 146

5.6.2 Coffee Machine Use Cases 147
5.6.3 SysML Use Case Diagrams 147
5.6.4 Modeling the Coffee Machine Use Cases 148
5.7 Analyze Use Cases 152
5.7.1 Use Case Analysis Pitfalls 154
5.7.2 Use Case Analysis with SysML 154
5.7.3 The "Make Coffee" Use Case 157
5.7.4 Modeling the "Make Coffee" Use Case 158
5.7.5 The "Clean and Descale Coffee Machine" Use Case 174
5.7.6 Modeling the "Clean and Descale Coffee Machine" Use Case 176
5.8 Analyze System States 184
5.8.1 System State Analysis Pitfalls 185
5.8.2 SysML State Machine Diagrams 186
5.8.3 Coffee Machine States and Transitions 186
5.8.4 Modeling the Coffee Machine States 187
5.9 Analyze and Decompose System Functions 210
5.9.1 Functional Analysis Pitfalls 211
5.9.2 Coffee Machine Functional Decomposition 213
5.10 Define Initial System Concept 220
5.10.1 Coffee Machine Components 222
5.10.2 Modeling the Initial System Concept of the Coffee Machine 223
5.11 Allocate Functions to System Components 224
5.11.1 Modeling the Allocation of Functions to Components of the Coffee Machine 226
5.12 Define System Internal Architecture 232
5.12.1 Modeling the Coffee Machine Internal Architecture 233
5.13 Model Parametrics 244
5.13.1 Modeling Parametrics for the Coffee Machine 246

Chapter 6 - Conclusion 259
6.1 What to Do Next? 259

Appendix A - Modeling Guidelines 263
MWP Modeling Guidelines 265

Appendix B - Installation 267

Appendix C - Collaboration 273
Publishing a Papyrus Model Using Git 274

Loading a Papyrus Model From Git 280

Appendix D - Profiles and Stereotypes 283
SysML Stereotypes 283
SysML Profiles 283
Creating the MWP Profile 284

Acronyms and Abbreviations 299

References 301

Index 305

Foreword

Although several graph-based modeling languages have been proposed for creating discipline-neutral system models, none of them has received as much attention as the Systems Modeling Language (SysML), and none of the alternative languages can be used in conjunction with so many different modeling tools. There are several reasons why SysML may currently be considered as the number one modeling language for Model-based Systems Engineering (MBSE). First, the language specification is freely available for download from the website of the Object Management Group (OMG) and may be used without charge, even for commercial purposes. Second, SysML is a specialized extension of the Unified Modeling Language (UML) reusing a subset of the UML 2, so many professional modeling tools that were available for the UML have been extended to support SysML modeling. Most of the other modeling languages can only be used with a unique tool that was developed for the specific language. Third, it is relatively easy to define custom specializations and extensions of SysML. Consequently, most SysML modeling tools permit the generation of domain-specific language extensions (or profiles) making it possible to use special modeling strategies and methods. Fourth, many books, videos, trainings & workshops, and other sources of information are available to learn about SysML and its practical use with professional modeling tools.

Although there are many well-established commercial SysML modeling tools on the market, they are not always the ideal choice for learning how to use the language and getting started with MBSE. To use a commercial SysML modeling tool, you need an expensive professional license. Trial licenses may be available free of charge, but they either limit the functionality of the tool considerably, or they expire after a few weeks. Consequently, free trial licenses are unsuitable for longer-term deep-dive MBSE experiences, as in higher education or research projects.

If no budget is available, free software is needed. Open-source modeling tools like Eclipse Papyrus can fill the gap and enable users to get a hands-on experience with SysML modeling. But how can you motivate learners to experiment with MBSE if they do not know how to create a good SysML model with the tool at hand? When the going gets tough, the tough get going – but what if learners do not want to be tough and refuse to spend their precious time on trial and error? This was the motivation for Philipp Helle to write this book, and I am grateful that he has come up with such a well-structured and easily readable introduction to MBSE with Eclipse Papyrus.

To guide you through the process of learning how to use SysML, he has chosen a simplified but universal modeling methodology which he calls MBSE with Papyrus (MWP). It is based on the approach I have been using in my Systems Engineering classes for more than a decade. So far, all my students have successfully managed to follow this approach in their semester-long projects, exploring and refining their ideas of innovative mobility systems in small teams. It has given them a good overview of MBSE with SysML, irrespective of the kind of modeling tool they were able to use.

I hope that this book inspires you to experiment with SysML and to identify your favorite use cases for SysML modeling. I am confident that the hands-on learning approach will help you to gain valuable insights, irrespective of whether you carry on using Eclipse Papyrus or eventually move on to other MBSE tools and/or languages.

— Dr. Jutta Abulawi
Professor of Systems Engineering at Hamburg University of Applied Sciences (Germany)

About MBSE4U

MBSE4U is a publishing organization for systems engineering books with a lean publishing process to support authors and publish books faster on the pulse of the markets with the goal of providing valuable knowledge for the community.

MBSE4U has published the following books so far:

Helle, P. (2024) *MBSE with SysML and Eclipse Papyrus.* MBSE4U.

Weilkiens, T., Vinarcik, M. and Fischer, C. (2023) *The Craft of MBSE.* MBSE4U.

Selic, B., Gullekson, G. and Ward, P.T. (2023) *Real-Time Object-Oriented Modeling.* MBSE4U.

Weilkiens, T. (2019) The New Engineering Game. MBSE4U.

Weinert, B. (2018) Ein Framework zur Architekturbeschreibung von sozio-technischen maritimen Systemen. MBSE4U.

Neureiter, C. (2017) A Domain-Specific, Model Driven Engineering Approach for Systems Engineering in the Smart Grid. MBSE4U.

Weilkiens, T. (2016) Variant Modeling with SysML. MBSE4U.

The current catalog can be found at www.mbse4u.com[1].

MBSE4U is also the home of *The MBSE Podcast* (mbse-podcast.rocks[2]).

[1] https://www.mbse4u.com
[2] https://mbse-podcast.rocks

About Philipp

Philipp Helle is a scientist in the Airbus *Central Research & Technology* organization. He studied linguistics, media culture, and computer science and received an MA from the University of Hamburg. Philipp joined Airbus in 2003 and is an Airbus Expert for "Digital Technologies for MBSE".

For more than 15 years, Philipp has researched and published in the field of MBSE and supported its deployment in the Airbus Group business units, modeling on all levels of abstraction from complete systems of systems down to individual control units.

Philipp is a guest lecturer at the Hamburg University of Applied Sciences (HAW Hamburg) for the module "Systems Engineering" in the master program "Aeronautical Engineering" training the next generation of MBSE enthusiasts. He is a member of GfSE, the German INCOSE chapter, and a certified Project Management Professional (PMP).

Acknowledgments

Manfred Sieber and Chokri Mraidha, thank you for providing valuable feedback.

Tim Weilkiens, thank you for your support in the development of this book and the opportunity to publish it.

Jutta Abulawi, thank you for providing me the opportunity to teach at the HAW Hamburg, which ultimately led to this book.

Chapter 1 - Introduction

In the intricate landscape of systems engineering, the demand for more efficient and coherent methodologies has led to the evolution of Model-Based Systems Engineering (MBSE). Unlike traditional document-centric approaches, MBSE leverages models as primary artifacts throughout the engineering lifecycle. This paradigm shift facilitates a more comprehensive and dynamic understanding of complex systems, ultimately enhancing the ability to manage and control them effectively.

> Systems Engineering is a transdisciplinary and integrative approach to enable the successful realization, use, and retirement of engineered systems, using systems principles and concepts, and scientific, technological, and management methods.–(INCOSE, 2019)

Systems engineering is an iterative process that involves the following key steps:

- **Requirements elicitation**: Identifying and defining the needs and expectations of stakeholders.
- **System architecture**: Designing the system's overall structure, functionality, and behavior.
- **Detailed design**: Specifying the individual components, interfaces, and operations of the system.
- **Verification and validation**: Ensuring that the system meets its requirements and is ready for deployment.
- **Operation and maintenance**: Managing the system throughout its operational life cycle, including monitoring, maintenance, and upgrades.

MBSE is an extension of systems engineering that incorporates the use of models to represent and analyze complex systems. Models are representations of the system's structure, behavior, and requirements that are used to communicate, analyze, and design the system. MBSE has become increasingly popular due to its ability to improve communication, reduce errors, and accelerate development.

The Systems Modeling Language (SysML) is a general-purpose modeling language that is specifically tailored for systems engineering. It provides a comprehensive set of graphical elements and diagrams that can be used to model various aspects of a system, including its structure, behavior, and requirements. SysML is a widely used language in industry and academia and is promoted by various organizations, including the International Council on Systems Engineering (INCOSE). It is an international standard maintained by the Object Management Group (OMG) and has been adopted as an International Organization for Standardization (ISO) standard.

The use of MBSE with SysML offers several benefits, including:

- **Improved communication**: Models provide a common language for stakeholders to communicate about the system, reducing misunderstandings and ensuring that everyone is on the same page.
- **Reduced errors**: Models can be used to analyze and identify potential problems early in the development process before they become expensive to fix.
- **Accelerated development**: Models can be used to automate and streamline the development process, leading to shorter development cycles and faster product delivery.
- **Enhanced decision-making**: Models provide a structured way to evaluate different design options and make informed decisions.

Eclipse Papyrus is an open-source modeling tool that provides support for various modeling standards, including UML and SysML. As a free, open-source tool, it does not require an upfront investment and is, therefore, a good fit for taking the first steps into the world of MBSE. It is built on the Eclipse platform and is designed to offer a flexible and extensible environment for model-driven development.

1.1 About this book

Unlike the official documentation (accessed: 17 April 2024)[1] of Eclipse Papyrus, which primarily focuses on the tool's functionality and installation and the use of the UML, this book takes a step into the practical application of Eclipse Papyrus for modeling complex systems using SysML. It provides a comprehensive guide for engineers, covering the entire modeling process and focuses on hands-on practical application, guiding readers through the creation of SysML models.

[1] https://help.eclipse.org/2022-12/index.jsp?topic=%2Forg.eclipse.papyrus.uml.doc%2Ftarget%2Fgenerated-eclipse-help%2FPapyrusStarterGuide.html&cp%3D66_0_0

The book addresses the limitations of the official documentation by:

- **Complementing the Papyrus documentation**: The official Eclipse Papyrus documentation is sparse and focused on UML rather than SysML. It is often incomplete or difficult to understand. Users who are new to SysML may find it difficult to learn how to use the tool effectively, and they may need to rely on external resources such as online tutorials, books, and forums to puzzle together an understanding of the tool.
- **Step-by-step modeling tutorial**: The book provides a step-by-step tutorial that walks readers through the process of creating a SysML model for various system aspects, such as requirements, behavior, and architecture.
- **Coffee machine case study**: The book showcases the power of Eclipse Papyrus by presenting a continuous coffee machine case study that illustrates how a SysML model can be used to successfully support systems development.

The book is not written as an in-depth SysML guide nor as a thorough guide to systems engineering. A lot of better books already exist for this purpose.

Please consider having a look at the following books if you want to learn more about SysML:

- *SysML Distilled: A Brief Guide to the Systems Modeling Language* (Delligatti, 2014)
- *A Practical Guide to SysML: the Systems Modeling Language* (Friedenthal, Moore and Steiner, 2015)
- *SysML for Systems Engineering: A Model-Based Approach* (Holt and Perry, 2018)
- *Model-Based System Architecture* (Weilkiens et. al., 2022)
- *Systems Engineering with SysML/UML: Modeling, Analysis, Design* (Weilkiens, 2007)
- *SYSMOD - The Systems Modeling Toolbox – Pragmatic MBSE with SysML* (Weilkiens, 2020)

And, of course, the official SysML specification, the current version is SysML v1.6 (OMG, 2019).

The following books provide solid foundations for systems engineering:

- *Systems Engineering Principles and Practice* (Kossiakoff et al, 2020)
- *Systems Engineering: Fundamentals and Applications* (Haberfellner et al, 2019)
- *System Engineering Analysis, Design, and Development: Concepts, Principles, and Practices* (Wasson and Augustine, 2016)
- *NASA Systems Engineering Handbook* (Kapurch, 2010)
- *INCOSE Systems Engineering Handbook: A Guide for System Life Cycle Processes and Activities* (Walden and INCOSE, 2023)

Additionally, the ISO/IEC/IEEE 15288:2023 standard *Systems and Software Engineering: System Lifecycle Processes.*

If you are interested more in systems thinking and systems theory take a look at:

- *Thinking in Systems: A Primer* (Meadows, 2008)
- *Systems Thinking, Systems Practice* (Checkland, 1981)
- *General System Theory: Foundations, Development, Applications* (Bertalanffy, 2009)

Overall, this book serves as a comprehensive guide for engineers seeking to effectively utilize Eclipse Papyrus for SysML modeling of complex systems. It fills the gap in the official documentation by providing in-depth practical knowledge using a hands-on tutorial, enabling engineers to confidently create and analyze SysML models with Eclipse Papyrus.

1.1.1 Target Audience

This book is written for anyone interested in learning about MBSE, from beginners to experienced practitioners. Primarily, the book will serve as a hands-on guide for individuals new to Eclipse Papyrus and/or SysML modeling. It will equip them with the necessary practical knowledge and skills to effectively utilize Papyrus for modeling complex systems using SysML. The book's focus on practical application by providing a step-by-step tutorial with a real-world example, and in-depth coverage of specific aspects of modeling will empower readers to quickly grasp the concepts and apply them to their projects.

For new users of Eclipse Papyrus:

- Provides a clear and concise introduction to the Eclipse Papyrus platform, including its key features, installation, and user interface navigation.
- Guides users through the creation of basic SysML models, covering core concepts like creating diagrams, elements, and relationships.
- Introduces essential modeling practices for various system aspects, such as requirements, architecture, and behavior.
- Shows how SysML can be extended and adapted to user needs using the UML profiling mechanism.

For new users of SysML modeling:

- Explains the fundamental purpose and benefits of SysML modeling in the context of systems engineering.

- Introduces the various SysML diagrams and their roles in representing different aspects of a system.
- Provides practical guidance on creating and using SysML models effectively to capture system requirements, architecture, and behavior.

Overall, the book aims to bridge the gap between theory and practice, enabling readers to transition from a basic understanding of SysML to hands-on modeling capabilities with the Eclipse Papyrus tool.

1.1.2 Using This Book

This book is organized as follows: Chapter 2 provides an introduction to the fundamental concepts and principles of systems engineering, modeling, and MBSE. It talks about the origins of systems engineering and its key characteristics. It introduces MBSE as a comprehensive approach that integrates modeling throughout the systems engineering lifecycle, enhancing efficiency, consistency, and communication, highlighting its benefits in representing, analyzing, and communicating complex system designs.

Chapter 3 introduces SysML. It provides an overview of the type of diagrams that SysML defines. Common misconceptions about the language and some common pitfalls that may be encountered and that can hinder the effective use of SysML are discussed.

Chapter 4 covers Eclipse Papyrus, the open-source modeling tool that will be used throughout this book to model using SysML.

The core of this book is the following Chapter 5: MBSE with Papyrus. In this chapter, a simple modeling methodology following the MBSE paradigm will be used to illustrate how Eclipse Papyrus and SysML can be used to build a model that covers the most important aspects of a system, such as requirements, architecture, and behavior. The book provides a step-by-step tutorial that walks readers through the process of creating a SysML model.

The depth of coverage of the Papyrus tool will change gradually as the book progresses. In the early sections, comprehensive step-by-step instructions for using the tool will be provided, covering every mouse click that needs to be made. No step will be skipped or left as an exercise for the reader. As the book progresses, it will be assumed that the earlier material and basic concepts of the tool have already been mastered. Explanations of mundane operations will not be repeated. Therefore, it is recommended to follow the examples in the book in the order that they are presented. This ensures a solid understanding of each concept before moving on to the next.

Finally, the book ends with a concluding chapter that also provides an outlook and pointers to where the reader might look next on their journey to becoming a MBSE professional.

1.1.3 Downloading the Examples

The coffee machine model and the MWP profile discussed in this book can be downloaded from MBSE4U.com[2].

1.1.4 Tool Version

The screen captures in this book were produced with Eclipse Papyrus version 6.6.0 2023-12 on Windows 10.

1.1.5 Permitted Use of the Examples

1. The book and examples are provided "As Is", without warranty of any kind, express or implied.
2. The book and examples can be used freely for all types of individual study and classroom instruction with the following restriction: Copyright laws are respected. Students must purchase or rent individual copies of the book. The book and examples cannot be bundled or redistributed in any fashion without the express written permission of the author.
3. Consultants and instructors from both private and public educational institutions are encouraged to incorporate this book and its accompanying example files into their classes, even those delivered for a fee, under the stipulated conditions. Consultants are permitted to integrate the book and examples as part of their paid services. However, they need to transparently communicate the source of the materials to their clients. Additionally, consultants should encourage their clients to independently purchase the book to ensure compliance with the specified terms of use. This approach ensures a fair and respectful utilization of the provided resources in educational settings.

1.2 Running Example - The Coffee Machine

A coffee machine, as illustrated by Figure 1.1, is a sophisticated kitchen appliance designed to provide users with a convenient and efficient means of brewing coffee. It integrates various components and functionalities to deliver a seamless user experience. The primary goal of a coffee machine is to transform coffee-related inputs, such as water and coffee pads, into a brewed coffee output that meets user preferences.

[2]https://mbse4u.com/mbse-sysml-papyrus

Figure 1.1. Running Example - The Coffee Machine

The coffee machine was chosen as an example for several reasons. First, everyone has personal experience using a coffee machine, which makes it a familiar and relatable subject. It exists in many homes and offices. This makes it relatable to readers and helps them connect the theoretical concepts of MBSE to a practical application. By understanding how MBSE can be applied to the coffee machine, readers can gain a better understanding of how it can be applied to other real-world systems. This familiarity will help readers better understand the concepts of systems engineering and MBSE as they are applied to the coffee machine system.

Second, the coffee machine system has sufficient complexity to illustrate the MBSE modeling with SysML but is not overly complex. This makes it an ideal example for introducing these concepts without overwhelming readers. The coffee machine system includes multiple interacting components, requires careful consideration of both functional and non-functional requirements, and can be modeled using SysML diagrams.

Finally, the coffee machine system is a system that can be easily understood without extensive technical domain knowledge. This makes it an accessible example for readers with varying levels of technical expertise. By using a relatively simple system like the coffee machine, readers can focus on the core concepts of MBSE without getting bogged down in complex technical details.

Chapter 2 - Background and Foundations

This chapter introduces the core concepts and principles that underpin systems engineering, modeling, and MBSE. It talks about the key characteristics and the history of systems engineering and explores the theory of modeling. Finally, it introduces MBSE, a comprehensive approach that integrates modeling throughout the systems engineering lifecycle, enhancing efficiency, consistency, and communication.

2.1 Systems Engineering

This introductory section lays the foundation for our exploration of systems engineering, emphasizing its role as a guiding framework for addressing the challenges posed by complex systems. Systems engineering is not merely a discipline; it is a paradigm that equips engineers with the mindset and tools needed to conquer the frontiers of innovation in an ever-evolving technological landscape.

A system is defined as a collection of interacting components that, together, form a larger entity with well-defined boundaries and functions to achieve one or more purposes. These components can be physical, conceptual, or a combination of both. Systems can exist at various scales, ranging from small, individual devices to complex, interconnected systems that span multiple domains.

Systems engineering plays a crucial role in the success of complex technical projects by ensuring effective integration and optimization of various components. It involves a holistic approach to design and development that takes into account the entire system and its interactions.

One of the key reasons why systems engineering is important is its ability to manage complexity. In today's world, technical projects are becoming increasingly complex with multiple subsystems and dependencies. Systems engineering provides a structured framework for understanding these complexities and addressing them systematically.

Another significant aspect of systems engineering is risk management. By identifying potential risks early on in the project lifecycle, engineers can develop strategies to mitigate or eliminate them. This proactive approach helps minimize costly delays or failures down the line.

Furthermore, systems engineering promotes collaboration among different disciplines involved in a project. It encourages effective communication and coordination between engineers from various backgrounds such as mechanical, electrical, software, and more. This cross-disciplinary collaboration leads to better decision-making and ultimately improves the overall performance of the system.

Additionally, systems engineering is also transdisciplinary. Transdisciplinarity goes beyond mere integration and seeks to create a new synthesis that transcends the boundaries of individual disciplines. It aims to develop a holistic and integrated understanding of complex problems that cannot be adequately addressed within the confines of a single discipline (Madni, 2018).

Systems engineering recognizes that complex systems cannot be fully understood by breaking them down into their constituent parts and analyzing them in isolation. Instead, it takes a holistic view, considering the interactions and interdependencies between different components and their environment. This transdisciplinary approach is essential for systems engineering because it allows for a deeper understanding of the system's behavior, its impact on its environment, and the trade-offs associated with different design and implementation options. By transcending the boundaries of individual disciplines, systems engineers can develop more comprehensive, innovative, and sustainable solutions to complex engineering problems.

Systems engineering operates within a structured process that guides engineers through the development lifecycle of a system. This process typically involves stages such as requirements analysis, system design, implementation, testing, and maintenance. Each stage is interconnected, and iterative, emphasizing feedback loops to refine and optimize the system as it evolves.

Systems engineering, systems theory, and systems thinking are interrelated fields that share a common focus on understanding and managing complex systems.

Systems theory is a broad interdisciplinary field that studies the structure, behavior, and interactions of complex systems. It provides a framework for understanding how systems function and how they can be designed and managed effectively.

Systems thinking is a way of viewing the world that focuses on the interconnectedness of parts and the overall patterns of behavior that emerge from those interactions. It emphasizes the importance of holistic thinking and understanding the context in which systems operate.

Systems engineering is based on the principles of systems theory. Systems theory provides the underlying framework for understanding the complexity of systems and the principles that govern their behavior. Systems engineering applies these principles to the design, development, and operation of real-world systems.

Systems thinking informs the practice of systems engineering. Systems thinking helps systems engineers understand the complex interactions and dependencies between the different components of a system. It also helps them to identify and manage potential risks and issues early in the development process.

Systems engineering provides a practical application for systems theory. It is a framework for applying the principles of systems theory to real-world problems and provides a set of tools and techniques that can be used to design, develop, and operate complex systems effectively.

2.1.1 Origins of Systems Engineering

The roots of systems engineering can be traced back to the early 20th century when the increasing complexity of engineered systems demanded new approaches to their development and management. The catalyst for the formalization of systems engineering can be traced to the demands of World War II. The war effort necessitated the coordination of diverse technologies, from radar and communications to weaponry and logistics. Engineers found themselves grappling with the challenges of integrating complex systems, leading to the recognition of the need for a more comprehensive and interdisciplinary methodology.

The term "Systems Engineering" was coined during the 1940s at Bell Telephone Laboratories (Schlager, 1956). Engineers and managers at Bell Labs were dealing with intricate telecommunications projects. They recognized the necessity of a systematic approach that considered the entire system rather than isolated components. Systems Engineering began to emerge as a separate branch of engineering in the late 1950s in response to the escalating arms race and the urgent need to develop and deploy increasingly sophisticated military technologies, including nuclear-armed missiles and satellites. The development of large-scale projects like the Apollo program further demonstrated the effectiveness of systems engineering principles in ensuring the success of ambitious engineering endeavors. The establishment of the Systems Engineering Group at the Massachusetts Institute of Technology (MIT) in 1958 further solidified the discipline's presence (Hossain, 2020).

As systems engineering matured, its applications extended beyond telecommunications and defense. It found relevance in diverse industries such as aerospace, automotive, healthcare, and information technology. The methodology's adaptability and effectiveness in addressing complex challenges led to its widespread adoption. As technology continues to advance, systems engineering remains a dynamic and essential field, providing a systematic and holistic approach to the design, development, and management of intricate systems.

2.1.2 Systems Thinking

Central to systems engineering is the adoption of a systems thinking mindset. This paradigm encourages engineers to view a system as a whole, considering not only its constituent parts but also the dynamic relationships and interdependencies between them. Systems thinking prompts a shift from linear, reductionist problem-solving to a more interconnected and comprehensive approach, enabling engineers to address the complexities inherent in modern systems.

Systems thinking is a holistic approach to problem-solving and decision-making that focuses on understanding the interconnections and relationships between various components of a system. It involves analyzing how different parts of a system interact with each other and how they contribute to the overall behavior and functioning of the system. Central tenets of systems thinking include:

- **Interconnections**: Systems thinking emphasizes the interconnectedness of different elements within a system. It recognizes that changes in one part of the system can have ripple effects on other parts.
- **Relationships**: Systems thinking looks at the relationships between different components of a system, such as cause-and-effect relationships or feedback loops, i.e., the process by which information about the output or outcome of a system is fed back into the system, influencing its future behavior. Understanding these relationships helps identify potential leverage points for intervention or improvement.
- **Emergent properties**: Systems thinking recognizes that systems often exhibit emergent properties - characteristics or behaviors that arise from interactions between individual components but are not directly attributable to any single component alone.
- **Complexity**: Systems thinking acknowledges that many real-world problems are complex, involving multiple interacting factors and uncertainties. It encourages considering multiple perspectives and taking into account both short-term and long-term consequences.

While there are several additional aspects of systems thinking introduced by multiple contributors, the following list summarizes the essential properties of a systems thinker (Waters Foundation, 2013).

A systems thinker

- seeks to understand the big picture.
- observes how elements within the system change over time, generating patterns and trends.
- recognizes that a system's structure (elements and their interactions) generates behavior.
- identifies the circular nature of complex cause-and-effect relationships.
- surfaces and tests assumptions.
- changes perspective to increase understanding.
- considers an issue fully and resists the urge to come to a quick conclusion.
- considers how mental models affect current reality and the future.
- uses understanding of system structure to identify possible leverage actions.
- considers both short- and long-term consequences of actions.
- finds where unintended consequences emerge.
- recognizes the impact of time delays when exploring cause-and-effect relationships.
- checks results and changes actions if needed: "successive approximation".

By adopting systems thinking, individuals and organizations can gain a deeper understanding of complex issues, identify unintended consequences, anticipate potential impacts, and develop more effective strategies for problem-solving and decision-making.

2.1.3 Key Concepts of Systems Engineering

At the heart of systems engineering lies the principle of transdisciplinary collaboration. Complex systems invariably involve diverse components, each requiring expertise from different engineering domains. The interdisciplinary approach fosters collaboration between specialists in various fields, such as mechanical, electrical, and software engineering. By integrating these diverse perspectives, systems engineering ensures a comprehensive understanding of the system and facilitates effective communication among team members with distinct expertise.

Trade-off analysis is a cornerstone of decision-making in systems engineering. As engineers strive to optimize a system for performance, cost, and other competing objectives, trade-offs become inevitable. This concept involves evaluating and balancing conflicting factors to arrive at the most favorable solution. Trade-off analysis guides engineers in making informed decisions, considering the implications of each trade-off on the overall system, and aligning choices with the project's goals and constraints.

In systems engineering, feedback loops play a crucial role in refining and optimizing the system throughout its lifecycle. Feedback mechanisms enable the continuous assessment of system performance against defined requirements. This iterative process ensures that any discrepancies or inefficiencies are identified and addressed promptly. The integration of feedback loops aligns with the adaptive nature of systems engineering, allowing for continuous improvement and adaptation to changing requirements.

Systems engineering operates within a structured lifecycle that guides the development, implementation, and maintenance of a system. The lifecycle typically includes phases such as conceptualization, requirements analysis, design, implementation, testing, deployment, and maintenance. Each phase is interconnected, with feedback loops facilitating iteration and refinement. The systems lifecycle provides a roadmap for engineers, ensuring a systematic and comprehensive approach to system development.

The ultimate goal of systems engineering is the holistic optimization of a system. This involves balancing conflicting objectives, managing emergent properties, and ensuring effective collaboration to create a system that meets its intended purpose efficiently and reliably. Holistic optimization extends beyond individual components, considering the system's impact on its environment and stakeholders. Systems engineers strive to achieve a balance that maximizes overall system performance while adhering to project constraints.

2.1.4 Recurring Themes in Systems Engineering

Changing the level of abstraction, decomposition, and iteration are fundamental and recurring themes in systems engineering, reflecting the dynamic nature of designing and developing complex systems. These themes are used in conjunction to achieve the overall systems engineering goals:

- **Adaptive nature**: These themes are interconnected and support the adaptive nature of systems engineering. As new information emerges, or requirements change, engineers can adjust the level of abstraction, decompose further, and iterate on existing designs.
- **Feedback Loops**: The iterative nature of systems engineering is reinforced by feedback loops between different abstraction levels and decomposition stages. Feedback from lower levels can inform changes at higher levels and vice versa.
- **Risk Mitigation**: Changing the level of abstraction, decomposition, and iteration also serve as risk mitigation strategies. They allow engineers to catch and address issues early, reducing the likelihood of costly errors or design flaws.

2.1.4.1 Changing the Level of Abstraction

Systems engineers frequently transition between different levels of abstraction, moving from a high-level overview to detailed specifications and back again. This allows them to balance the need for conceptual understanding with the precision required for implementation.

At a high level of abstraction, systems engineers focus on the overall purpose and function of the system without getting bogged down in specific details. This provides a holistic view of the system and helps to identify key requirements and constraints. As the design process progresses, systems engineers go deeper into the system's architecture and components, gradually increasing the level of detail. This involves defining interfaces, refining specifications, and making trade-offs between competing requirements.

The ability to switch between different levels of abstraction is crucial for systems engineers because it allows them to consider the system from multiple perspectives and make informed decisions at each stage of the development process.

Strongly related to changing the level of abstraction is the differentiation between viewing the systems as a black box versus viewing the system as a white box. Black box view and white box view are two perspectives used to analyze a system.

A black box view focuses on the inputs and outputs of a system without considering its internal workings. It treats the system as a "black box" whose inner workings are not visible or relevant to the analysis. This simplified view is particularly useful for understanding the overall behavior and interactions of the system.

A white box view, on the other hand, goes deeper into the internal structure and mechanisms of a system. It exposes the detailed components, their interconnection, and the processes that transform the inputs into outputs. This detailed view is crucial for understanding the system's design, implementation, and troubleshooting.

2.1.4.2 Decomposition

Decomposition is a fundamental process of breaking down a complex system into smaller, more manageable components. This hierarchical approach allows systems engineers to focus on the details of each individual component without losing sight of the overall system architecture.

Decomposition allows systems engineers to tackle large and complex systems in a manageable way, breaking them down into smaller, more manageable pieces that can be analyzed, designed, and implemented independently. This approach promotes modularity and makes it easier to identify and isolate potential problems and make changes to the system.

Decomposition also facilitates reuse, as components can be designed and tested independently and then reused in different contexts. This helps to reduce development time and improve the overall efficiency of the systems engineering process.

Decomposition is also a prerequisite for working in separate teams in large enterprises. It enables parallel development of decomposed modules and accelerates the overall development process.

Decomposition is a powerful tool in systems engineering that can be applied to both the functions and the structure of a system.

Function decomposition involves breaking down a system's overall functionality into smaller, more manageable tasks or functions. This approach helps in understanding the system's purpose, identifying its essential building blocks, and ensuring that each function contributes to the overall goals.

Consider a smartphone's camera function. It can be decomposed into subfunctions such as image capture, image processing, and image storage. These subfunctions can further be broken down into even more granular tasks, such as controlling the focus, sharpening an image, and compressing the data.

Structural decomposition involves breaking down a system into its physical components or subsystems. This approach focuses on the organization and arrangement of the system's elements, their relationships, and how they interact with each other.

Consider a car's engine system. It can be decomposed into the engine block, cylinder head, crankshaft, and other components. These components are arranged in a specific manner to enable the engine to convert fuel into power.

In practice, both function and structure decomposition are most often used together to provide a comprehensive understanding of a system. This approach allows engineers to grasp the system's purpose, its internal workings, and how its functions are implemented within its structure.

Returning to the smartphone camera example, function decomposition would reveal the various camera functions, while structural decomposition would show the physical components that perform these functions. By combining both approaches, engineers can understand the overall camera system and how it operates.

Decomposition is a versatile tool that can be applied to a wide range of systems, from simple devices to complex multinational projects. By breaking down systems into manageable components, engineers can gain a deeper understanding, facilitate parallel work, and optimize testing and maintenance, ensuring the successful development and operation of complex systems.

2.1.4.3 Iteration

Iteration is a recurring theme in systems engineering because it allows for continuous improvement and refinement throughout the development process. Systems engineers often go back and forth between different stages of the development process, making adjustments and refining their designs based on feedback and new information.

Too often, the system definition is viewed as a linear, sequential, single pass through the processes. However, valuable information and insight need to be exchanged between the processes to ensure a good system definition that effectively and efficiently meets the mission or business needs.

To foster effective communication and accommodate ongoing learning and decision-making, the incorporation of iteration and recursion into the life cycle processes becomes essential. Introducing appropriate feedback loops ensures that knowledge gained from subsequent analyses and process applications is considered, facilitating the evolution of the technical solution over time. This iterative approach not only enhances the adaptability of the system definition but also allows for the dynamic integration of insights gleaned from evolving circumstances and evolving requirements.

The step-by-step tutorial provided by this book can make it seem that all the method steps need to be completed before moving on and that there is a strict sequence. This, however, is misleading and largely due to the sequential nature of books. In reality, frequent iterations between methods steps are occurring often and should be considered a natural part of the systems engineering process.

2.2 Modeling

In the realm of science, mathematics, and engineering, models play a pivotal role in our understanding and manipulation of the world around us. A model, in its essence, is a representation that encapsulates the essential features of a system, phenomenon, or concept while discarding unnecessary details. This enables deeper understanding, analysis, and prediction. This section explores the general concept of models, their role in scientific inquiry, and their practical application in engineering endeavors.

The following quotes, highlighted by the author, from Rothenberg et. al. (Rothenberg et. al., 1989) paint an overall picture of models:

> "Modeling, in the broadest sense, is the *cost-effective* use of *something in place of something else* for some *cognitive* purpose."
>
> "It allows us to use something that is *simpler, safer, or cheaper* than reality instead of reality for *some purpose*."
>
> "A model represents reality for the *given purpose*; the model is an *abstraction of reality* in the sense that it cannot represent all aspects of reality."
>
> "This allows us to deal with the world in a *simplified manner, avoiding the complexity, danger, and irreversibility of reality*."–Rothenberg et. al.

Stachowiak (Stachowiak, 1973) lists three fundamental properties of models:

- **Mapping**: Models are always models of something, i.e., mappings from representations of natural or artificial originals that can be models themselves.
- **Reduction**: Models, in general, capture not all attributes of the original represented by them, but rather only those seeming relevant to their model creators and/ or model users.
- **Pragmatism**: Models are not uniquely assigned to their originals per se. They fulfill their replacement function a) for particular – cognitive and/ or acting models using subjects, b) within particular time intervals, and c) restricted to particular mental or actual operations.

Scientific models are fundamental tools for organizing and expressing our understanding of the natural world. They range from simple conceptual frameworks to complex mathematical representations. Models in science serve various purposes, such as explaining observed phenomena, making predictions, and facilitating communication among scientists. Whether depicting the motion of celestial bodies, the behavior of subatomic particles, or the dynamics of ecosystems, scientific models provide a structured means for scientists to explore, test hypotheses, and refine their understanding of the universe.

Scientific models take diverse forms, each tailored to the specific needs of a scientific inquiry. Some common types include:

- **Conceptual models**: Descriptive representations that aid in visualizing abstract concepts or relationships.
- **Mathematical models**: Formulas or equations that express quantitative relationships and enable precise predictions.

- **Physical models**: Physical replicas or simulations that mimic the behavior of real-world phenomena.
- **Computational models**: Computer-based simulations that allow for the dynamic analysis of complex systems.

A good model aligns with its intended purpose, whether for analysis, communication, or decision support. A good model is one that effectively captures the essential features of a real-world system, provides accurate predictions, and can be readily understood and interpreted. It should be simple enough to handle without sacrificing accuracy and should be verifiable through real-world data or experiments.

It is crucial to recognize that models are not perfect representations of reality. They are simplifications, abstractions, and approximations that aim to capture the essence of a system while omitting irrelevant details. Models are tools, not perfect mirrors of reality.

Generally, all models should be used with caution, recognizing their limitations and potential biases. They should not be blindly accepted as infallible truths but rather used to generate hypotheses, guide investigations, and inform decisions. The interplay between models and reality is a dynamic process where models are constantly refined and adapted to better reflect the actual world.

2.2.1 Models in Engineering

In engineering, models are tools that bridge theoretical concepts with practical applications. Engineering models serve a myriad of purposes throughout the design and development lifecycle. They aid in conceptualization, design validation, performance analysis, and decision-making. Engineering models range from schematic diagrams and 3D renderings to sophisticated simulations and prototypes, each contributing to a comprehensive understanding of the system under consideration.

Models are employed in engineering for various purposes, including:

- **Problem formulation and understanding**: Models help engineers formulate problems clearly, identify key factors and relationships, and gain a deeper understanding of the system's behavior and constraints.
- **Design exploration and optimization**: Models enable engineers to explore various design alternatives, evaluate their feasibility, and optimize the system's performance, cost, and other objectives.
- **Simulation and prediction**: Models can be used to simulate the system's behavior under various conditions, predict its performance, and identify potential issues or challenges before the system is physically built.

- **Communication and collaboration:** Models serve as a common language for engineers to communicate design ideas, share insights, and collaborate effectively across different teams.
- **Testing and validation:** Models can be used to develop test cases and evaluate the performance of the system's implementation, ensuring that the system meets its requirements and design specifications.

Models are indispensable tools for engineers, enabling them to tackle complex engineering challenges effectively. By carefully selecting, validating, and utilizing models, engineers can make informed design decisions, optimize system performance, and ensure the successful development and operation of engineered systems.

2.2.2 Process, Method, and Tool

Understanding the difference between process, method, and tool is essential in the frame of MBSE. While these terms are often used interchangeably, they have distinct meanings and applications.

A *process* refers to a series of actions or steps that are taken in a specific order to achieve a particular outcome or goal. It involves a systematic approach that outlines the sequence of activities required to complete a task or reach an objective. Processes provide structure and guidance by defining the overall workflow and ensuring consistency in how tasks are performed. In systems engineering, processes define the activities that are carried out throughout the life cycle of a system, from concept to disposal. These processes are typically defined in a lifecycle model, which outlines the phases and activities involved in each stage of the system's development.

On the other hand, a *method* is a systematic approach to carrying out a step within a process. It is more specific than a process and focuses on the techniques or strategies employed to accomplish a desired result. Methods can vary based on factors such as industry standards, best practices, personal preferences, or unique circumstances. In systems engineering, methods provide more detailed guidelines and techniques for performing specific tasks within a process.

Lastly, a *tool* is an instrument or resource that aids in performing tasks within a process according to a method. Tools can be physical objects or software applications designed to facilitate specific functions. They assist by automating processes, enhancing efficiency, improving accuracy, or enabling new capabilities. In systems engineering, tools can range from simple notepads and flowcharts to sophisticated modeling and simulation software.

Tools can be used to automate tasks, collect data, analyze results, and communicate information.

To summarize:

- **Process**: A series of actions or steps taken to achieve an outcome.
- **Method**: A particular way or approach used within a process.
- **Tool**: An instrument or resource that aids in performing tasks within a process/method.

A *methodology* is a framework that combines processes, methods, and tools to provide a structured approach to solving a specific problem or achieving a particular goal. In systems engineering, methodologies provide a standardized way of approaching complex systems development, ensuring that critical steps are not overlooked and that best practices are followed. Understanding these distinctions helps individuals and organizations effectively plan and execute their work by selecting appropriate methods and utilizing suitable tools within defined processes.

2.3 Model-Based Systems Engineering

MBSE is an approach to system design and development that utilizes models to represent the various aspects of a complex system. It is a model-centric approach, where models serve as the primary artifacts to describe, analyze, and manage system specifications. In MBSE, models serve as a central repository of information that can be shared and accessed by all stakeholders involved in the system's lifecycle. These models can also provide a visual representation of the system's components, their relationships, and their behavior over time.

In traditional systems engineering, models are employed for various purposes, such as requirements analysis, system architecture design, and validation. However, the transition to MBSE elevates the role of models to a central position, emphasizing their dynamic and interconnected nature throughout the systems engineering process. Models in MBSE encapsulate not only the structural aspects of a system but also its behavior, interactions, and requirements in a comprehensive and integrated manner.

The key idea behind MBSE is to enable a more holistic understanding of the system by capturing its complexity in a structured manner. By using models, engineers can analyze different operational scenarios, identify potential issues or conflicts early in the design process, make informed decisions based on accurate and up-to-date information, and, depending on the language and the tool, even simulate system behavior.

> Contrary to popular myth, models are not new to systems engineering. Models are the way engineers analyze both problems and solutions, so system models are as old as systems engineering itself. With the traditional focus on written specifications as the "source of truth," models were secondary and descriptive - sometimes reflected as simple sketches, sometimes shown in formal diagrams, partially captured in analysis packages, and often trapped in the mind of the chief engineer. The transformation from document-centric to model-centric practices is not about the introduction of models. It is about making models explicit and moving them to the foreground, where they serve as the authoritative tool for design, analysis, communication, and system specification.–David Long in (Weilkiens et al., 2022)

Advantages of MBSE:

- **Enhanced communication**: MBSE facilitates clearer and more effective communication among stakeholders by providing visual and intuitive representations of system concepts.
- **Improved collaboration**: Stakeholders from diverse disciplines can collaborate seamlessly through a shared model, fostering interdisciplinary understanding and cooperation.
- **Early validation**: MBSE enables early validation of system specifications through simulations, reducing the likelihood of costly errors in later stages of development.

MBSE has gained traction in industries such as aerospace, automotive, defense, and healthcare, where complex systems need to be designed with precision. While the adoption of MBSE has seen significant progress in recent years, several challenges and factors contribute to its limited widespread use. Understanding these barriers is crucial for addressing them and promoting the broader implementation of MBSE across various industries.

- **Cultural resistance and legacy practices**: Many organizations have well-established document-centric practices deeply ingrained in their workflows. Transitioning to a model-centric approach requires a cultural shift, and there may be resistance to abandoning familiar methods.
- **Training and skill gaps**: The successful implementation of MBSE relies on a workforce with expertise in modeling languages and associated tools. The shortage of skilled professionals who are proficient in both systems engineering and advanced modeling techniques can hinder adoption.

- **Perceived complexity and learning curve**: MBSE involves the creation and management of complex models that capture various aspects of a system. Organizations may perceive the learning curve associated with MBSE as a barrier, particularly when transitioning from traditional engineering practices.
- **Tool integration challenges and interoperability issues**: In some cases, the interoperability between different modeling tools can be a challenge. Organizations may face difficulties in seamlessly integrating models across various disciplines, leading to information silos.
- **Initial investment**: Implementing MBSE may require an initial investment in new tools, training, and infrastructure. Some organizations may be hesitant to allocate resources for these changes, especially if they perceive their current methods as sufficient.
- **Risk aversion and fear of disruption**: Organizations may be risk-averse, fearing that the transition to MBSE could disrupt ongoing projects or lead to inefficiencies during the learning curve. This fear can impede the willingness to embrace change.
- **Industry-specific challenges**: In industries with stringent regulatory requirements, such as aerospace and healthcare, adapting established processes to comply with MBSE standards may pose challenges. Ensuring that MBSE practices align with regulatory expectations is crucial for adoption.
- **Inadequate standardization**: Although there are efforts to standardize MBSE practices and languages, a lack of universally accepted standards can create confusion and hinder widespread adoption. Standardization is crucial for ensuring consistency and interoperability across different applications and industries.
- **Limited awareness and educational gaps**: Some professionals and organizations may not be fully aware of the benefits and capabilities of MBSE. Educational gaps and insufficient awareness campaigns may contribute to a lack of interest or understanding.
- **Perceived applicability for simpler systems**: MBSE has been proven to be beneficial for large and complex systems where more traditional methods are not working anymore. For smaller projects or organizations dealing with less intricate systems, the perceived benefits of a complex MBSE methodology may not outweigh the associated costs and efforts.

Addressing these challenges requires a concerted effort from industry stakeholders, including ongoing education, standardization efforts, tool improvements, and organizational support for transitioning to MBSE practices. As awareness grows and best practices emerge, the barriers to MBSE adoption are likely to diminish, paving the way for its increased use in various engineering domains.

Overall, MBSE offers numerous benefits, including improved productivity, reduced errors or rework costs, enhanced traceability between requirements and design elements, and better decision-making capabilities based on data-driven insights from simulations or analyses. As technology continues to advance rapidly, MBSE is expected to play an increasingly vital role in shaping the future of systems engineering.

2.3.1 MBSE Methodology

Defining a methodology is crucial when adopting MBSE because it provides a structured approach to planning, organizing, and executing system development activities. A well-defined methodology ensures that MBSE is applied consistently and effectively, maximizing its benefits and minimizing potential challenges.

Who defines the methodology depends heavily on the scope of the deployment and application. It can range from an individual user who is defining their methodology as they are progressing in the modeling effort to a dedicated MBSE deployment team in a large organization that can take years to prepare and document a method, adapt and tailor the language and possibly a chosen tool, and train the engineers that are supposed to use the method.

The reasons for and the benefits of a well-defined MBSE methodology include:

- **Establishing consistency and standardization**: A methodology provides a common framework for MBSE activities across the organization, ensuring that teams follow consistent practices and terminology. This fosters collaboration, reduces misunderstandings, and promotes knowledge sharing.
- **Managing complexity and scope**: Complex systems often involve numerous stakeholders, diverse requirements, and intricate interactions. A methodology provides a structured approach to defining, organizing, and managing the system's complexity, ensuring that all aspects are considered and addressed effectively.
- **Enhancing communication and collaboration**: Models serve as a common language for communicating system requirements, design decisions, and performance analysis. A methodology promotes the effective exchange of information among stakeholders, ensuring that everyone is aligned and working towards common goals. A methodology establishes a standardized approach to organizing and presenting models, making them more effective for communication and collaboration among stakeholders. It defines conventions for naming, labeling, and documenting models, enhancing their clarity and accessibility.

- **Promoting reusability and knowledge management**: Models developed during MBSE can be reused in subsequent projects or phases of the same project, reducing development time and ensuring consistency across the system's lifecycle. A methodology encourages the organization and management of these reusable assets.
- **Aligning with organizational processes**: MBSE should be integrated with existing organizational processes, such as project management methodologies and quality assurance practices. A methodology facilitates this alignment, ensuring that MBSE is not a standalone approach but rather an integral part of the organization's overall development process. MBSE methodologies often incorporate elements from other engineering disciplines, such as requirements engineering, architecture design, and verification and validation (V&V). This integration ensures that modeling is aligned with the overall systems engineering process and the specific needs of the project.
- **Continuous improvement**: A methodology facilitates continuous improvement in the MBSE practice by providing a framework for evaluating the effectiveness of modeling activities and identifying areas for improvement. It encourages feedback loops and iterative refinement of the modeling process.
- **Tool selection and usage**: A methodology guides the selection and utilization of appropriate MBSE tools. It ensures that tools are used effectively to support the modeling process and to integrate with other project tools and workflows.
- **Standardization and conformance**: MBSE methodologies can align with industry standards and guidelines, ensuring that models and modeling practices are compatible with broader standards and expectations. This standardization can facilitate collaboration with external stakeholders and promote the interoperability of models across different projects and organizations.
- **Knowledge sharing and transferability**: A well-defined methodology serves as a valuable knowledge base for systems engineers and other stakeholders. This knowledge can be readily transferred across projects and teams, promoting consistency and efficiency in modeling practices.

2.3.2 Existing MBSE Methods and Methodologies

Several MBSE methods and methodologies have been published over the years. Estefan and Weilkiens recently published an overview (Estefan, J.A., Weilkiens, T., 2022).

Some of them are promoted by tool vendors, such as Harmony SE from IBM or MagicGrid from Dassault Systèmes. While notionally tool-independent these methodologies are often tightly linked to the tool from the same vendor, which has potential advantages, e.g., the methodology works well with the tool but also disadvantages, e.g., a potential vendor lock-in.

Typically for these methodologies, the tool existed before the methodology was developed which means that the tool drove the methodology.

Others are completely tool-independent, such as OOSEM (Object-Oriented Systems Engineering Method) or SYSMOD (Systems Modeling Process). The focus of these methods is often on the correctness of the method in an ideal usage of the language as it is specified. Since the implementation of the language inside the tools can differ between the tools and the official specification, actually using these methods with a tool can lead to the identification of some gaps and necessary adaptations of the method or the tool.

Then there is Arcadia, an MBSE method developed by the Thales Group, which comes hand in hand with Cappella, an open-source modeling tool specifically designed to support the Arcadia method. In this case, the method existed before the tool was developed which means that the method drove the tool development.

Then there almost countless other methodologies that have been developed inside companies or as part of a Ph.D. thesis and which are only partially documented or published in the open.

2.3.3 MBSE Vision and Current State of Practice

Distinguishing between the vision for MBSE and its current state of practice is crucial for several reasons. The ultimate vision for MBSE is to revolutionize the way complex systems are conceived, designed, and managed throughout their entire lifecycle. At its core, MBSE seeks to enhance collaboration, communication, and efficiency by leveraging visual models that comprehensively represent system characteristics, behaviors, and interactions. Understanding the disparities between the desired future state and the present reality helps guide strategic efforts, set realistic expectations, and drive continuous improvement within the field of systems engineering. If the current limitations of MBSE are not communicated clearly, unrealistic expectations for what it can achieve may be set, which can lead to frustration and disappointment among users and other stakeholders. By understanding the gaps between the current state of practice and the vision for MBSE, the areas on which to focus the efforts to make MBSE more prevalent and effective can be identified. The vision for MBSE includes:

- **Holistic representation**: To have a unified and holistic representation of systems that captures not only their physical aspects but also their requirements, behaviors, and relationships in a single, integrated model.
- **Interdisciplinary collaboration**: Facilitate seamless collaboration between multidisciplinary teams, including engineers, analysts, and stakeholders, by providing a common visual language that transcends traditional disciplinary boundaries.

- **Lifecycle integration**: Extend the use of models beyond the design phase to cover the entire lifecycle, including requirements analysis, V&V, maintenance, and evolution.
- **Automated analysis and simulation**: Enable automated analysis and simulation directly from the models, allowing for early detection of issues, validation of requirements, and optimization of system performance.
- **Improved decision support**: Enhance decision-making by providing decision-makers with intuitive visualizations and analysis tools derived directly from the model, supporting more informed and timely choices.
- **Traceability and transparency**: Ensure comprehensive traceability, where every element in the model is traceable to its source and its impact on other elements is transparent, facilitating impact analysis and change management.

While significant strides have been made in realizing the vision of MBSE, certain challenges and gaps persist in the current state of practice:

- **Tool interoperability**: Interoperability between different MBSE tools and between MBSE tools and other systems engineering tools remains a hurdle. Seamless integration and data exchange between tools is essential for a truly collaborative and interoperable MBSE environment.
- **Standardization**: The absence of universal standards hampers the seamless exchange of models and information. Establishing and adhering to standardized modeling languages and methodologies is crucial for achieving a more cohesive MBSE ecosystem.
- **Cultural shift**: Embracing a paradigm shift towards model-centric practices requires a cultural change. Organizations need to invest in training and change management to ensure widespread adoption of MBSE principles.
- **Scalability**: Applying MBSE to large, complex systems presents scalability challenges. Ensuring that MBSE methodologies and tools can effectively handle the intricacies of massive systems is an ongoing area of improvement.
- **Model maintenance and evolution**: Managing the evolution of models over the system's lifecycle requires robust version control, configuration management, and tools to handle changes efficiently.
- **Integration with legacy systems**: Integrating MBSE practices with existing legacy systems and engineering processes is often complex. Legacy systems may not seamlessly align with model-centric approaches, necessitating careful integration strategies.

In summary, the ultimate vision for MBSE is to create a paradigm where models serve as the backbone for system engineering throughout the entire lifecycle. While significant progress has been made, ongoing efforts are essential to address current challenges and propel MBSE towards its full potential. As the field continues to mature, a collaborative and concerted effort from the industry, academia, and standards organizations will play a pivotal role in realizing the vision of MBSE.

2.3.4 Misconceptions About MBSE and Clarifications

Several misconceptions persist regarding the capabilities and limitations of MBSE. This section addresses common misconceptions and provides clarifications to enhance understanding and dispel any associated myths.

- **MBSE is only for complex systems.** While MBSE is indeed beneficial for large and complex systems, it is also scalable and adaptable to projects of varying sizes. Just like systems engineering itself, MBSE can and should be tailored to the envisaged application.
- **MBSE is a replacement for traditional systems engineering methods.** MBSE is, at its core, still systems engineering and can also be dubbed systems engineering with more modern methods. It is an evolution rather than a replacement. MBSE does encourage a shift from documents to models and has a model-centric view, but it does not entirely eliminate documentation. Documentation is still relevant, and models can be accompanied by textual descriptions, ensuring a balance between visual representation and detailed documentation. Additionally, in document generation feature of MBSE tools allows automatically creating reports and other documents directly from the system model.
- **MBSE is too complex and difficult to learn.** MBSE is a powerful tool, but it is not overly complex. The basic principles of MBSE can be learned relatively quickly, and there are many resources available to help people learn the language and tools of MBSE. While there is a significant initial learning curve and adaptation period, the long-term benefits of MBSE often outweigh the upfront investment. Once teams become proficient, the efficiency gains in communication, collaboration, and system understanding can result in time savings.
- **MBSE is only about drawing diagrams.** MBSE is about using models to represent and analyze complex systems. These models can be used to communicate with stakeholders, identify and mitigate risks, and make better design decisions. While diagrams are a significant aspect of MBSE, it is not solely about drawing pictures. MBSE emphasizes the integration of models, ensuring that different aspects of a system are represented in a coherent and interconnected manner to support decision-making. Models are more than just diagrams.
- **MBSE is not worth the investment:** MBSE can save money and time by improving the efficiency and effectiveness of systems engineering projects. It can also help to reduce the risk of project failure.

- **MBSE is just a collection of modeling tools.** MBSE involves not just the use of modeling tools but a methodological shift in how systems are engineered. It includes defining modeling languages, establishing modeling practices, and ensuring that models are integrated into the engineering workflow.
- **MBSE requires a specific modeling language.** While specific modeling languages like SysML are commonly associated with MBSE, the essence of MBSE lies in the model-centric approach. Different projects and domains may use varying modeling languages or combinations to suit their specific needs.
- **MBSE is only applicable in certain industries.** MBSE is versatile and applicable across a wide range of industries, including aerospace, automotive, healthcare, and more. Its principles of model-centric engineering and collaborative modeling can be adapted to various domains.
- **MBSE is all about technology, not people.** MBSE is a methodology that involves both technology and people. While modeling tools play a role, the success of MBSE also depends on effective collaboration, communication, and the active involvement of stakeholders throughout the engineering process.
- **MBSE will dramatically speed up the development process.** MBSE offers long-term benefits in terms of enhanced understanding, traceability, and collaboration. However, the front-loading of effort, including learning curves and robust modeling, typically will initially impact the speed of development projects. The realization of a dramatic speedup and overall project cost reductions often becomes more apparent in subsequent phases as the benefits of the upfront investment in modeling become evident, leading to increased efficiency, reduced errors, and improved decision-making throughout the project lifecycle.

Understanding and addressing these misconceptions is crucial for organizations considering the adoption of MBSE. By recognizing the breadth and flexibility of MBSE, teams can better appreciate its potential benefits and overcome any barriers to successful implementation.

Chapter 3 - The Systems Modeling Language (SysML)

SysML is a general-purpose modeling language used in systems engineering to describe, analyze, and design complex systems. SysML provides a standardized notation and semantics for representing system requirements, structures, behaviors, and interactions. It supports modeling of both hardware and software components as well as their interactions within a system.

SysML is based on the Unified Modeling Language (UML) and extends it with specific constructs for system modeling. It allows engineers to capture the various aspects of a system in a visual and structured manner, facilitating communication between different stakeholders involved in the development process.

The relation between UML and SysML is depicted in Figure 3.1. SysML re-uses a subset of UML 2, which is called UML4SysML. SysML complements this subset with additional modeling elements, which do not exist in the UML. These are defined in the SysML profile (see Appendix D - Profiles and Stereotypes to learn about profiles). The remaining part of the UML is not used by SysML.

SysML is not a methodology, but rather a language. This means that it does not provide a set of processes or steps for developing systems. Instead, SysML provides a set of diagrams and notations that can be used to represent systems in a standardized and well-defined way. The SysML specification serves as a foundational framework for systems modeling, providing a rich set of modeling elements and diagrams to represent the various aspects of a system. However, it does not dictate a rigid methodology or prescriptive steps for using these elements.

SysML empowers the user with the flexibility to adapt and tailor their modeling approach to suit the unique circumstances and requirements of their project or organization depending on the scope of the MBSE deployment. Rather than imposing a one-size-fits-all approach, SysML encourages the users to make informed choices about when and how to use its modeling elements, ensuring that the modeling process is effective and tailored to the specific needs of the project or the organization.

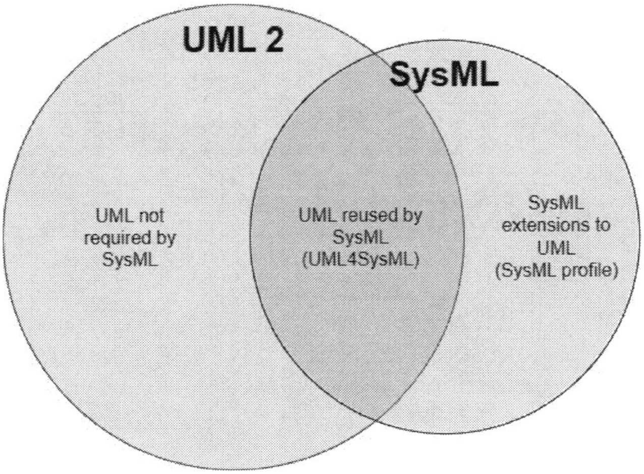

Figure 3.1. Overview of SysML/UML Interrelationship (OMG, 2019)

For instance, the decision of when to create a use case diagram is not defined by the SysML specification. Similarly, the choice of using an activity diagram to elaborate a use case is not a mandatory step. The method team can evaluate whether an activity diagram is the most suitable method for representing the flow of activities within a use case or whether another modeling element might be more effective. Furthermore, SysML does not mandate the number of internal block diagrams (IBDs) that must be created. The project team should assess whether IBDs are necessary for representing the system architecture and, if so, determine the appropriate level of detail and the number of diagrams required. All these are methodological decisions. They fall outside the scope of the SysML specification

By embracing this flexibility, the SysML user can effectively leverage the power of SysML to create meaningful models that support the successful development and implementation of their system. They can adapt their modeling approach to address unique challenges, optimize modeling efforts, and ensure that the resulting models are directly relevant to the project's goals.

3.1 A Brief History of SysML

UML, introduced in the late 1990s and adopted by the OMG as a standard in 1997, revolutionized the software development industry by providing a standardized graphical language for modeling software systems. Its comprehensive set of diagrams and notations facilitated communication, understanding, and analysis of complex software architectures. However, as UML gained popularity, its use expanded beyond software systems, encompassing broader systems engineering endeavors.

Systems engineering faced challenges in employing UML effectively. While UML provided valuable tools for modeling software components and data flows, it lacked the necessary constructs for representing essential aspects of systems engineering, such as requirements, processes, and physical systems.

Recognizing the need for a more specialized modeling language tailored to systems engineering, a group of systems engineering experts from the INCOSE and the OMG as maintainers of the UML specification formed the OMG Systems Engineering Domain Special Interest Group (SE DSIG). Their goal was to develop an extension and modification of UML to address the specific needs of systems engineering.

In 2007, the OMG released the SysML v1.0 specification. This marked a significant milestone, establishing SysML as a standardized modeling language for systems engineering. The current version of OMG SysML is v1.6, which was issued by the OMG in December 2019. In addition, SysML was published by the International Organization for Standardization (ISO) in 2017 as a full International Standard, ISO/IEC 19514:2017. The SysML v1.7 specification has already been published as a Beta version in 2022 and is currently being finalized. It will likely be the last v1.x standard of SysML. The OMG has been working on the next generation of SysML and issued a Request for Proposals (RFP) for version 2 on December 8, 2017, following its open standardization process. In 2023, the OMG approved the SysML v2 beta specification, and the final adoption of the standard is currently expected to happen in 2024. SysML v2 is a completely new language independent from UML and is expected to be a major boost for MBSE implementation in the industry.

3.2 The SysML Diagrams

SysML provides a variety of diagram types to represent different aspects of a system separated into the four pillars of SysML:

1. Structure
2. Behavior
3. Requirements
4. Parametrics

In total, nine different diagrams are defined by SysML as shown in Figure 3.2.

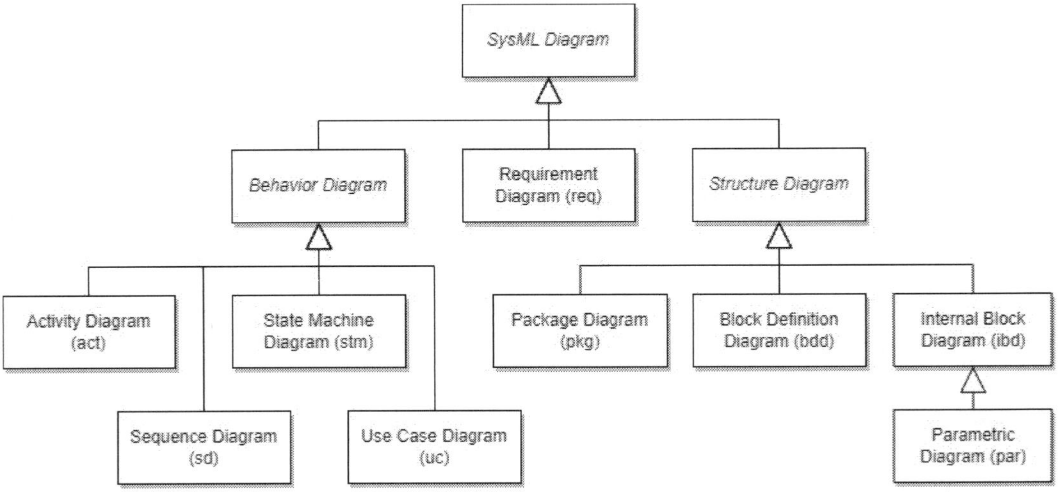

Figure 3.2. SysML diagrams

SysML's structure pillar is concerned with how a system is composed of its constituent parts and how these parts are organized and interconnected. It provides a graphical and textual notation for representing the system's architecture, including blocks, interfaces, ports, and connectors. This pillar is essential for understanding the physical and logical organization of the system.

The *Package Diagram* is a static structural diagram that shows the relationships among packages and their contents. It organizes the various SysML elements into packages, providing a way to manage and structure a complex model.

The *Block Definition Diagram (BDD)* is a static structural diagram that shows system components, their contents (Properties, Behaviors, Constraints), Interfaces, and relationships.

The *Internal Block Diagram (IBD)* is a static structural diagram owned by a particular Block that shows its encapsulated structural contents: Parts, Properties, Connectors, Ports, and Interfaces. Stated otherwise, an IBD is a white-box perspective of an encapsulated "black-box" Block.

SysML's behavior pillar focuses on how the system functions and how it reacts to external stimuli. It provides a notation for modeling the system's interactions, workflows, and state transitions. This pillar is essential for understanding how the system behaves under different conditions.

The *Use Case Diagram (UC)* shows communications among system Use Cases and external users in the context of a system boundary. Use Case diagrams are typically used to provide a high-level view of the subject system and convey the top-level system requirements in non-technical terms for all stakeholders, including customers and project managers as well as architects and engineers.

The *Activity Diagram (AD)* shows system dynamic behavior using a combined Control Flow and Object (data) Flow model.

The *Sequence Diagram (SD)* is a dynamic behavioral diagram that shows interactions (collaborations) among distributed objects or services via sequences of messages exchanged, along with corresponding (optional) events. It illustrates the interactions between components or objects over time, emphasizing the order of events.

The *State Machine Diagram (STM)* is a dynamic behavioral diagram that models the different states that a system or object can be in and how it transitions between those states in response to events.

SysML's requirements pillar is concerned with capturing and managing the system's functional and non-functional requirements. It provides a notation for defining, refining, and analyzing requirements, ensuring that they are clear, measurable, and traceable throughout the system lifecycle.

The *Requirement Diagram (REQ)* provides a modeling construct for textual requirements and the relationship between requirements and other model elements that satisfy or verify them.

SysML's parametrics pillar is concerned with representing the system's quantitative attributes and their relationships. It provides a notation for defining parameters, constraints, and equations, enabling engineers to model the system's behavior in terms of quantifiable parameters.

The *Parametric Diagram (PD)* is a new SysML diagram type, i.e., it does not exist in the UML, that describes the constraints among the properties associated with blocks. This diagram is used to integrate behavior and structure models with engineering analysis models such as performance, reliability, and mass property models.

The four pillars of SysML are interdependent and work together to support the comprehensive modeling of complex systems.

- Structure provides the foundation for understanding the system's physical and logical architecture, while behavior models how the system interacts and functions within that architecture.
- Requirements define the system's intended behavior and performance, while parametrics provide a quantitative framework for analyzing and optimizing the system's performance.

- The interplay of these pillars enables engineers to create a holistic model that captures the system's structure, behavior, requirements, and parametric relationships.

By integrating these four pillars, SysML provides a powerful framework for systems modeling, enabling engineers to create, manage, and utilize models that support the successful development and operation of complex systems.

3.2.1 The Model versus the Diagrams

While diagrams are a major part of SysML, it is important to note that the diagrams are not the model. Instead, a diagram is a view of the model and typically contains only a fraction of the overall model. A diagram is normally built for a specific purpose to communicate a chosen aspect of the system model. Model elements can be used on one, several, or even no diagram at all.

> Deleting an element from a diagram is not the same as deleting an element from the model. All SysML tools have two types of deletion when an element is deleted from a diagram: delete from the diagram only or delete from the model, which, of course, also leads to a removal of the element from the diagram.

SysML diagrams are not independent entities but rather visualizations or graphical representations of the underlying model elements. Changes made in one diagram impact the underlying model, and vice versa. The relationship between a SysML diagram and the model it represents is dynamic and bidirectional.

When you create elements (blocks, connectors, etc.) in a SysML diagram, you are essentially adding graphical representations of those elements to the view presented by the diagram. These graphical representations provide a way for modelers to interact with and understand the model without having to deal directly with the complexity of the underlying model.

SysML encourages the use of multiple diagrams to present different aspects of the system. This approach allows modelers to focus on specific details without being overwhelmed by the entire model. Each diagram acts as a view tailored to communicate particular information about the system. The same model elements can appear in multiple diagrams, each providing a specialized view. For example, a block representing a subsystem might appear in a Block Definition Diagram (BDD) and be further detailed in an Internal Block Diagram (IBD).

The SysML standard ensures that there is consistency and coherence between the diagrams and the underlying model. Changes made in one view are reflected in the others to maintain a synchronized representation of the system.

SysML tools often provide cross-referencing capabilities, enabling users to navigate between the graphical representation in a diagram and the detailed information in the model. This facilitates efficient navigation and understanding of the relationships between different parts of the system.

3.3 Common Misconceptions about SysML

Several common misconceptions about SysML can hinder its adoption and effective use. Here are some of the most prevalent misconceptions:

- **SysML is too complex and difficult to learn**. SysML is a powerful tool, but it is not overly complex. The basic principles of SysML can be learned relatively quickly, and there are many resources available to help people learn the language and tools of SysML.
- **SysML is just a drawing tool**. SysML is about using models to represent and analyze complex systems. These models can be used to communicate with stakeholders, identify and mitigate risks, and make better design decisions. The diagrams are just one part of SysML. A model is more than a collection of diagrams.
- **SysML is only for software engineers**. SysML is not only for software engineers. It can be used by a variety of disciplines, including mechanical engineers, electrical engineers, systems engineers, and architects. SysML provides a common language for communicating about complex systems across different disciplines.
- **SysML is not practical for real-world projects**. SysML is a mature and well-established language that has been used on a wide range of successful projects. There are several case studies and testimonials that demonstrate the effectiveness of SysML in real-world applications.
- **SysML is a silver bullet for solving all problems**. SysML is a tool that can be used to improve the way that we model and analyze complex systems. However, it is not a silver bullet and must be used in conjunction with other processes and tools to be effective.
- **SysML is too expensive to implement**. SysML tools are available for all levels of budget. There are open-source tools, such as Papyrus, which is covered by this book, that are free to use, and there are also commercial tools that can be purchased at a variety of price points.
- **SysML is not the right tool for my industry**. SysML is a general-purpose language that can be used for developing systems in any industry. It has been widely adopted in a variety of industries, including aerospace and defense, automotive, medical devices,

and software. SysML is a versatile modeling language that can be adapted and customized for a wide range of purposes. This flexibility is achieved through the use of profiles, which are extensions of SysML that provide additional concepts and notation.
- **SysML aims to replace all the existing domain engineering tools**. SysML complements the specialized tools of each specialist. While each discipline has its own set of tools tailored to its specific needs, SysML serves as a common language and a single source of truth to facilitate collaboration and integration. The mechanical engineer relies on various computer-aided design (CAD) tools to design and analyze physical components. Software engineers utilize a range of programming languages and design tools to create software modules. SysML does not replace these specialized tools but rather provides a unifying framework to connect them. SysML's role is to create a shared understanding of the system's requirements, architecture, and behavior. By establishing a single source of truth, SysML ensures that each specialist has a consistent and up-to-date view of the system, enabling them to use their specialized tools effectively and efficiently. This collaboration leads to a more cohesive and integrated system that meets the needs of all stakeholders.

By understanding and addressing these misconceptions, we can more effectively promote the use of SysML and its benefits in systems engineering.

3.4 SysML Pitfalls

SysML is a powerful tool for modeling complex systems, but like any modeling language, it comes with its set of challenges. Several common pitfalls can hinder the effective use of SysML and lead to problems in the systems engineering process. Here are some of the most prevalent pitfalls and some ideas for mitigating them:

- **Over-complexity in models**: Creating overly complex models with excessive detail can make them difficult to understand, maintain, and communicate to stakeholders. Focus on capturing essential aspects of the system. Use modularization and abstraction to manage complexity, and avoid unnecessary details in the early stages.
- **Lack of consistency**: Inconsistencies in terminology, notations, or conventions within the SysML model can lead to confusion and misinterpretation. Establish and adhere to a consistent modeling approach. Define naming conventions, notations, and guidelines for the entire modeling team to follow.
- **Inadequate training and knowledge**: Insufficient training or knowledge of SysML among team members can lead to modeling errors, misinterpretations, and inefficiencies. Invest in training for team members to ensure a solid understanding of SysML

concepts, language constructs, and best practices. Encourage ongoing learning and knowledge-sharing within the team.

- **Overlooking model maintenance**: Neglecting to update SysML models in response to changes in requirements, design, or system understanding can result in outdated and inaccurate representations. Establish a process for model maintenance and updates. Regularly review and revise SysML models to reflect the current state of the system.
- **Over-reliance on tool features**: Depending too heavily on specific features of modeling tools without understanding the underlying SysML concepts can limit flexibility and hinder collaboration. Ensure that team members have a solid understanding of SysML independent of the tool being used. Use tools as aids, not substitutes, for understanding the modeling language.
- **Not involving stakeholders in the modeling process**: Identify the key stakeholders who will be using the models and ensure that they are involved in the modeling process. Conduct workshops with stakeholders to gather their requirements and input into the modeling process.

By being aware of these common pitfalls and implementing mitigation, practitioners can enhance the effectiveness of SysML modeling, fostering clearer communication, better collaboration, and more accurate representation of complex systems.

Chapter 4 - Eclipse Papyrus

Eclipse Papyrus is an open-source modeling tool that provides support for various modeling standards, including UML and SysML. It is built on the Eclipse platform and is designed to offer a flexible and extensible environment for model-driven development.

The main characteristics and features of Papyrus are:

- **Modeling language support**: Papyrus supports multiple modeling languages, including UML, SysML, and more. This makes it versatile for different domains and purposes, such as software engineering, systems engineering, and business modeling.
- **Diagram editors**: Papyrus provides graphical diagram editors for creating and editing models. Users can create various types of diagrams, such as class diagrams, activity diagrams, sequence diagrams, and more, depending on the modeling language being used.
- **Model repository**: Papyrus provides a central repository for storing and managing models. This repository allows multiple users to collaborate on a project by sharing and versioning models. It supports model versioning, change tracking, and collaborative modeling.
- **Code generation**: Papyrus supports code generation from models. This feature allows users to generate code from UML or SysML models, facilitating model-driven development. It helps maintain consistency between the model and the code implementation.
- **Customization and extensibility**: Papyrus is highly extensible and customizable. Users can define their modeling profiles, stereotypes, and customizations to tailor the tool to the specific needs of their domain or organization. This extensibility makes Papyrus suitable for a wide range of applications.
- **Verification and validation**: Papyrus includes features for model V&V. It helps to ensure that models adhere to defined rules, constraints, and standards. This aids in improving the quality of models and the resulting systems.
- **Integration with other tools**: Papyrus can be integrated with other development tools and platforms. It supports interoperability with tools that are part of the Eclipse ecosystem, as well as with external tools through standard interfaces.
- **Profile editor**: Papyrus includes a profile editor that allows users to define custom stereotypes, tagged values, and constraints. This is particularly useful in extending the modeling languages to meet specific modeling needs.

- **Simulation and analysis**: Papyrus supports simulation and analysis capabilities for models. This is especially valuable in system engineering where dynamic behavior and system interactions need to be modeled and analyzed.
- **Documentation generation**: Papyrus can generate documentation from models, helping to maintain comprehensive and up-to-date documentation about the system.

Papyrus has a solid foundation in the form of the Eclipse Rich Client Platform (RCP) and the Eclipse Modeling Framework (EMF). This makes it generally a reliable tool, but it is still an open-source software product. As such, it is subject to the same potential issues as any other open-source software, including bugs and compatibility issues. To mitigate these risks, users of Papyrus need to adopt a proactive approach to data management. This includes regularly saving their work and maintaining backups of their data and models. This will help to ensure that users are not affected by any unexpected issues that may arise with the software. Specific recommendations for saving and backing up data in Papyrus include:

- **Saving the work frequently.** This is especially important when working on complex models or when making significant changes.
- **Using a version control system.** A version control system, such as Git (see Appendix C - Collaboration), can be used to track changes to the models over time. This will provide an additional layer of protection in case of data loss or corruption.
- **Regularly back up the entire Eclipse Papyrus workspace.** This will ensure that you have a complete copy of all of your models and data in case of any major problems with the software.

By following these recommendations, users of Papyrus can help protect their data and models from potential issues and ensure that they can continue to work on their projects without interruption.

4.1 Eclipse Rich-Client Platform

To account for some peculiarities of Papyrus, e.g., the menu structure, how properties are displayed and edited in the Properties View, and the nested preferences settings, it is helpful to understand that Papyrus is built on top of the Eclipse RCP.

The Eclipse RCP is a robust and extensible framework for building modular and customizable desktop applications. It serves as the foundation for the Eclipse IDE and allows developers to create feature-rich, platform-independent applications. A key strength of RCP lies in its support for extension mechanisms through the use of plugins.

In RCP, a plugin is a modular unit of functionality that contributes to the overall application. Each plugin encapsulates specific features, views, editors, or other components. The modular nature of plugins promotes code organization, reusability, and maintainability.

An Eclipse product is a packaged and distributable version of an Eclipse-based application designed for standalone use by end-users. It provides a convenient way to bundle, configure, and distribute Eclipse applications with a specific focus or set of features. An Eclipse product specifies the set of plugins and features that constitute the application. This ensures that the product includes the necessary functionality and extensions. Products can be branded to give them a distinct look and feel. Branding includes customizing logos, splash screens, and other visual elements to provide a unique identity for the application.

Reasons and benefits for building Papyrus as a RCP product include:

- **Modularity and extensibility**: RCP is designed to support modular development through the concept of plugins. Eclipse Papyrus, being a complex modeling tool, benefits from RCP's modularity by organizing its functionality into discrete plugins. This modularity makes it easier to manage and extend Papyrus with additional features or domain-specific modeling capabilities.
- **Plug-in extension points**: Papyrus, as a modeling tool, often requires extensibility to support various modeling domains and notations. RCP's extension point mechanism allows Papyrus to define specific extension points, enabling users and third-party developers to contribute additional features, modeling languages, and viewpoints. This aligns with the extensibility needs of a diverse user base. Examples of this specialization or customization include Papyrus RT[1] for Real-time Systems modeling or Papyrus for Robotics[2], a Papyrus-based modeling environment dedicated to robotics.
- **Customizable user interface (UI)**: RCP provides a flexible and customizable UI framework. Eclipse Papyrus can take advantage of this feature to offer a user interface that is tailored to the specific needs of different modeling activities. Users can arrange and customize views, editors, and perspectives based on their modeling requirements.
- **Rich set of widgets and components**: RCP comes with a rich set of standard UI components and widgets. Papyrus, as a modeling tool, can utilize these components to create a consistent and user-friendly interface for creating, editing, and visualizing models. This helps in providing a standardized look and feel across different modeling perspectives.
- **Eclipse Modeling Framework (EMF) integration**: Papyrus often works closely with the EMF, which is seamlessly integrated into the Eclipse platform. EMF provides a

[1] https://projects.eclipse.org/projects/modeling.papyrus-rt
[2] https://eclipse.dev/papyrus/components/robotics/

comprehensive set of tools for modeling and code generation. Leveraging RCP ensures that Papyrus can easily integrate with EMF and take advantage of its capabilities.

- **Rich tooling support**: RCP includes a wealth of tools and utilities for building robust desktop applications. Papyrus can leverage these tools for tasks such as debugging, profiling, testing, and version control. This contributes to the overall robustness and quality of the modeling tool.
- **Community and ecosystem**: The Eclipse community and ecosystem provide a wealth of resources, plugins, and expertise that Eclipse Papyrus can tap into. By being part of the Eclipse ecosystem and using RCP, Papyrus benefits from the collaborative development efforts and contributions from the Eclipse community.
- **Cross-platform compatibility**: RCP ensures that applications built on it, including Eclipse Papyrus, are inherently cross-platform. This means that Papyrus users can run the modeling tool on various operating systems, ensuring broad accessibility and usability.

4.2 Papyrus GUI Overview

Perspectives are a fundamental aspect of the Eclipse RCP to enable customization of the graphical user interface (GUI). They provide a structured way to organize and arrange tools, views, and editors into workspaces tailored to specific tasks or workflows. Perspectives enable users to customize the arrangement of tools, views, and editors within the workspace, catering to their individual preferences and workflows. They are centrally managed within the Eclipse framework, allowing users to easily switch between different perspectives to adapt to different tasks or work contexts.

When Papyrus is started with the Papyrus perspective with the default settings and a project is open, the GUI consists of seven major parts, as shown in Figure 4.1.

1. Menu Bar: This contains the standard Eclipse menus, such as the "File" menu for opening and saving projects. In the Papyrus perspective, the Menu Bar also contains a "Papyrus" menu, which allows the creation of new diagrams or has some possibilities for working with diagrams.
2. Project Explorer: This shows the Eclipse projects in the current workspace. The list of projects is not limited to Papyrus projects but can show all other types of Eclipse projects as well.
3. Model Explorer: This shows the Papyrus model in a hierarchical view and allows the creation of new model elements and diagrams.

4. Outline: This shows a small overview of the currently opened diagram, which enables navigation on very large diagrams.
5. View Area: This shows the currently opened view, i.e., a diagram or a table, and enables working with it, e.g., creating new elements, and moving elements around. If multiple views are open at the same time, they will all appear as tabs at the bottom of the View Area to enable switching between views, similar to how web browsers work.
6. Diagram Palette: This shows all the elements that can be added to the currently opened diagram depending on its type.
7. Properties View: Shows detailed information about the currently selected element. The element can either be selected in the Model Explorer or in the View Area. The Properties View consists of several tabs, to show and manipulate different aspects of the chosen element.

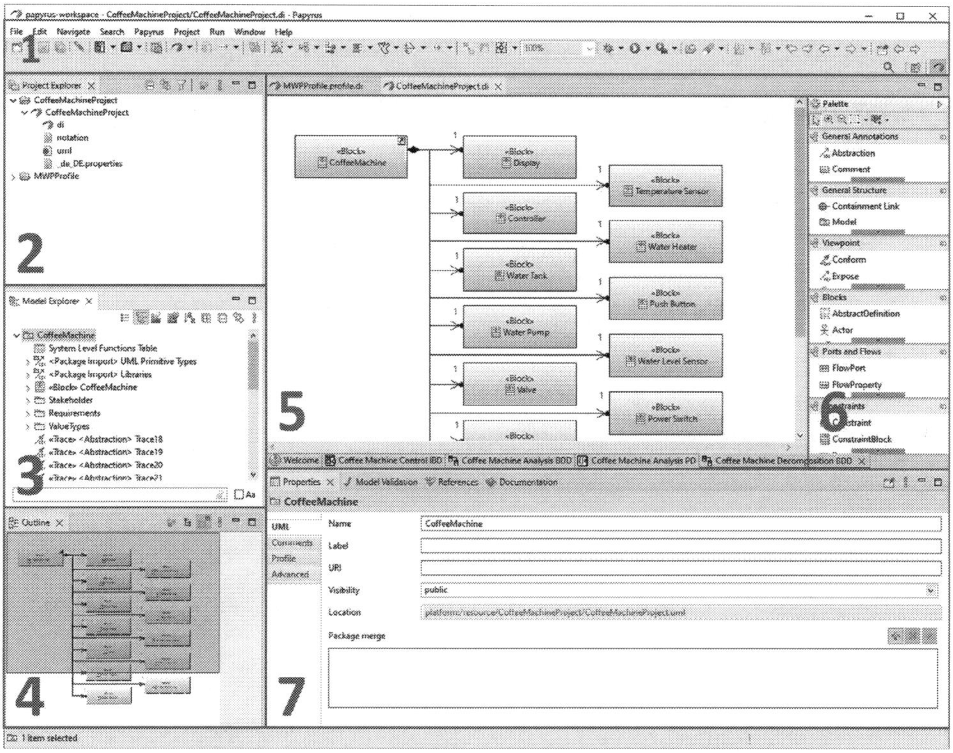

Figure 4.1. Papyrus GUI Overview

The individual windows that comprise the Papyrus perspective, such as the Project Explorer or the Model Explorer, can be moved around or closed by the user depending on their personal preference or current steps in the workflow. To reset the Papyrus perspective to its initial state, click on the button with the small Papyrus logo on the right side of the menu bar and select "Reset," as shown in Figure 4.2.

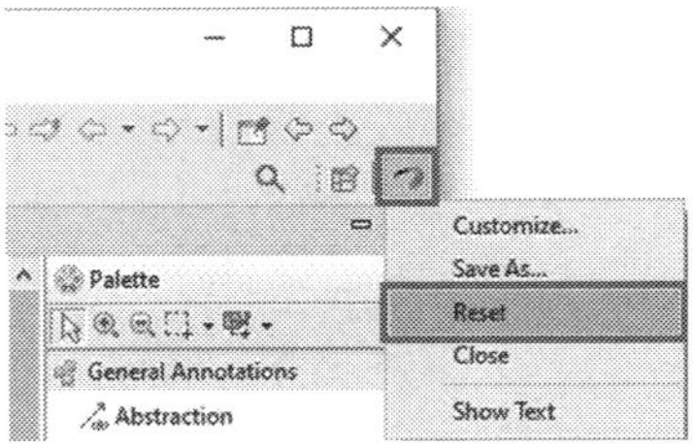

Figure 4.2. Reset the Papyrus perspective

4.3 Papyrus Preferences and Default Settings

Many of the Papyrus default settings, e.g., the font type and size, can be customized by the user. You can find the Papyrus settings in the Eclipse preferences. To open the Eclipse preferences, open the Window menu and select "Preferences," as depicted in Figure 4.3.

Figure 4.3. Open the Eclipse preferences

The Eclipse preferences dialog shows a list of plugins that can be adapted on the left side. Expand the "Papyrus" item to see all the topics for which Papyrus offers customization options, as shown in Figure 4.4.

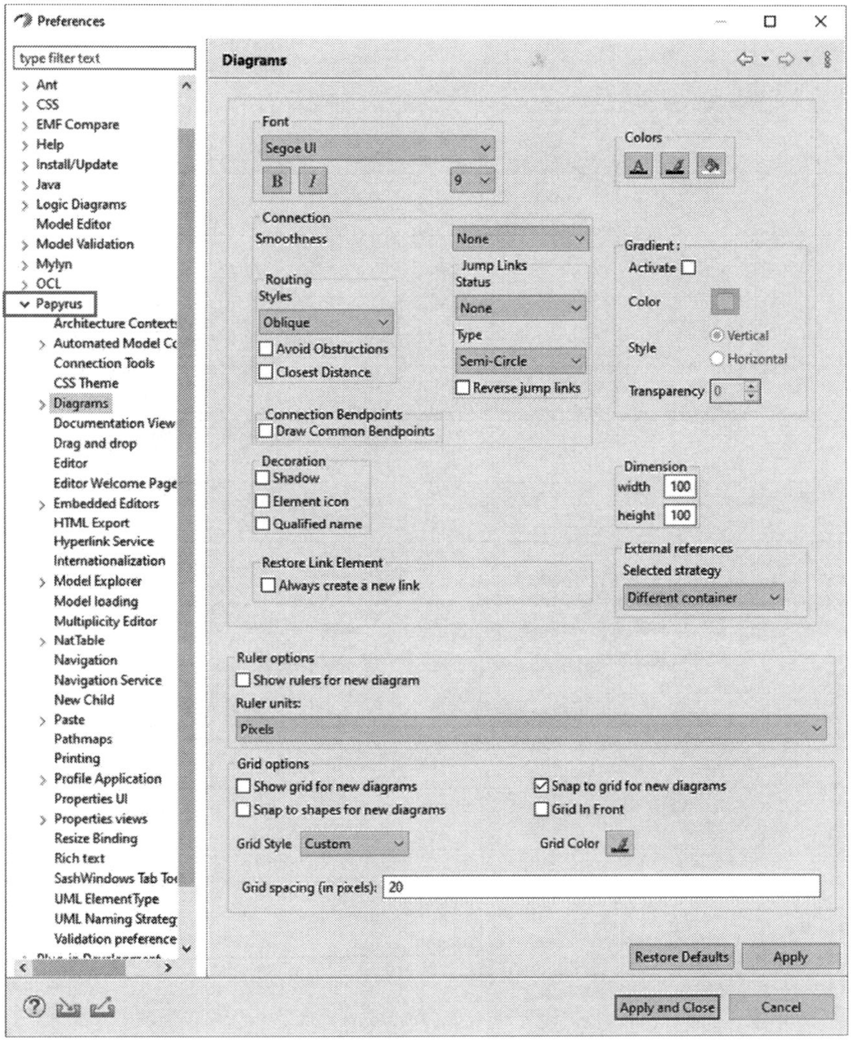

Figure 4.4. Papyrus preferences

Chapter 5 - MBSE with Papyrus

Enabling MBSE requires a foundation of three things: a method, a language, and a tool, as shown in Figure 5.1. The method defines the purpose and direction. A chosen modeling tool and language are just a collection of tools in a toolbox that are required for building a model. Actually building a model requires guidance in the form of method steps that prescribe for each type of information about the system what kind of modeling elements should be used to represent it.

Additionally, while method, language, and tool form the foundation of MBSE, the success of MBSE ultimately relies on the people who employ these elements. It hinges on the ability of individuals to understand and use the tool and the language effectively. Systems engineers must understand the MBSE method and utilize it to guide the development process.

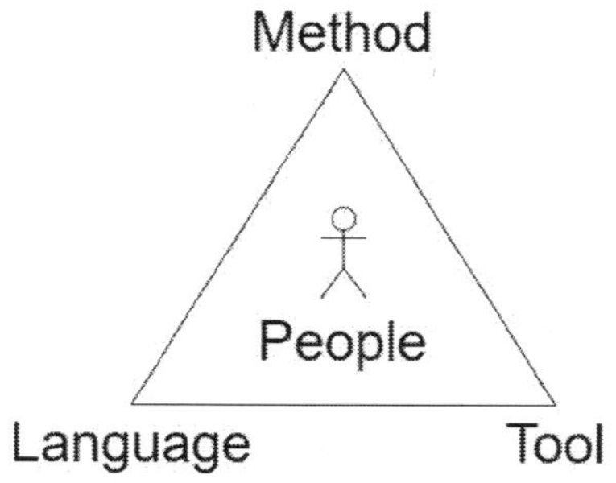

Figure 5.1. The MBSE triangle

In a typical MBSE deployment effort in a large organization, the language and the tool should not be a given. Instead, the purpose of the overall modeling approach drives the development or selection of the method and the selection of the language and the tool. Of course, for this book, the language and the tool are already selected. What is missing is the method. To learn SysML modeling with Eclipse Papyrus, a simple modeling methodology following the MBSE paradigm will be used to cover the most important aspects of a system: requirements, behavior, and architecture. This method is aptly called MBSE with Papyrus (MWP) method. The MWP method is used for teaching MBSE to students of aeronautical engineering at the Hamburg University for Applied Sciences Hamburg (HAW Hamburg). Students are expected to learn SysML and apply it to a system of their choosing using Papyrus in small teams. The MWP method is based on the work by Jutta Abulawi (Abulawi 2018). Figure 5.2 depicts an overview of the MWP method.

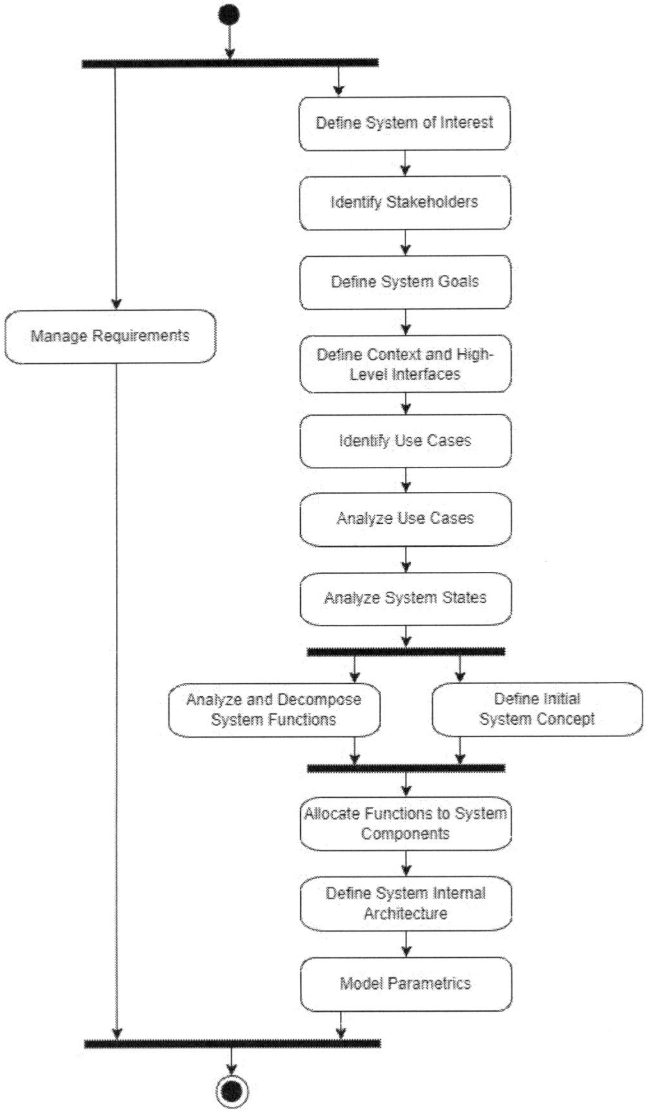

Figure 5.2. MBSE with Papyrus process

This book uses the MWP method to teach the usage of Eclipse Papyrus, but, of course, Papyrus can also be used to model following other methods and the skills acquired by following the modeling instructions in this book can be used when another modeling method is used.

While systems engineering covers the whole system development lifecycle from initial conception to eventual retirement and disposal, the MWP method has a smaller scope focusing on the concept phase of system development. This includes the identification of the specific needs and requirements that the system should address, the definition of the system's scope, functionalities, measures of effectiveness, and constraints, and the preliminary design for the system's architecture, components, and interfaces. The goal of the MWP method is to build a descriptive model. Simulation as well as V&V are out of the scope of the method, although Papyrus can be used to support both.

The MWP method can be separated into two phases, a black box phase and a white box phase. Initially, the system is treated as a black box, focusing on its relation to stakeholders and external systems, its goals, high-level interfaces and the use cases, which a user might have in mind when using the system. This provides a holistic understanding of the system's purpose and capabilities and a solid basis for the actual design of the system. The method steps during the black box phase are:

- **Define System of Interest**: The boundaries of the system of interest are established to clearly define what is included in the system and what is considered external or outside its scope.
- **Identify Stakeholders**: Stakeholders, i.e., individuals, groups, or entities, that have an interest, influence, or concern in the system are identified.
- **Define System Goals**: Measurable goals that the user needs to achieve to fulfill the stakeholder requirements are established.
- **Define Context and High-Level Interfaces**: The context of the system is defined, and the interfaces it has to external units are developed.
- **Identify Use Cases**: Use cases, i.e., interactions between the users or other external entities and the system, are identified.
- **Analyze Use Cases**: The identified use cases are systematically analyzed to detail the interactions between users, the system, and potentially other external systems to achieve a specific goal. The main goal of the use case analysis is to identify system-level functions.
- **Analyze System States**: The system's behavior as it interacts with its environment and responds to various stimuli is defined.

In the next phase, the system is treated as a white box, penetrating the system's inner workings, revealing its internal structure and connections between its components. It dissects the system's elements to understand the mechanisms that drive its behavior. Where the focus of the black-box phase is on understanding the boundary conditions of the system, the white-box phase is focused on design and implementation. The method steps during the white box phase are:

- **Analyze and Decompose System Functions**: The system-level functions identified during the black-box phase are broken down into smaller, more manageable component-level functions.
- **Define Initial System Concept**: The hierarchy of the system is defined. The system components are identified and it is defined how many of each component are used and what the role of each component is in the system architecture.
- **Allocate Functions to System Components**: The component-level functions are allocated to the system components to define how the system components interact to fulfill the system-level functions.
- **Define Internal System Architecture**: The internal architecture of the system is defined that describes how the system's components are connected, to each other and the external interfaces of the system.
- **Model Parametrics**: Constraints of the system are identified and related to the characteristics of the system and its components.

Additionally, the management of requirements is an additional ongoing process step that is concurrent with the rest of the MWP method and is performed throughout the project lifecycle.

- **Manage Requirements**: Textual requirements are identified, elicited and related to other elements of the system to ensure traceability.

The following sections of the book explain all the method's steps in more detail and comprise a step-by-step tutorial that walks readers through the process of creating a SysML model using the Papyrus tool. Note that, as the description of the method progresses, the scope of the description will narrow. While the description of the stakeholder analysis covers a wide range range of system stakeholders, the description of the parametric modeling will only focus on a single aspect of the system. The book exemplifies the method step and widening the scope at each method step is left as an exercise to the reader.

5.1 Manage Requirements

In MBSE, requirements engineering and management are concurrent activities throughout the systems engineering process. This means that requirements are identified, elicited, and analyzed at various stages of the MBSE methodology, and these requirements are reflected in the system models. Requirements management is not a linear process that occurs in isolation at the beginning of the project. The identification and elicitation of requirements is an ongoing process throughout the development lifecycle. This allows for a more iterative and flexible approach to requirements management, as requirements can be identified during the various method steps and later be easily updated as the system design evolves.

- **Requirement identification**: Requirement identification is the process of capturing requirements systematically, giving them a unique name and identifier.
- **Requirement elicitation**: Once requirements are identified, they must be elicited in a clear, unambiguous, and complete statement. Requirement elicitation may also involve evaluating if a requirement is achievable.
- **Traceability establishment**: Traceability links are established to track the relationships between higher-level requirements and their decomposed lower-level requirements and also between requirements and other elements in the model. This ensures that changes at one level are reflected and managed throughout the hierarchy. Traceability is the basis for impact analysis. Changes to requirements can be analyzed for their impact on other requirements, the system functions and architecture, and other aspects captured in the model. Visualization capabilities in MBSE tools enable stakeholders to understand requirements more intuitively, as they can be represented graphically alongside other system model elements to which they are traced.

As requirement management is a continuous activity in the MWP method, identifying new requirements, creating requirements in the model and establishing traceability between requirements and other requirements and model elements are covered in the description of the other method steps.

5.1.1 Requirement Management Pitfalls

There are common pitfalls that teams may encounter during requirement engineering and management. Recognizing these pitfalls is crucial for conducting a thorough and effective analysis. Some common pitfalls in requirement engineering and management include:

- **Unclear or vague requirements**: Unclear or vague requirements provide little direction to engineers, opening up the possibility of misinterpretations and misunderstandings. Ensure that requirements are clear, concise, and unambiguous. Use clear language, define terminology, and provide examples to illustrate requirements. There are many books and guidelines such as the *INCOSE Guide to Writing Requirements* (INCOSE, 2023) that provide guidance for writing clear, concise and unambiguous requirements.
- **Unrealistic or unachievable requirements**: Unrealistic or unattainable requirements can place an excessive burden on the system, often exceeding its capabilities and resources. This can lead to performance bottlenecks, quality compromises, and cost overruns as engineers struggle to meet unrealistic specifications. Moreover, unrealistic requirements can hinder the development process, causing delays and missed deadlines as engineers struggle to reconcile unattainable goals with practical constraints. Assess the feasibility of requirements throughout the development process to identify and address potential unrealistic expectations. This involves conducting trade-off studies, evaluating resource constraints, and involving stakeholders in requirement prioritization.
- **Lack of requirements traceability**: Without a clear linkage between requirements, design elements, and implementation artifacts, it becomes challenging to track the impact of changes throughout the development process. This can lead to costly rework as changes cascade through the system, introducing inconsistencies and missed defects. Moreover, the lack of traceability impedes effective V&V efforts, making it difficult to ensure that the system meets its specified requirements. Establish a clear and consistent traceability mechanism to link requirements to design elements and implementation artifacts. The integration of the requirements engineering and management process into the overall MBSE approach makes it possible to link the requirements directly to other related model artifacts.
- **Ineffective Change Management**: Ineffective change management processes can lead to an uncontrolled influx of changes, often without proper assessment or authorization. This can introduce inconsistencies, compromise the system's integrity, and erode the trust of stakeholders who may perceive the project as disorganized and unreliable. Make sure to assess the impact of a change in a requirement on the rest of the system before you make the change. In a professional environment, implement a structured change management process to handle changes to requirements in a controlled and organized manner. This involves defining change control procedures, assessing the impact of changes, and involving stakeholders in the change approval process.

By addressing these common pitfalls, engineers can ensure that requirements are effectively captured, managed, and validated throughout the system lifecycle, leading to the development of systems that meet stakeholder needs and achieve their intended objectives.

5.2 Define System of Interest

In systems engineering, the *System of Interest (SOI)* refers to the specific, well-defined portion of the world that is the focus of the engineering effort. This system is the primary subject of analysis, design, and optimization within the discipline. The boundaries of the system of interest are established to clearly define what is included in the system and what is considered external or outside its scope. Key characteristics of the system of interest include:

- **Boundaries**: Clearly defined boundaries separate the system of interest from its external environment. These boundaries help determine what is included within the system and what falls outside of it.
- **Components**: The system is composed of various interacting and interrelated components or elements. These components work together to achieve the system's defined objectives or functions.
- **Purpose or objectives**: The system of interest has a specific purpose, objectives, or functions it is designed to fulfill. These objectives guide the engineering efforts to ensure that the system meets its intended goals.
- **Interactions**: The components within the system of interest interact with each other and, in some cases, with elements in the external environment. Understanding these interactions is crucial for effective systems engineering.
- **Interfaces**: The interfaces between the system and its external environment or between internal components are carefully considered. Managing interfaces is essential to ensure proper communication and integration.
- **Environment**: The system of interest exists within a broader context, known as its environment. This environment includes external factors, conditions, and influences that may impact the system's behavior and performance.
- **Constraints**: The system of interest operates within certain constraints, which could include limitations on resources, time, budget, or other factors that may impact its design and operation.
- **Lifecycle**: Systems engineering considers the entire lifecycle of the system, from conceptualization and design to implementation, operation, and eventual retirement or decommissioning.

The definition of the system of interest is a critical step in systems engineering, as it sets the stage for the entire engineering process. Engineers must clearly articulate the boundaries and characteristics of the system to ensure a focused and effective approach to designing, analyzing, and managing the complex interactions within the defined system. The system of interest may vary depending on the specific project or application, ranging from a simple mechanical device to a large-scale socio-technical system.

It is important to understand that the perception and definition of a particular system, its architecture, and its system elements depend on a stakeholder's interests and responsibilities. One stakeholder's SOI can be viewed as a system element in another stakeholder's SOI. Furthermore, an SOI can be viewed as being part of the environment for another stakeholder's SOI.

5.2.1 SysML Blocks

In SysML, blocks are the fundamental building blocks of a system model. They represent physical or conceptual elements that have a well-defined purpose, structure, and behavior. Blocks can be used to model a wide range of system entities, such as:

- **Hardware components**: Blocks can represent physical hardware components such as CPUs, memory modules, and sensors. They can be used to model the structure, behavior, and interactions of these components.
- **Software modules**: Blocks can represent software modules such as classes, functions, and interfaces. They can be used to model the structure, behavior, and dependencies of these modules.
- **Processes**: Blocks can represent processes or workflows within a system. They can be used to model the sequence of activities, data flows, and decision points in a process.
- **Organizations**: Blocks can represent organizations or teams within a system. They can be used to model the structure, roles, and responsibilities of these organizations.
- **Items**: Blocks can represent, very generally, any item - representing a real-world entity or an abstract concept - that is used by the system.

Various relationships can exist between blocks.

- **Composition**: Blocks can be composed of other blocks, which allows the creation of hierarchical structures that represent the organization of a system. Composition is primarily used to model a whole-part relationship between blocks. It indicates that one block is composed of, or includes, other blocks as its parts. The parts are integral components of the whole and have a lifecycle that is dependent on the lifecycle of the whole.

- **Aggregation:** Aggregation is similar to Composition but is a more loose association between blocks, where the aggregating block does not own the sub-blocks. Unlike composition, the destruction of the whole does not necessarily lead to the destruction of its constituents. The constituents in an aggregation relationship can have an independent lifecycle.
- **Generalization and specialization:** Blocks can inherit properties and behaviors from other blocks. This allows the creation of reusable blocks that can be customized to meet the specific needs of a particular system. Inheritance is useful for promoting model element reuse and for ensuring that blocks maintain a consistent structure and behavior. Generalization represents an "is-a" relationship between blocks, indicating that one block is a specialized version of another. In a generalization relationship, the more general block (the superclass or parent) can be defined, and one or more blocks (the subclasses or children) can specialize and inherit characteristics from the general block.

Blocks are a powerful and versatile modeling construct that can be used to represent a wide range of system entities. Their ability to be abstract, encapsulated, composed, and related to each other makes them essential for creating modular, reusable, and analyzable system models.

5.2.2 Model the System of Interest for the Coffee Machine

The first step is to create a new Papyrus project for the model. To do this, open the File menu and select *New->Papyrus Project.*

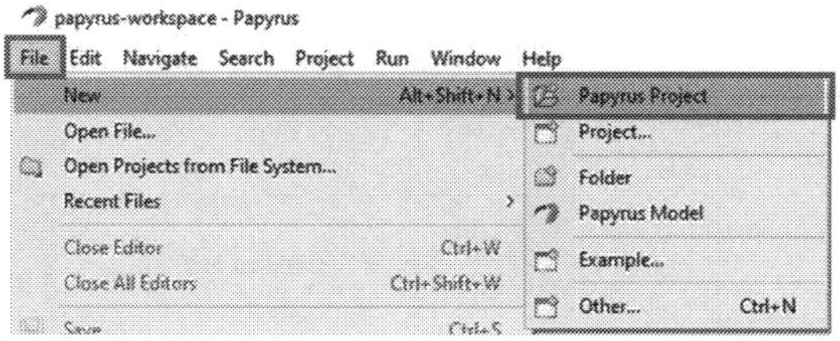

Figure 5.3. Create a new project

To specify that the model uses SysML and not UML as the modeling language, select SysML 1.6 in the architecture context selection dialog.

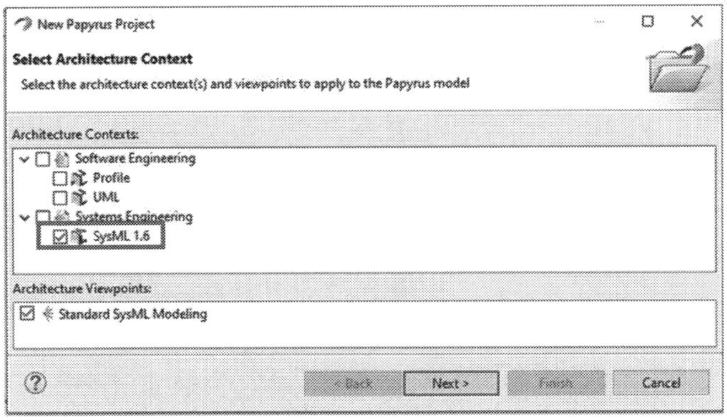

Figure 5.4. Select the SysML 1.6 Architecture Context

Next, provide a name for the modeling project. By default, the project will be created in the current workspace directory (see Appendix B - Installation) but it is possible to provide a different location as well.

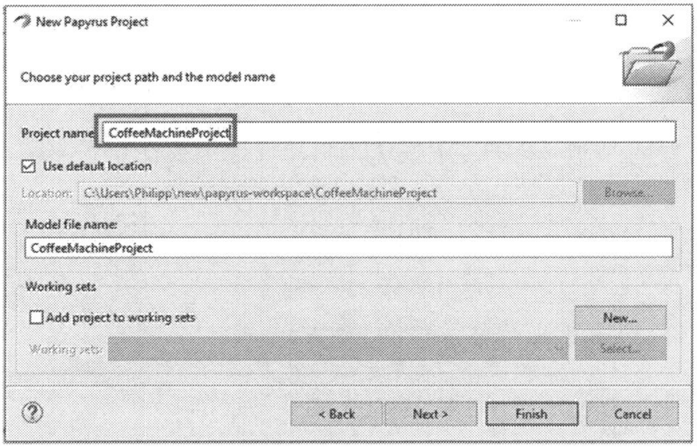

Figure 5.5. Provide the project name

On completion, the new project will be added to the Project Explorer and immediately be opened. Note that the model, as seen in the Model Explorer, is not empty. Some pre-defined packages have already been added to the project based on the selection of the architecture context during project creation.

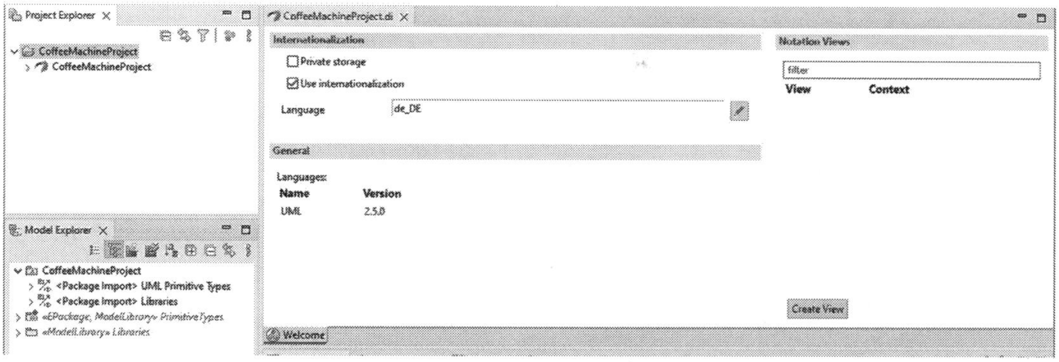

Figure 5.6. The new project has been opened in Papyrus

The MWP profile (see Appendix D - Profiles and Stereotypes) is a customization of SysML that provides some extensions specific to the MWP method used in this book. The MWP profile needs to be added to the workspace to use in a modeling project. To do this, open the File menu and select *Import*. A pre-requisite for the next step is that the MWP profile has been downloaded (see Section 1.1.3 - Downloading the Examples).

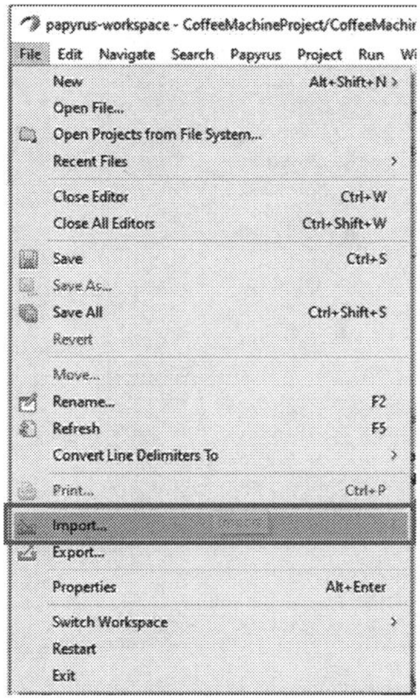

Figure 5.7. Open the "Import" dialog

Then, expand the General section and select Existing Projects Into Workspace.

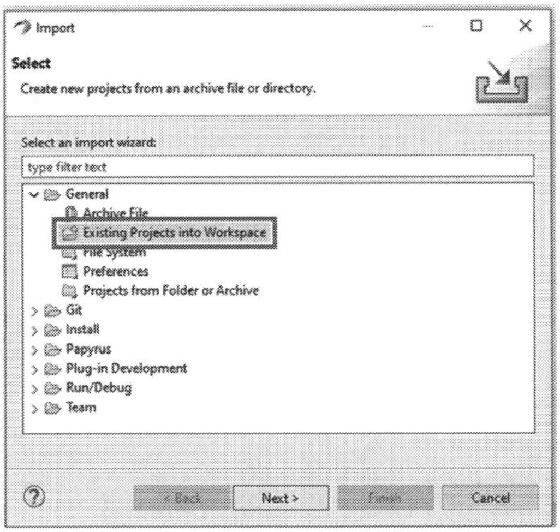

Figure 5.8. Choose the import wizard

Click on Browse next to Select Root Directory, to open a dialog for proving the directory where the MWP profile was stored.

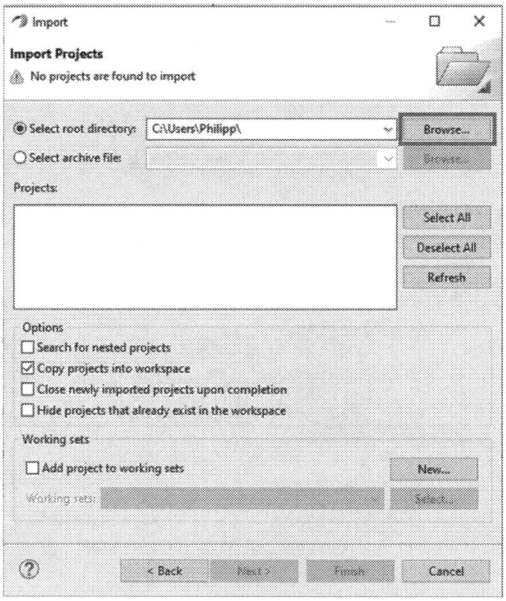

Figure 5.9. Import the MWP profile into the new project

Provide the location of the MWP profile.

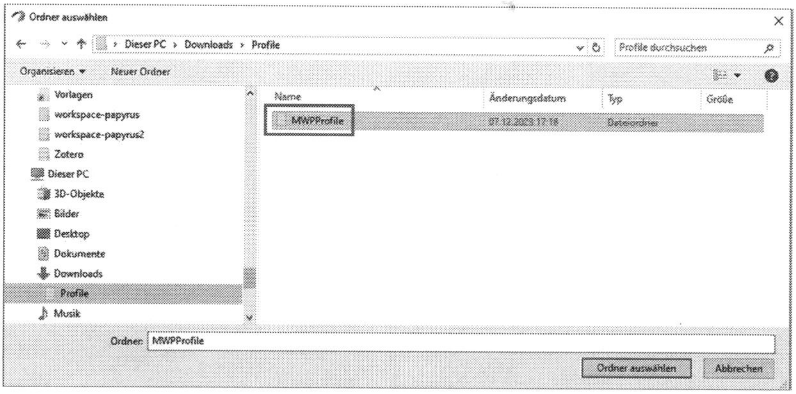

Figure 5.10. Select the MWPProfile project

Back in the "Import Projects" dialog, make sure that MWPProfile is selected in the Projects section.

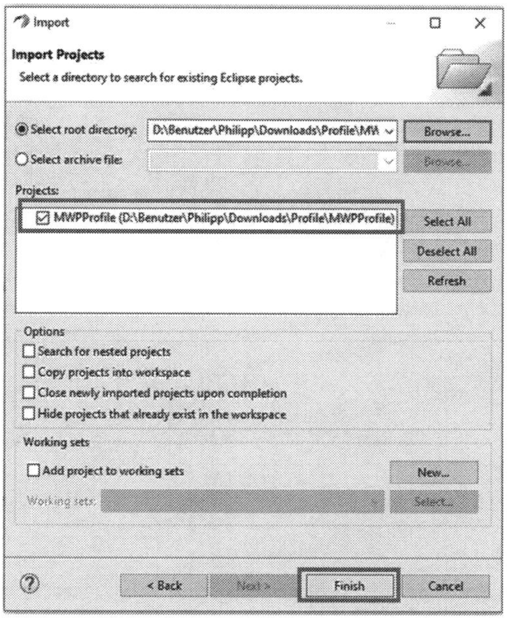

Figure 5.11. Finish importing the MWP profile into the new project

Once the import is finished, the MPW profile is added to the Project Explorer.

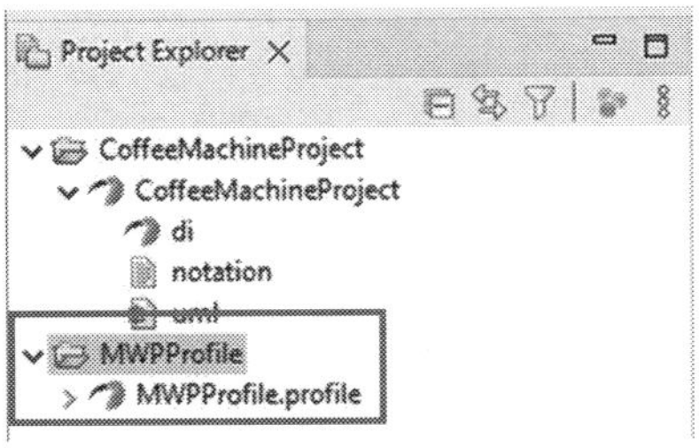

Figure 5.12. MWP profile in the Project Explorer after import

To use a profile, it needs to be applied to a modeling project. This makes all the stereotypes defined by the profile accessible from the modeling project. To achieve this, select the "CoffeeMachineProject" element, the root element in the Model Explorer, and in the Properties View select the Profile tab. The section Profile Applications shows that the SysML 1.6 profile has already been applied to the project. Then click on the button with the plus symbol.

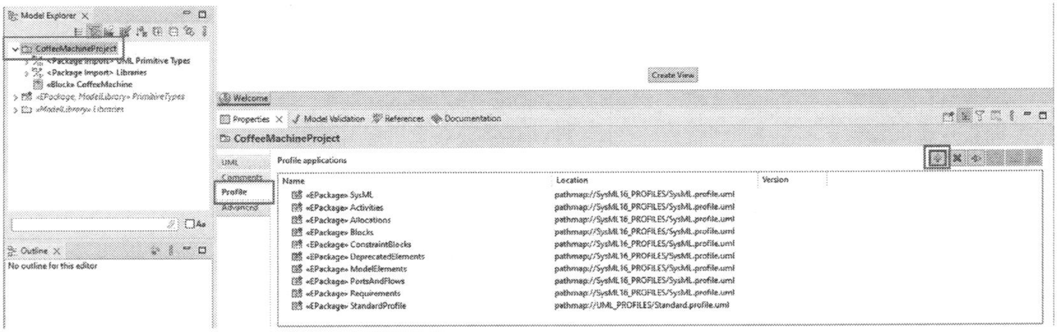

Figure 5.13. Apply MWP profile to the Coffee Machine Project

Select the MWP profile, in the form of the "MWPProfile.profile.uml" file, and move it from left to right by clicking on the button with the single arrow pointing from left to right. Close the dialog by clicking on OK.

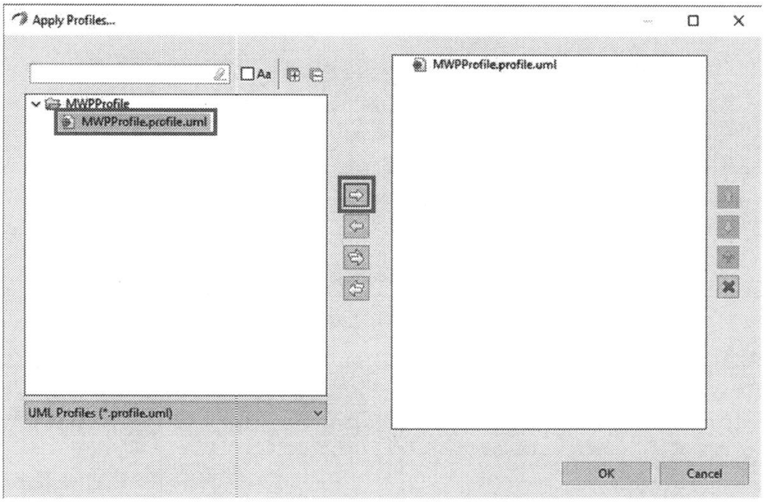

Figure 5.14. Apply MWP profile to the Coffee Machine Project

The MWP profile will be shown next to the SysML 1.6 profile in the section Profile Applications.

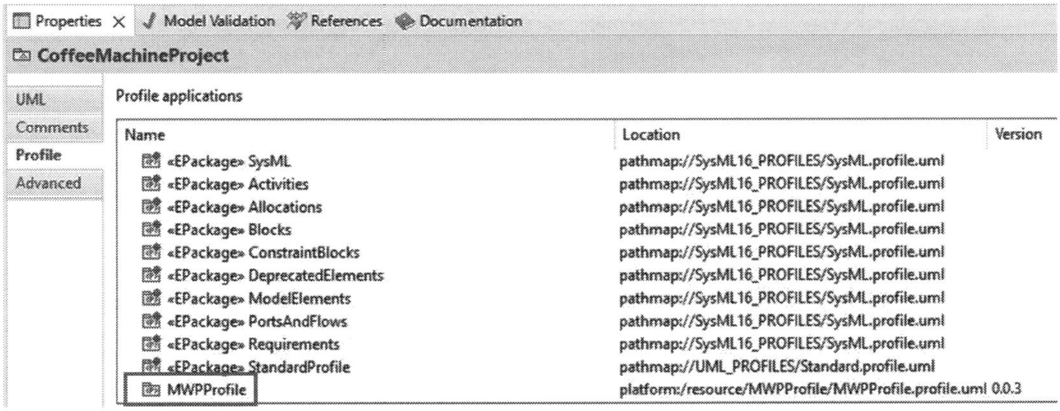

Figure 5.15. MWP profile applied to the Coffee Machine Project

The first step, in creating actual model content, is to create a block that represents the coffee machine as the system of interest. To create a new block, right-click where it should be created, in this case in the CoffeeMachineProject, then select *SysML 1.6 Child->Block*.

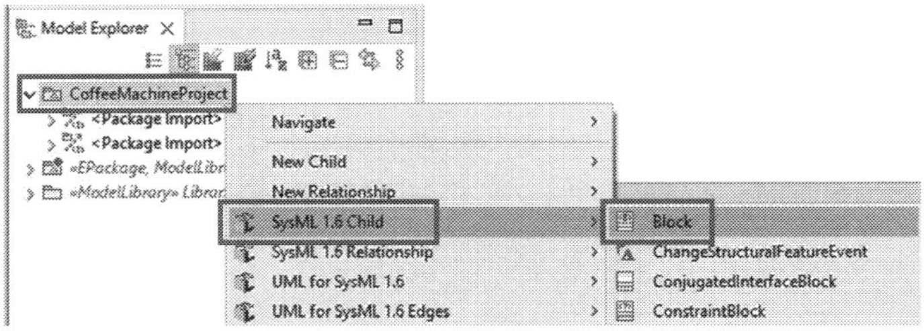

Figure 5.16. Create a new block

Name the block "CoffeeMachine". To mark this block as the system of interest, the MWP profile provides a stereotype "SystemOfInterest" that can be added to blocks. To add the stereotype, select the block "CoffeeMachine", go to the Profile tab in the Properties View, and click on the button with the plus symbol.

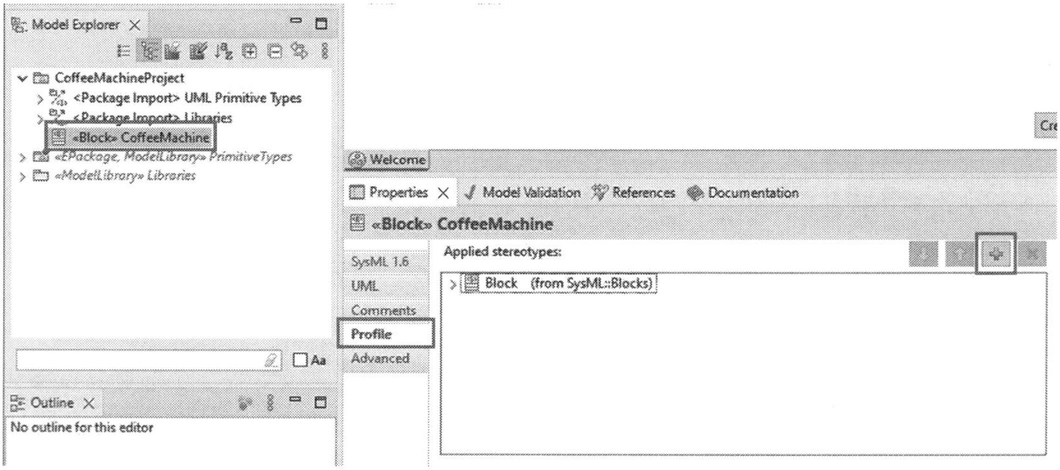

Figure 5.17. Add a new stereotype

In the "Apply Stereotypes" dialog select the "SystemOfInterest" stereotype and move it from left to right by clicking on the button with the single arrow pointing from left to right.

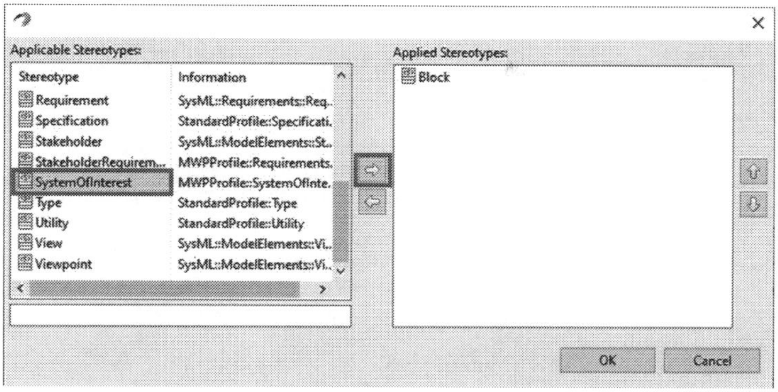

Figure 5.18. Select the new stereotype

The stereotype "SystemOfInterest" will be added to the list of applied stereotypes.

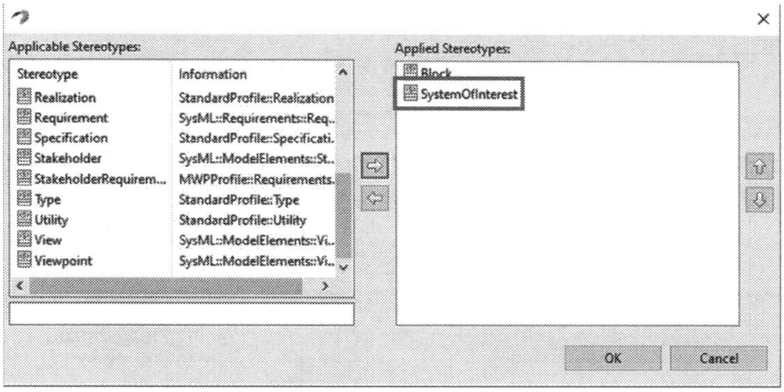

Figure 5.19. New stereotype in the list of applied stereotypes

In the MWP profile, the "SystemOfInterest" stereotype is a specialization of the "Block" stereotype (see Appendix D - Profiles and Stereotypes). Therefore keeping the "Block" stereotype is redundant. However, if the "Block" stereotype is removed from the "CoffeeMachine" block then Papyrus does not recognize the element correctly anymore and prevents the user from adding an internal block diagram to it.

The newly applied stereotype is now shown in the list of Applied stereotypes in the Profile tab in the Properties View. It is also shown in the Model Explorer.

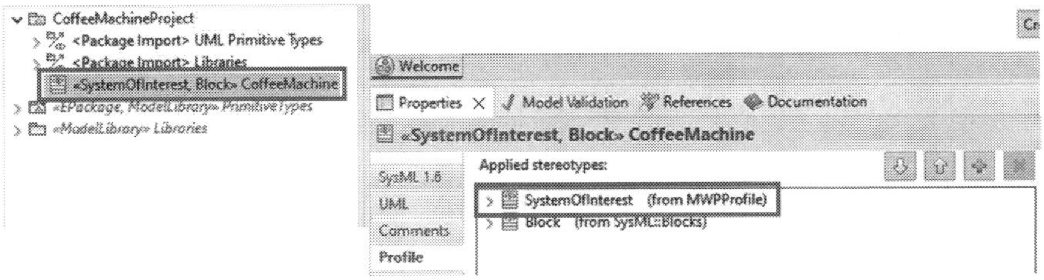

Figure 5.20. The new stereotype has been applied

5.3 Identify Stakeholders

A stakeholder is an individual, group, or entity that has an interest or concern in the development, operation, or outcomes of a system. Stakeholders can include a diverse range of entities, such as end-users, customers, project managers, regulatory bodies, suppliers, operators, maintainers, and other entities directly or indirectly affected by the system.

Stakeholders play a crucial role in influencing and being influenced by the system under consideration. The identification, engagement, and management of stakeholders are integral components of systems engineering, as they help ensure that the system meets its intended purpose and satisfies the needs and expectations of those involved or affected.

Successful projects depend on meeting the needs and requirements of the stakeholders throughout the life cycle. The stakeholder requirements are the primary reference against which the operational capability is validated.

Stakeholders have varying levels of interest and influence in the system. Their interests may be financial, operational, regulatory, or strategic. The degree of influence a stakeholder holds can significantly impact decisions related to the system. Stakeholders have expectations and needs that the system is designed to address. These expectations can encompass functionality, performance, reliability, safety, regulatory compliance, cost-effectiveness, and other relevant aspects.

Stakeholders can assume different roles and responsibilities throughout the systems engineering life cycle. For example, end-users may focus on usability and user experience, while regulatory bodies may be concerned with compliance and safety. When direct contact is not possible, systems engineers can find agents to represent the concerns of a class of stakeholders.

Identifying stakeholders is a critical step in the early stages of a project or initiative. It involves recognizing individuals, groups, or entities that have an interest, influence, or concern in the project's outcomes. Several methods can be employed to identify stakeholders in different contexts. Methods for stakeholder identification include:

- **Stakeholder identification workshops**: Conducting workshops with project team members, subject matter experts, and key decision-makers can be an effective way to brainstorm and identify stakeholders. These workshops encourage open discussions, knowledge sharing, and the generation of comprehensive stakeholder lists.
- **Document analysis**: Reviewing existing documentation, project plans, reports, and organizational charts can provide insights into potential stakeholders. Analyzing documents related to the project or the organization can help identify key individuals or groups that may be impacted by or have an impact on the project.
- **Surveys and questionnaires**: Distributing surveys or questionnaires to internal and external parties can help gather information about potential stakeholders. Questions may focus on roles, responsibilities, expectations, and levels of interest in the project. This method is particularly useful for larger projects or when dealing with a geographically dispersed audience.
- **Interviews**: Conducting one-on-one interviews with key individuals or representatives from different departments can provide in-depth insights into stakeholder concerns and expectations. Interviews allow for more personalized discussions and the opportunity to uncover stakeholders who may not be immediately apparent.
- **Focus groups**: Organizing focus group sessions with representatives from various stakeholder groups can facilitate open discussions about project impacts and expectations. This method encourages collaboration and helps identify stakeholders based on shared concerns or interests.
- **Reviewing organizational policies and procedures**: Examining organizational policies, procedures, and guidelines can reveal stakeholders involved in decision-making processes. This method is particularly useful for understanding formal roles and responsibilities within the organization.
- **Expert judgment**: Seeking input from subject matter experts or individuals with experience in similar projects can provide valuable insights into potential stakeholders. Experts can contribute their knowledge of the industry, project domain, or organizational dynamics to identify relevant stakeholders.
- **Publicly available information**: Utilizing publicly available information, such as press releases, news articles, or public records, can help identify stakeholders external to the organization. This method is particularly useful for projects with a public or regulatory dimension.
- **Brainstorming sessions**: Organizing brainstorming sessions with project team members and key stakeholders can be an informal and creative way to generate a com-

prehensive list of potential stakeholders. This method encourages free thinking and exploration of different perspectives.

It is often beneficial to use a combination of these methods to ensure a thorough and accurate identification of stakeholders. The stakeholder identification process is iterative, and as the project progresses, new stakeholders may emerge or existing ones may change their level of influence or interest. Regularly revisiting and updating the stakeholder list is important throughout the project life cycle.

Additional tips for the stakeholder analysis:

- Every human actor that interacts with the system during operation is a stakeholder.
- Not every stakeholder is an actor. Many stakeholders do not interact with the system but still provide very important requirements.
- The types of stakeholder and actor roles that exist depend heavily on the system and its marketing and operating concept.
- If a stakeholder does not have any requirements for the system, it is not a stakeholder for the system of interest.
- If different stakeholders belong to the same organization, their relation should be noted. The organization is the general stakeholder, and the stakeholders contained within it that belong to the organization are specializations.

5.3.1 Stakeholder Analysis Pitfalls

Stakeholder analysis is a critical step in the systems engineering process, helping identify and understand the individuals or groups who can significantly influence or be influenced by a system. However, there are common pitfalls that teams may encounter during stakeholder analysis. Recognizing these pitfalls is crucial for conducting a thorough and effective analysis. Some common pitfalls in stakeholder analysis include:

- **Incomplete stakeholder identification**: If not all stakeholders are identified, their needs, expectations, and potential impacts on the system may be overlooked, leading to incomplete requirements and potential conflicts later in the project. To avoid this, a variety of techniques, such as stakeholder brainstorming sessions, surveys, and interviews, should be used to ensure a comprehensive list.
- **Superficial analysis**: Superficial analysis may result in a limited understanding of stakeholders' concerns, potentially leading to inadequate system design and failure to address critical issues. To mitigate this, thorough interviews, surveys, and workshops to gain deeper insights into stakeholders' interests and concerns should be conducted.

Qualitative and quantitative analysis methods should be used to understand the depth of their influence.

- **Static analysis**: Stakeholder dynamics can change throughout the project lifecycle. Failing to revisit and update the stakeholder analysis can result in outdated information and missed opportunities for engagement. Regular reviews of the stakeholder analysis throughout the project lifecycle should be scheduled.
- **Ignoring internal stakeholders**: Internal stakeholders can significantly impact the success of a project. Neglecting their perspectives may lead to internal conflicts and hinder collaboration.
- **Stereotyping stakeholders**: It should not be assumed that stakeholders within a particular category share identical interests and concerns. Each stakeholder is unique, and making assumptions based on stereotypes can lead to misunderstandings and misalignments with their actual needs and expectations.
- **Excluding end-users**: Neglecting the perspectives of end-users or ultimate beneficiaries of the system. End-users play a crucial role in system acceptance and success. Excluding their input may result in a system that does not meet their practical needs.
- **Overlooking negative stakeholders**: It is easy to focus only on positive or supportive stakeholders and ignore those with potential objections or resistance. Negative stakeholders can pose risks to the project. Understanding their concerns allows for proactive management and mitigation of potential issues.

Regular reviews, ongoing communication, and flexibility in adapting to changing stakeholder dynamics are essential for effective stakeholder management throughout the systems engineering process.

Initiated in 2002, the NHS National Programme for IT in the United Kingdom (UK) was a massive government-led initiative aimed at modernizing and digitizing healthcare information systems across the National Health Service (NHS) in England. The project had ambitious goals, including the development of a centralized electronic health record system, a national broadband network for healthcare, and other IT infrastructure improvements. Clinicians and healthcare professionals were vocal about their concerns regarding the usability and functionality of the new systems. Their feedback on critical issues, such as the design of the electronic health record system and its impact on patient care, was not adequately considered. The lack of attention to clinician concerns resulted in a system that did not effectively integrate with existing clinical workflows. Healthcare professionals were burdened with systems that did not enhance patient care and, in some cases, disrupted established practices.

The NHS National Programme for IT faced severe consequences, including substantial delays, escalating costs, and, ultimately, project failure. In 2011, after spending billions of pounds, the UK government decided to dismantle the NPfIT, acknowledging that the project did not deliver the expected benefits and that its approach had been flawed from the outset. The failure of the NPfIT underscores the critical importance of engaging and considering the perspectives of all relevant stakeholders in a development project. The failure to involve healthcare professionals, clinicians, and patients in the planning and decision-making processes resulted in a system that did not meet the real-world needs of those it was intended to serve (Campion-Awwad, Hayton, Smith and Vuaran, 2014).

5.3.2 Stakeholders for the Coffee Machine

A comprehensive stakeholder analysis using a combination of the methods listed above resulted in the following list of relevant Stakeholders for the coffee machine:

User
Description: Individuals who interact with and use the coffee machine.
Interests: User-friendly interface, brewing options, reliability, and ease of maintenance.

Producer
Description: Companies or individuals responsible for designing, manufacturing, and developing the coffee machine.
Interests: Product design, performance, manufacturing efficiency, and cost-effectiveness.

Product Designer
Description: Professionals involved in the design and aesthetics of the coffee machine.
Interests: Ergonomics, user experience, and visual appeal.

Product Engineer
Description: Technical experts responsible for the engineering aspects of the coffee machine.
Interests: Technical specifications, functionality, and system performance.

Marketing
Description: Professionals involved in the marketing of the coffee machine.
Interests: Marketability, design, and unique selling points.

Quality Assurance
Description: Individuals or teams responsible for testing and ensuring the quality of the

coffee machine.
Interests: Identifying and rectifying defects, ensuring compliance with standards.

Customer Support
Description: Teams providing assistance to end users regarding the coffee machine.
Interests: Addressing user queries, providing technical support, and handling warranty or maintenance issues.

Distributors
Description: Entities involved in the distribution and sale of the coffee machine.
Interests: Marketing, sales, and customer satisfaction.

Regulatory Authorities
Description: Government agencies responsible for regulating household appliances.
Interests: Compliance with safety, environmental, and quality standards.

Environmental Organizations
Description: Groups focused on environmental conservation and sustainability.
Interests: Compliance with eco-friendly practices, and product lifecycle considerations.

Suppliers
Description: Companies or individuals providing components or materials for the coffee machine.
Interests: Quality of supplied materials, timely deliveries, and cost considerations.

Competitors
Description: Other companies that are producing similar coffee machines.
Interests: Market competition, product differentiators, and industry trends.

Understanding the perspectives and interests of these stakeholders is essential for developing a coffee machine that meets the diverse needs and expectations of the individuals and entities involved in its lifecycle.

5.3.3 Modeling the Coffee Machine Stakeholders

Create a new package that will contain all the stakeholders. To create a new package, right-click on the "CoffeeMachineProject" element in the Model Explorer and select *UML for SysML 1.6->Package.*

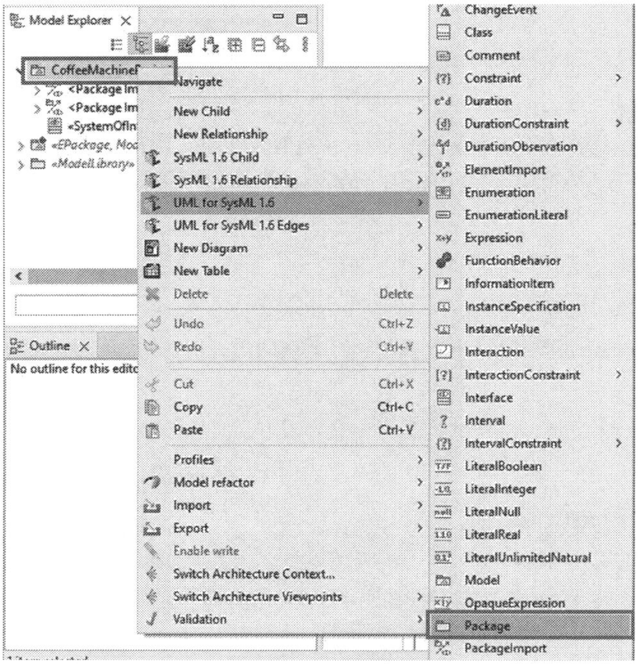

Figure 5.21. Create a new package

Then name that new package element "Stakeholder".

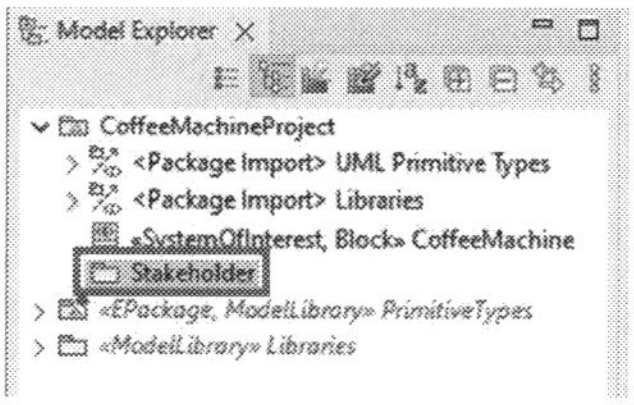

Figure 5.22. The new package in the Model Explorer

Next, add a stakeholder element to the newly created "Stakeholder" package, by right-clicking on the package and selecting *SysML 1.6 Child->Stakeholder.*

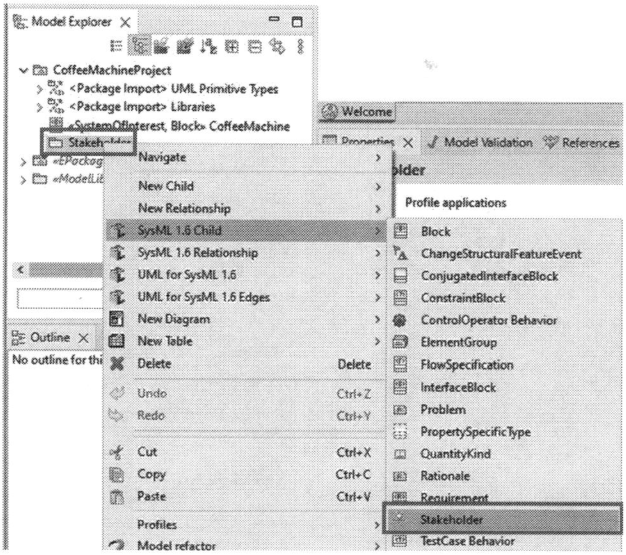

Figure 5.23. Create a new stakeholder element

To create a new element, we sometimes have to select *SysML 1.6 Child* and sometimes *UML for SysML 1.6*. The reason for that is the relation between UML and SysML as discussed in Chapter 3. If an element belongs to the UML4SysML subset, it will be found in the sub-menu UML for SysML 1.6, and if an element belongs to the SysML profile, then it will be found in the *SysML 1.6 Child* sub-menu.

A new dialog will appear, forcing you to select the concrete metaclass for the new stakeholder element.

This is necessary because the stereotype Stakeholder in the SysML 1.6 profile has Classifier as a base class. Classifier, however, is an abstract class. So, when a new stakeholder element is created it is necessary to specify the non-abstract sub-class of Classifier that is the base class for the new element.

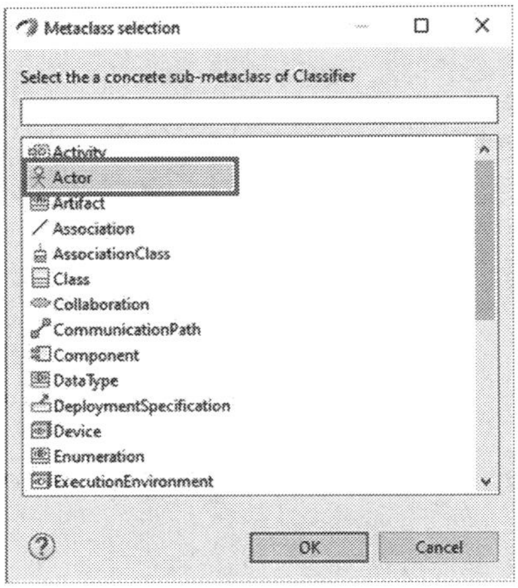

Figure 5.24. Select the metaclass "Actor"

Name that new stakeholder element "User".

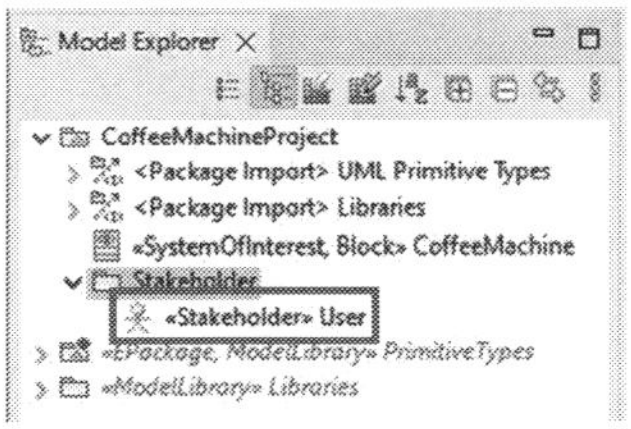

Figure 5.25. The new stakeholder in the Model Explorer

Apply the stereotype "ExtendedStakeholder" from the MWP profile to the new stakeholder called "User".

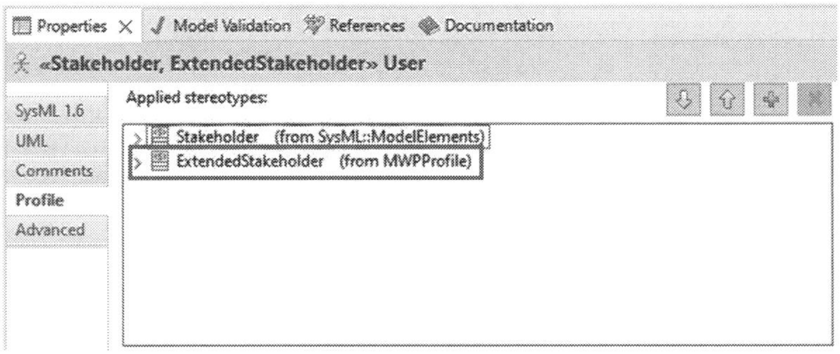

Figure 5.26. Applied stereotype "ExtendedStakeholder"

The “ExtendedStakeholder” stereotype is a specialization of the SysML stakeholder element and allows setting the additional properties for a stakeholder that is stereotyped as “ExtendedStakeholder”: relevance and interests. To set the values for the properties that are defined by a stereotype, expand the stereotype in the applied stereotypes section of the profile tab in the Properties View on an element, and select the property that you want to set.

Since “relevance” is typed by an enumeration with three values, LOW, MID, and HIGH, Papyrus will provide a drop-down list to select the value.

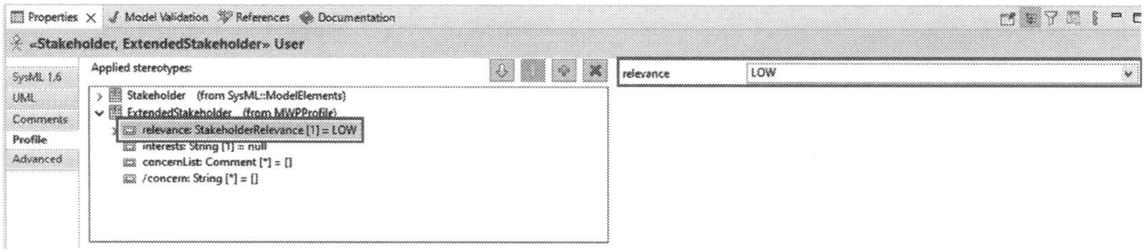

Figure 5.27. Set stakeholder relevance

Since “interests” is typed as a string, Papyrus will provide a text area to define the value.

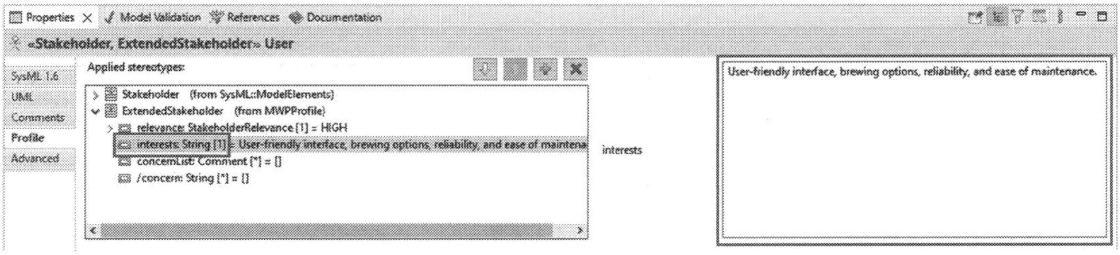

Figure 5.28. Set stakeholder interests

A textual description for any model element can be added by selecting the documentation view of an element.

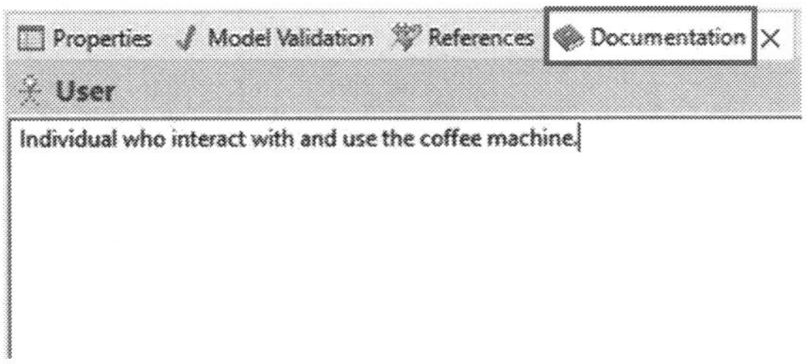

Figure 5.29. Provide textual description

The documentation will be added as a new model element in the form of a comment, that is owned by the element for which the documentation was added.

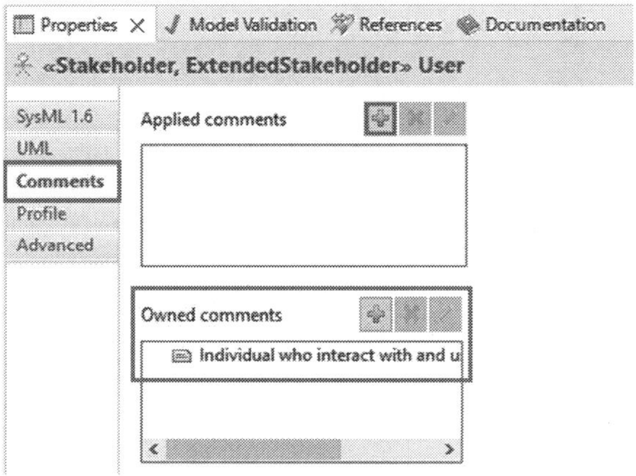

Figure 5.30. The documentation is a comment element

Next, we want to display the new “User” stakeholder on a diagram. For this, we choose a SysML Use Case diagram.

Normally, to display classifiers and their relations, Block Definition Diagrams are used. However, Papyrus does not display the names of actors and stakeholder elements correctly on BDDs, so a Use Case diagram is used.

To create a new use case diagram, right-click on the “Stakeholder” package and select *New Diagram->SysML 1.6 Use Case Diagram.*

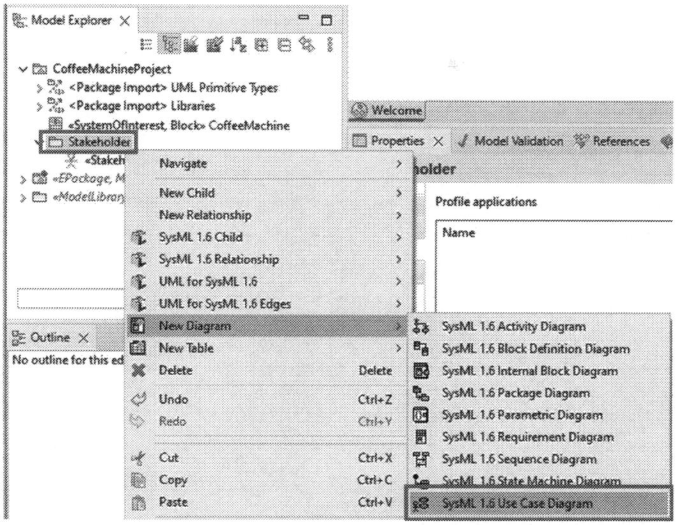

Figure 5.31. Create a use case diagram

Name the diagram "Stakeholder UD" (see Appendix A - Modeling Guidelines) in the dialog that appears.

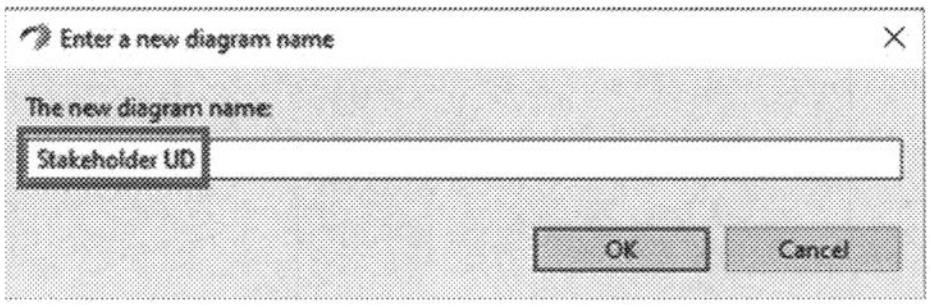

Figure 5.32. Name the use case diagram

The new diagram is added to the "Stakeholder" package and automatically opens in the View Area of Papyrus.

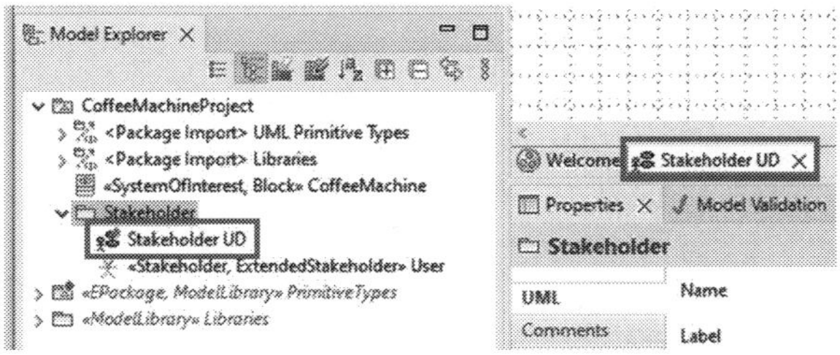

Figure 5.33. The use case diagram is added to the Model Explorer and opened in the View Area

Note the grid, which Papyrus shows on the new diagram. While it is useful for aligning elements on diagrams, it makes the diagrams harder to read. To toggle the displaying of the grid on a per-diagram basis open the "Papyrus" menu and select or deselect *Diagram->View->Grid.*

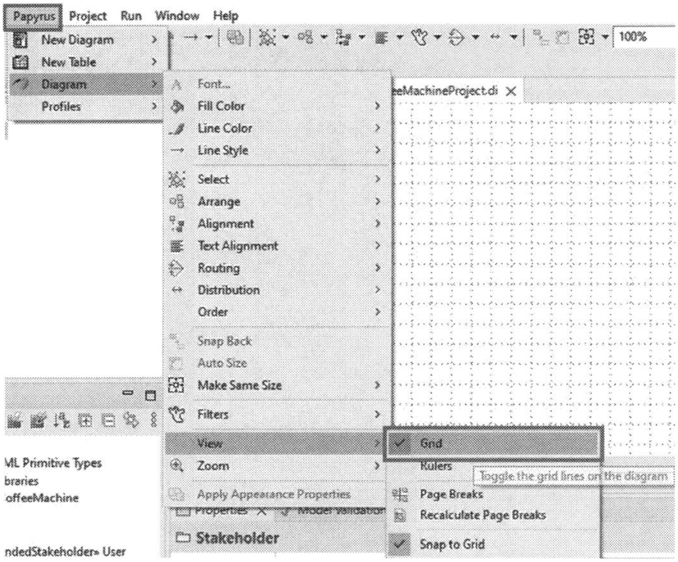

Figure 5.34. Toggle grid display on a per-diagram basis

To toggle the displaying of the grid for all newly created diagrams open the Papyrus preferences and select or deselect the "Show grid for new diagrams" option in the Diagrams settings.

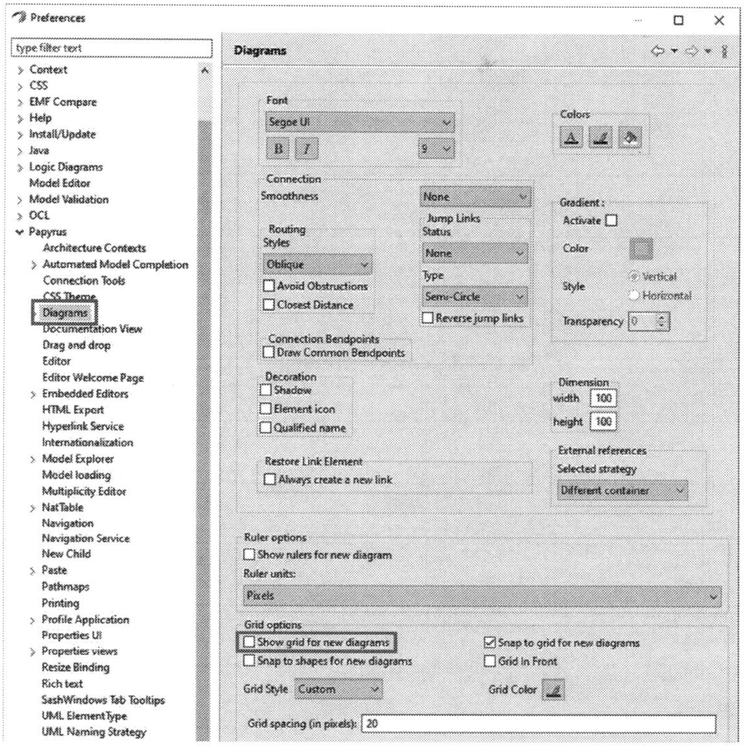

Figure 5.35. Toggle grid display default setting

To add the “User” stakeholder to the diagram, simply drag and drop it from the Model Explorer onto the diagram.

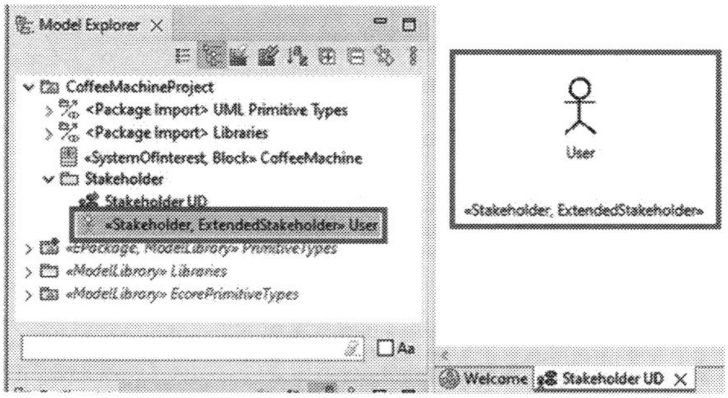

Figure 5.36. The “User” stakeholder on the diagram

Repeat these steps for creating and specifying new stakeholders for all the stakeholders that have been identified.

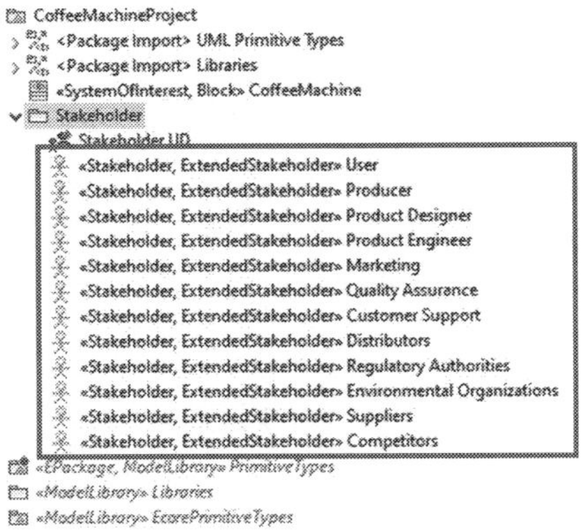

Figure 5.37. Complete list of stakeholders in the Model Explorer

Then, add them to the "Stakeholder UD" diagram.

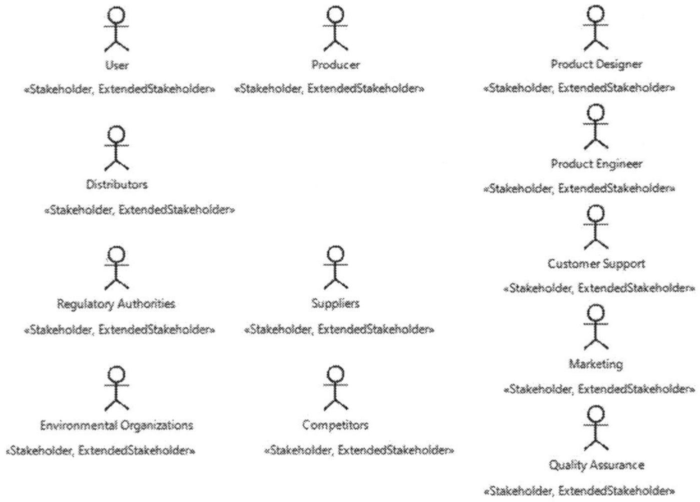

Figure 5.38. All the stakeholders on the diagram

In the next step, we add relations between the stakeholders. More specifically, we want to indicate that "Product Designer", "Product Engineer", "Marketing", "Quality Assurance", and "Customer Support" are a subtype of "Producer". They share concerns since they are all working together for the company that is producing the coffee machine. To model this kind of relationship we use a generalization relation.

Generalization is the process of identifying common characteristics and behaviors among different elements and representing them as a more abstract, generalized element called a supertype. This abstract representation captures the shared features, enabling reuse and reducing redundancy in the model. Specific elements that inherit from the supertype are called subtypes. They inherit the common features of their supertype and may also have additional unique features.

To create the generalization relationship between the "Product Designer" and the "Producer", select Generalization from the use case Diagram Palette, first left-click on the "Product Designer", then move the mouse onto the "Producer" and click again.

In SysML, generalization is represented using a directed line connecting the subtype to its supertype. The line at the supertype end has a closed arrow shape without a filling.

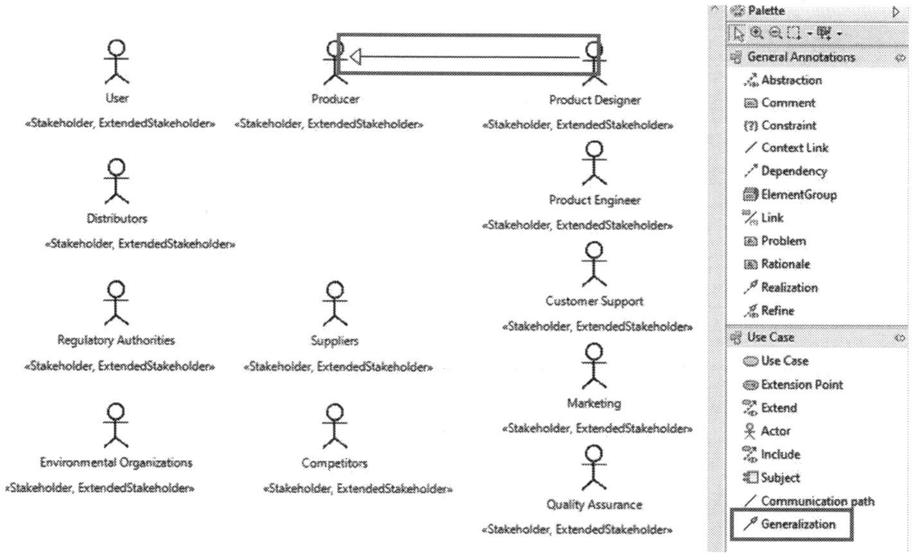

Figure 5.39. Modeling a generalization

Repeat the same for all the other stakeholders that are subtypes of the "Producer" stakeholder. Once all the relations are added, the stakeholder diagram is finished.

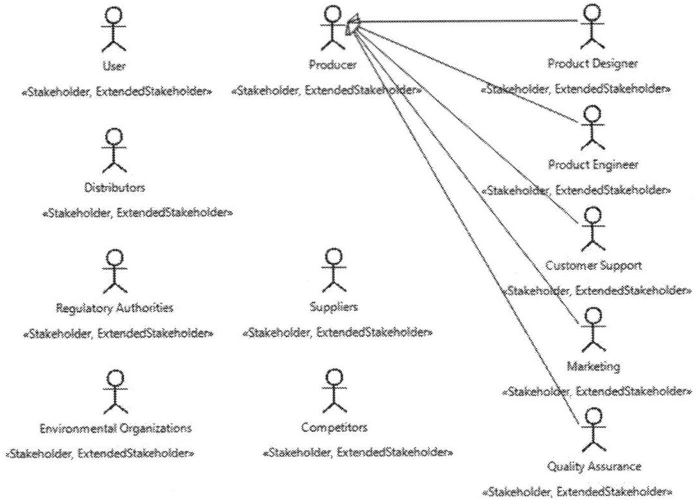

Figure 5.40. Finished stakeholder diagram

The generalization is not just an arrow on the diagram. In the Model Explorer, it can be seen that the generalization relation is a model element in its own right that belongs to the specialized elements.

Figure 5.41. Generalizations in the Model Explorer

After creating all the stakeholders and modeling the relations between them, the next step is to add them to a table. Tables are a fundamental feature of SysML, providing a versatile, structured, and organized way to represent information. They provide a clear and concise format for presenting structured data, making it easier to understand, analyze, and manage complex information.

To create a table for the stakeholders, right-click on the "Stakeholder" package and select *New Table->Generic Table.*

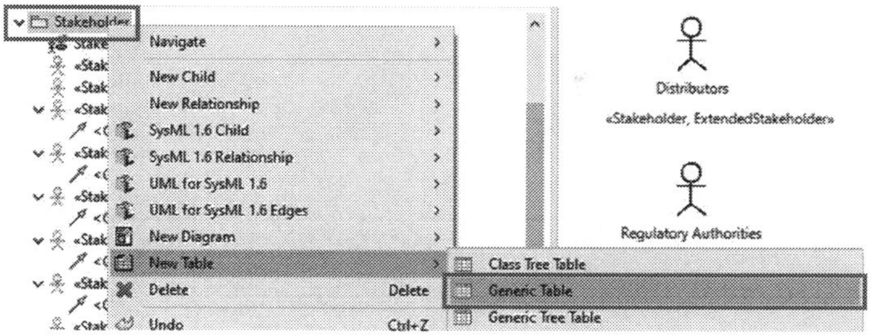

Figure 5.42. Create a generic table

In the next dialog, name the table "Stakeholder Table".

Figure 5.43. Name the table

The new table is added to the "Stakeholder" package and automatically opens in the View Area of Papyrus. The newly created table is empty.

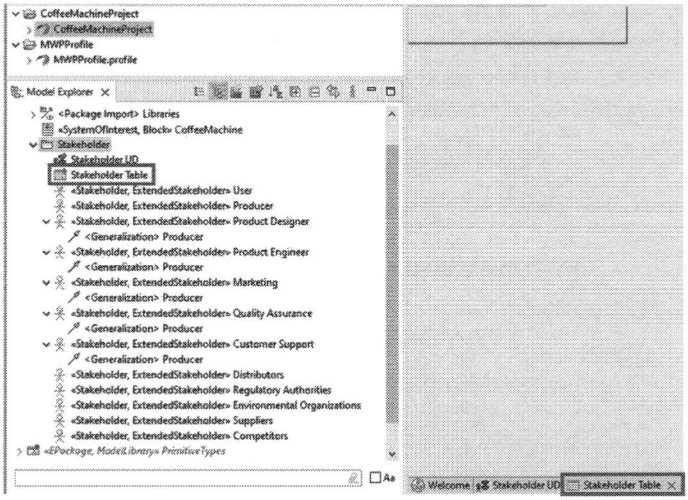

Figure 5.44. The table is added to the Model Explorer and opened in the View Area

To add a stakeholder to the table, drag and drop the "User" stakeholder from the Model

Explorer onto the empty table. Papyrus will automatically add all the properties that the newly added element has, as new columns to the table. Since most of the properties are empty and not relevant to our stakeholder table, the table needs to be customized.

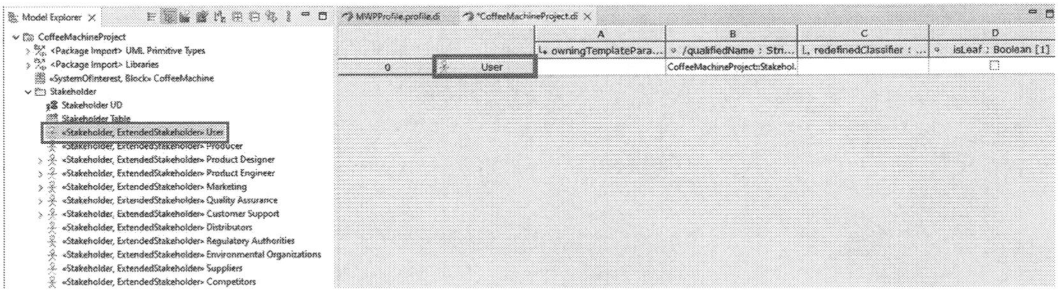

Figure 5.45. Add the "User" stakeholder to the table

The appearance tab in the Properties View of the table enables the user to define, how the columns and rows are presented in a table.

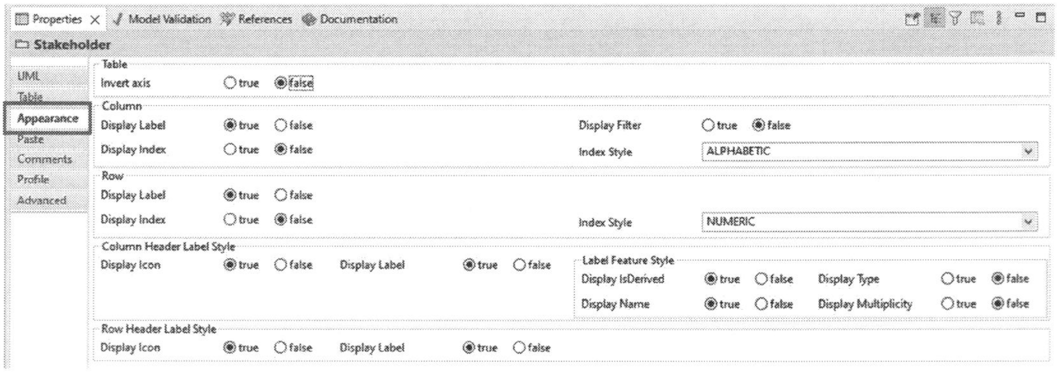

Figure 5.46. Set the display options for the table

The appearance settings in the screenshot above remove superfluous information, like the numbering of the rows and the columns from the table.

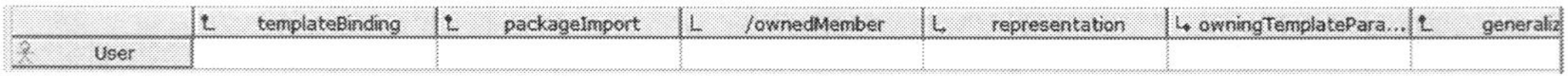

Figure 5.47. The table after changing the display options

To specify, which properties of the row elements should be shown in the table as columns, right-click on the table and select *Columns->Select/Destroy Columns.*

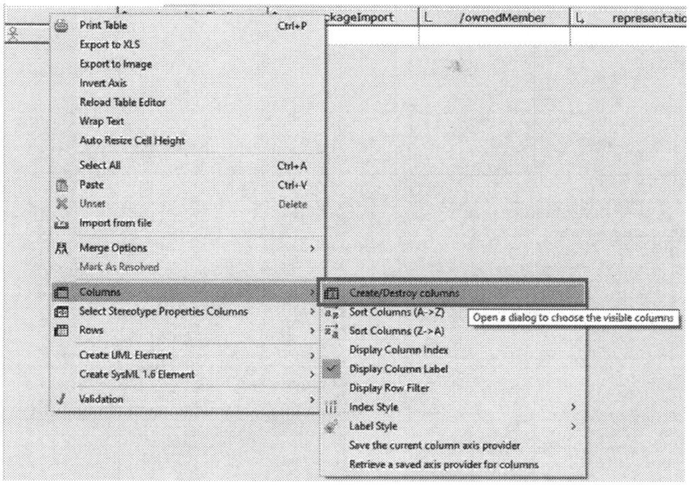

Figure 5.48. Open the “Select Columns” dialog

The “Select Columns” dialog has all available properties on the left side and all the display columns on the right side. Since only a few columns are needed for the stakeholder table, the first step is to remove all the currently selected properties by clicking on the double arrows pointing from right to left.

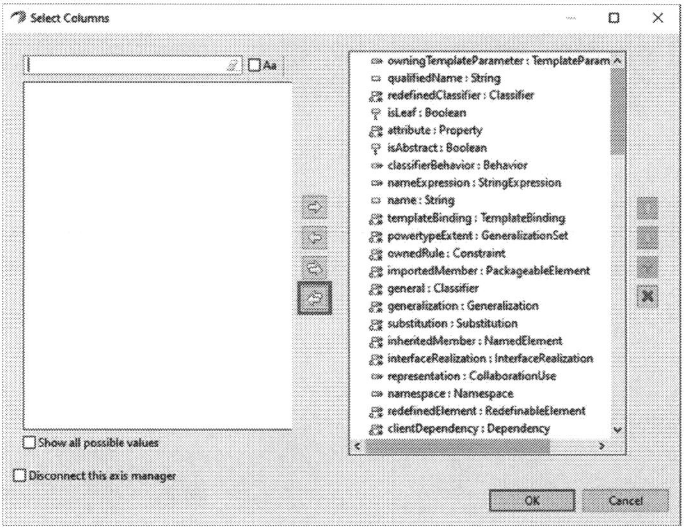

Figure 5.49. Remove all currently selected properties

Then, the properties that should be shown as columns can be selected on the left side and individually moved to the right side by clicking on the single arrow pointing from left to right.

Papyrus automatically analyzes the model elements that are currently displayed in the table and shows all their properties. Since the "User" stakeholder is stereotyped as an "ExtendedStakeholder", the properties "relevance" and "interests" of the "ExtendedStakeholder" stereotype are available for display.

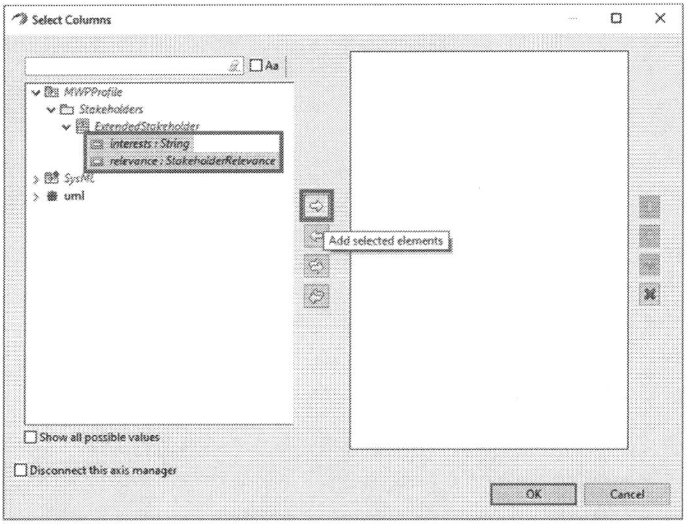

Figure 5.50. Move the properties that should appear in the table from left to right

Complement the "relevance" and "interests" properties from the "ExtendedStakeholder" stereotype with the properties "name", "allParents" and "ownedComment" from the actor metaclass.

Setting the "Disconnect the axis manager" property in this dialog ensures that the selected properties will not be changed when a new element is added to the table. If this option is not selected, then adding a new element to the table will add all of the properties of the newly added element to the table as columns again.

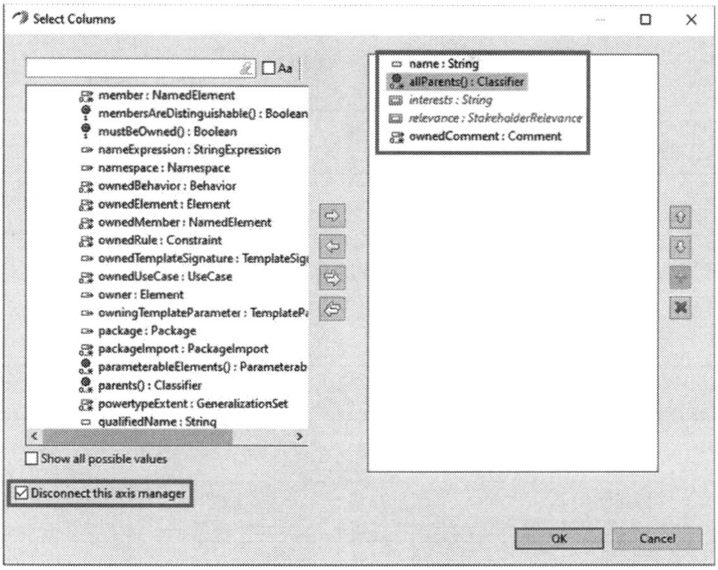

Figure 5.51. Finished column selection

Then, add the other stakeholders to the same table by drag and drop. The finished table provides a nice overview of all the stakeholders in the model.

	name	Generalizations	interests	relevance	Documentation
User	User		User-friendly interface, brewing .	HIGH	Individual who interact with and use th.
Producer	Producer		Product design, performance, ...	HIGH	Company or individuals responsible fo..
Product Designer	Product Designer	Producer	Ergonomics, user experience, a...	MID	Professionals involved in the design an..
Product Engineer	Product Engineer	Producer	Technical specifications, functio..	MID	Technical experts responsible for the e...
Marketing	Marketing	Producer	Identifying and rectifying defect.	LOW	Individuals or teams responsible for m...
Quality Assurance	Quality Assurance	Producer	Identifying and rectifying defect.	LOW	Individuals or teams responsible for tes.
Customer Support	Customer Support	Producer	Addressing user queries, providi..	MID	Teams providing assistance to end user.
Distributors	Distributors		Entities involved in the distributi.	HIGH	Entities involved in the distribution an...
Regulatory Authorities	Regulatory Authorities		Compliance with safety, enviro...	HIGH	Government agencies responsible for r..
Environmental Organizations	Environmental Organizations		Compliance with eco-friendly p..	LOW	Groups focused on environmental con..
Suppliers	Suppliers		Quality of supplied materials, ti...	MID	Companies or individuals providing co..
Competitors	Competitors		Market competition, product di...	LOW	Other companies producing similar co..

Figure 5.52. Finished stakeholder table

Note that tables do not provide a static view of the model. They can be used to interact with the model and make changes directly. Double-click on a property and the value of the property can be changed in the table.

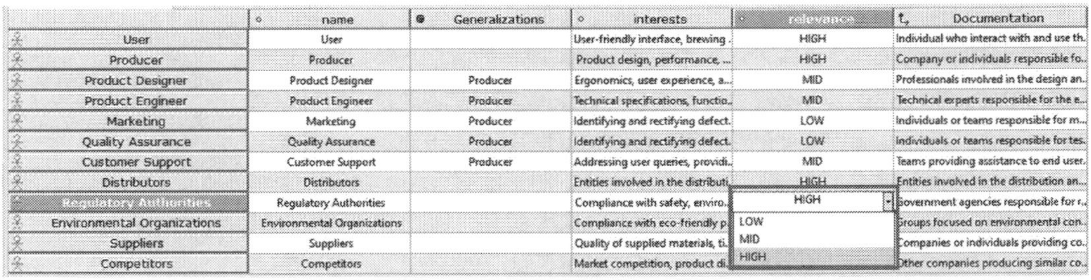

Figure 5.53. Tables are active elements that can be used to change values

Papyrus can export a model table into the XLS data format from Microsoft Excel. Exporting the data into a table makes sharing data with people who do not want to use the modeling tool easier and also provides a way to transfer the data into other tools. To export a table, right-click on the table and select *Export to XLS.*

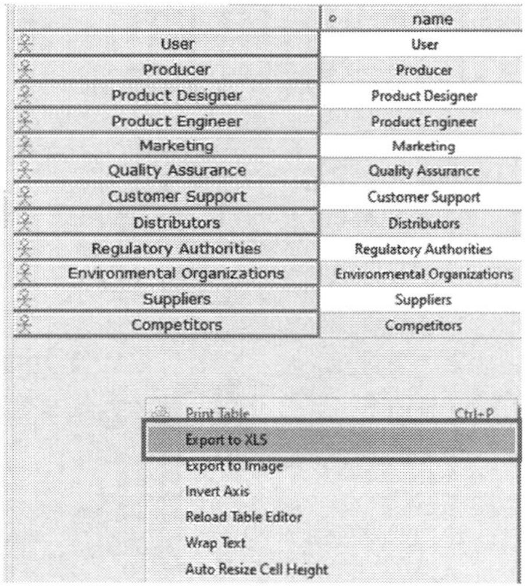

Figure 5.54. Export table to XLS

A new dialog appears that enables selecting the directory to which the table will be exported and its filename.

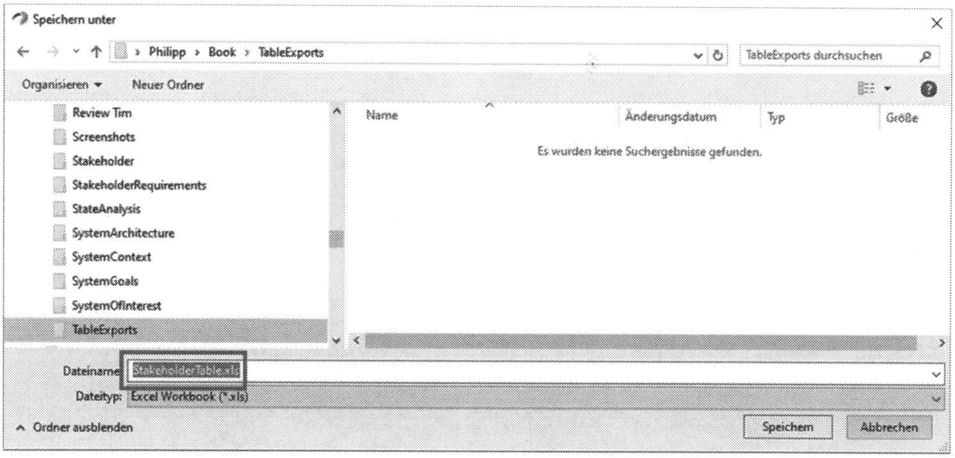

Figure 5.55. Select an export directory and provide a file name

Once the table is exported, it will be automatically opened with Microsoft Excel.

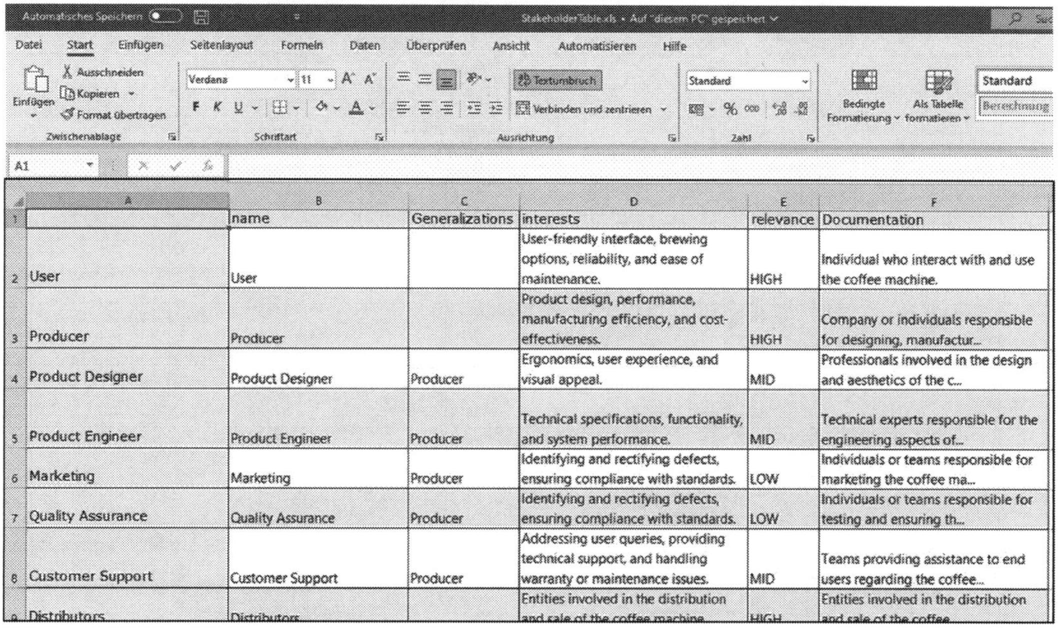

Figure 5.56. The exported table opened in Microsoft Excel

5.3.4 Stakeholder Requirements

Stakeholder requirements are the expressions of needs, expectations, and constraints that stakeholders have for a system. Stakeholder requirements are typically high-level and expressed in natural language or user-friendly terms. They provide a broad understanding of the desired functionality, features, and characteristics of the system from the perspective of various stakeholders. Stakeholder requirements serve as the foundation for the development of system requirements and play a crucial role in ensuring that the final system meets the needs and expectations of its intended users and other stakeholders. The process of capturing stakeholder requirements involves eliciting information from stakeholders through techniques such as interviews, surveys, workshops, and direct observation. It aims to understand the needs and expectations of different stakeholders.

5.3.5 Stakeholder Requirements for the Coffee Machine

A comprehensive stakeholder requirements analysis using a combination of the methods listed above resulted in the following list of relevant stakeholder requirements for the coffee machine:

SH-REQ-001
Name: Usability
Text: The coffee machine must provide a user-friendly interface with intuitive controls for selecting brewing options.
Stakeholder: User
Verification Method: Conduct user testing sessions with representative users to evaluate the ease of use and intuitiveness of the interface.

SH-REQ-002
Name: Cost Effective
Text: The coffee machine design should prioritize cost-effectiveness in manufacturing without compromising on quality and performance.
Stakeholder: Producer
Verification Method: Perform a cost-benefit analysis, comparing manufacturing costs with the desired quality and performance metrics.

SH-REQ-003
Name: Appealing Design
Text: The coffee machine design must incorporate a visually appealing and ergonomic layout, enhancing user experience.
Stakeholder: Product Designer

Verification Method: Assess design prototypes through user feedback and ergonomic evaluations to ensure visual appeal and usability.

SH-REQ-004
Name: Well Engineered
Text: The coffee machine must meet technical specifications, including precise temperature control during the brewing process.
Stakeholder: Product Engineer
Verification Method: Use testing equipment to measure and verify temperature control accuracy during multiple brewing cycles.

SH-REQ-005
Name: Marketable
Text: The coffee machine must be easy to position in the market.
Stakeholder: Marketing
Verification Method: Conduct market surveys and analyze competitor products to identify differentiating features.

SH-REQ-006
Name: Design For Test
Text: The coffee machine must be easy to test.
Stakeholder: Quality Assurance
Verification Method: Implement comprehensive testing protocols, including functionality, stress, and regression testing, and address identified defects.

SH-REQ-007
Name: User Manual
Text: The coffee machine must have a comprehensive user manual.
Stakeholder: Customer Support
Verification Method: Verify the completeness and clarity of the user manual through user feedback.

SH-REQ-008
Name: Facilitate Sales
Text: The coffee machine packaging should be attractive, and marketing materials should highlight key features to facilitate sales.
Stakeholder: Distributors
Verification Method: Conduct market surveys and feedback sessions with retailers to assess the effectiveness of packaging and marketing materials in driving sales.

SH-REQ-009
Name: Regulatory Standards

Text: The coffee machine must comply with safety, environmental, and quality standards set by regulatory authorities.
Stakeholder: Regulatory Authorities
Verification Method: Perform certification testing with recognized testing laboratories to ensure compliance with relevant standards and regulations.

SH-REQ-010
Name: Eco Friendly
Text: The coffee machine should consider eco-friendly practices, such as using recyclable materials and minimizing environmental impact.
Stakeholder: Environmental Organizations
Verification Method: Conduct a life cycle analysis to assess the environmental impact of the coffee machine, including material selection and energy consumption.

SH-REQ-011
Name: Industry Standards
Text: The system design and components shall comply with industry-standard specifications and regulations, ensuring compatibility with commonly used materials and technologies in the market.
Stakeholder: Suppliers
Verification Method: Analyze the system design to check and confirm compliance with relevant industry standards.

SH-REQ-012
Name: Preserve Rights
Text: The coffee machine must not violate the copyrights or infringe on the patents of the competitor and their products.
Stakeholder: Competitor
Verification Method: Perform IP analysis and assess the coffee machine design against it.

SH-REQ-013
Name: Rapid Heating
Text: The coffee machine shall minimize the time between turning on the machine and brewing the first cup, providing users with a quick and convenient coffee preparation experience.
Stakeholder: User
Verification Method: Testing the coffee machine to ensure it meets the specified time requirements and effectively minimizes the waiting time for the user.

SH-REQ-014
Name: Compact size

Text: The coffee machine shall be easily moveable by a single adult person.
Stakeholder: User
Verification Method: Conduct user testing sessions with representative users.

SH-REQ-015
Name: Reliable
Text: The coffee machine shall be reliable and long-lived.
Stakeholder: User
Verification Method: Extensive stress testing.

5.3.6 Modeling the Coffee Machine Stakeholder Requirements

Create a new package called "Requirements" that will contain all requirements, and inside that package, create a new package called "Stakeholder Requirements".

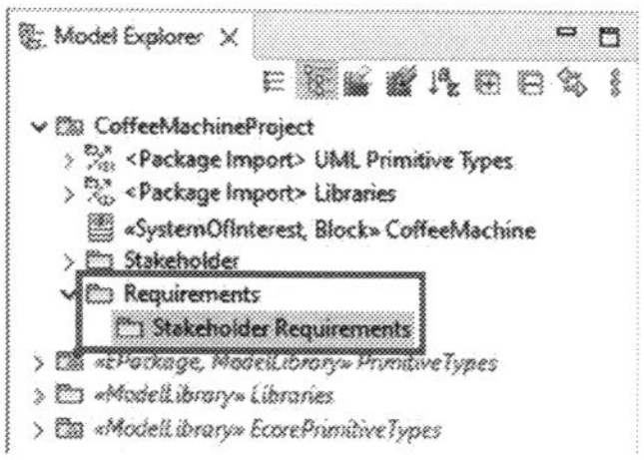

Figure 5.57. New packages for requirements

Next, add a requirement element to the newly created "Stakeholder Requirements" package, by right-clicking on the package and selecting *SysML 1.6 Child->Requirement.*

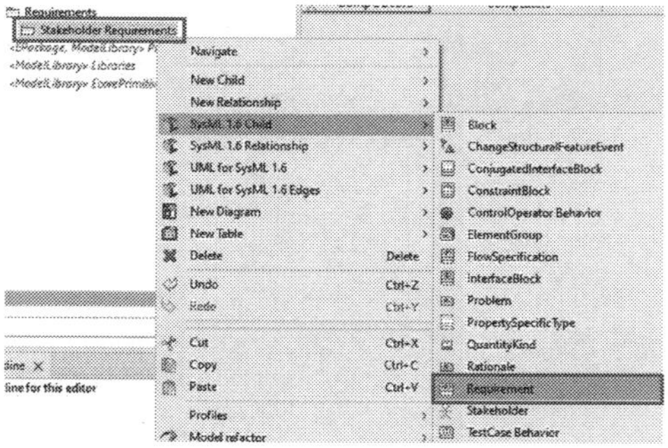

Figure 5.58. Create a new requirement element

To set the name, identifier, and text of the newly created requirement, use the SysML 1.6 tab of the Properties View of the newly created requirement.

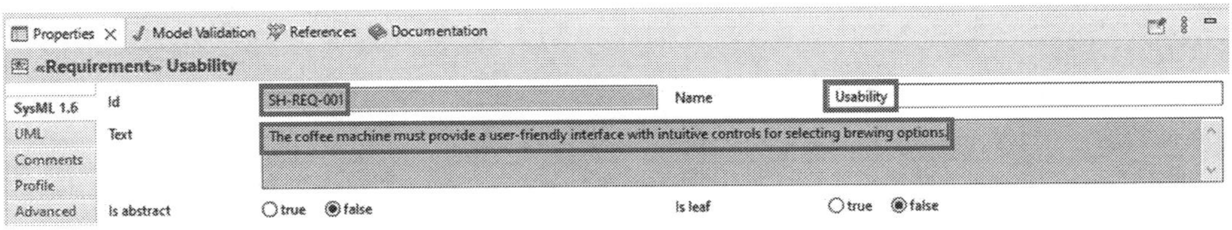

Figure 5.59. Requirement properties

Next, apply the "StakeholderRequirement" stereotype. To relate the newly created requirement to its stakeholder, select the "stakeholder" property of the "StakeholderRequirement" stereotype in the Profile tab of the Properties View of that requirement and click on the button with the three dots.

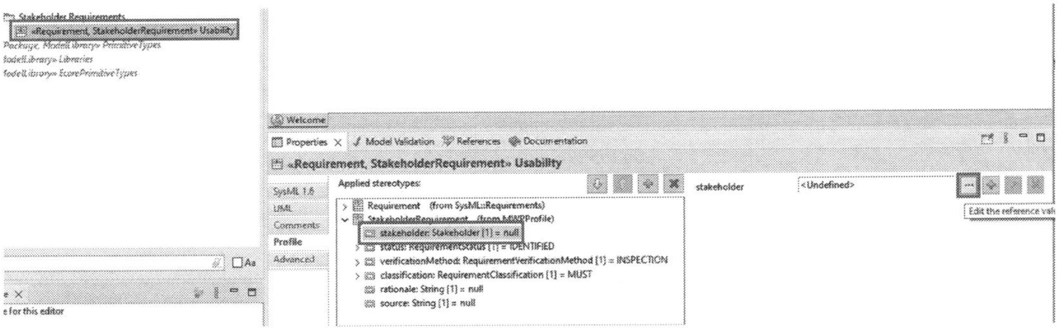

Figure 5.60. Name the requirement and edit the stakeholder property

A new dialog appears that enables selecting the stakeholder that belongs to the stakeholder requirement.

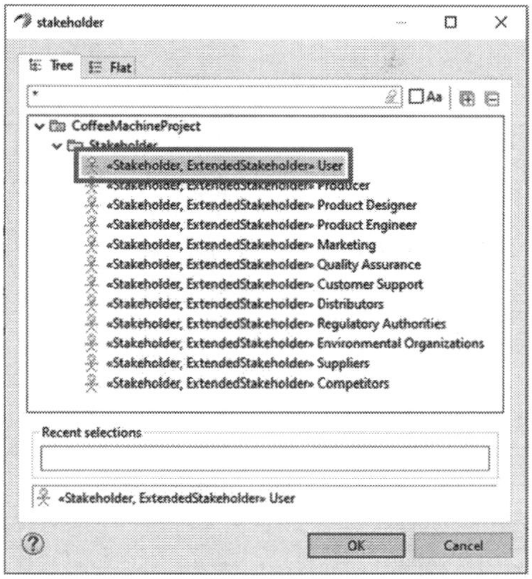

Figure 5.61. "Select stakeholder" dialog

Set the values for all the other stakeholder requirement stereotype properties.

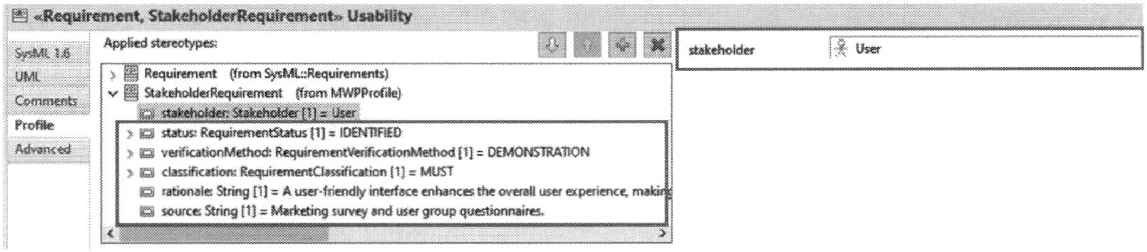

Figure 5.62. Stakeholder requirement stereotype properties

After creating the requirement, the next step is to add it to a table. To create a requirements table, right-click on the "StakeholderRequirements" package and select *New Table->SysML 1.6 Requirement Table.* Name the new table "Stakeholder Requirements Table".

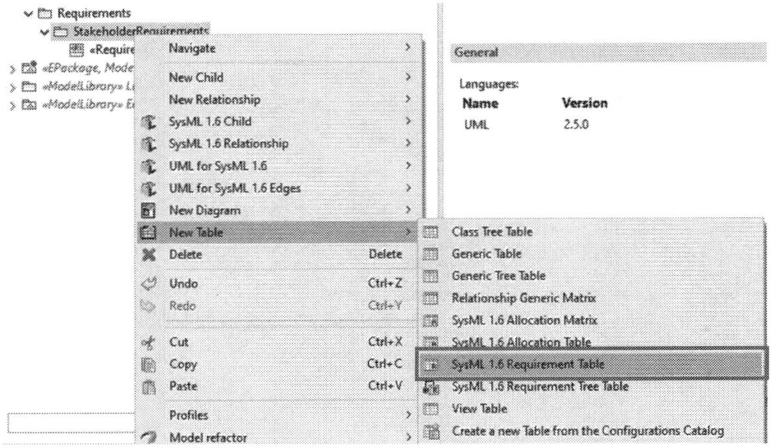

Figure 5.63. Create requirement table

The requirement table uses a pre-configured layout and automatically contains all requirement elements that are stored in the same package as the requirement table. There is no need to drag and drop the requirement onto the table, and the display options for the columns and the rows are already set.

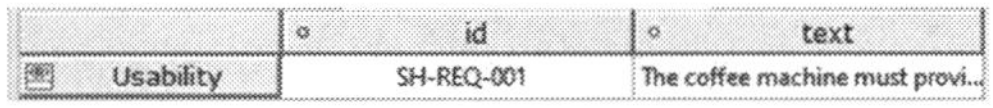

	id	text
Usability	SH-REQ-001	The coffee machine must provi...

Figure 5.64. Newly created requirement table

However, the requirement table does not know about the additional properties from the "StakeholderRequirement" stereotype. To specify the columns of the table, again right-click on the table and select *Columns->Select/Destroy Columns*. Then add all the properties that should be displayed as a column in the table.

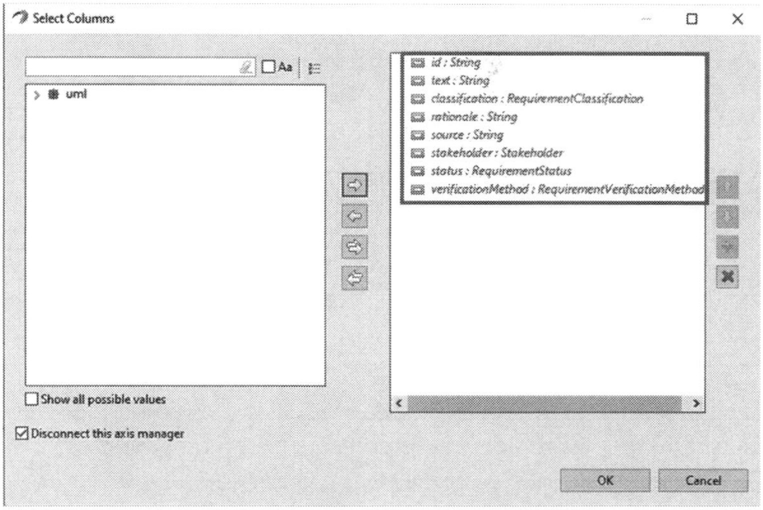

Figure 5.65. Finished column selection

As already stated before, tables are a way to directly interact with the underlying model. To show this, create the next requirement directly from the table by right-clicking on the table and selecting *Create SysML 1.6 Element->Create a new Requirement.*

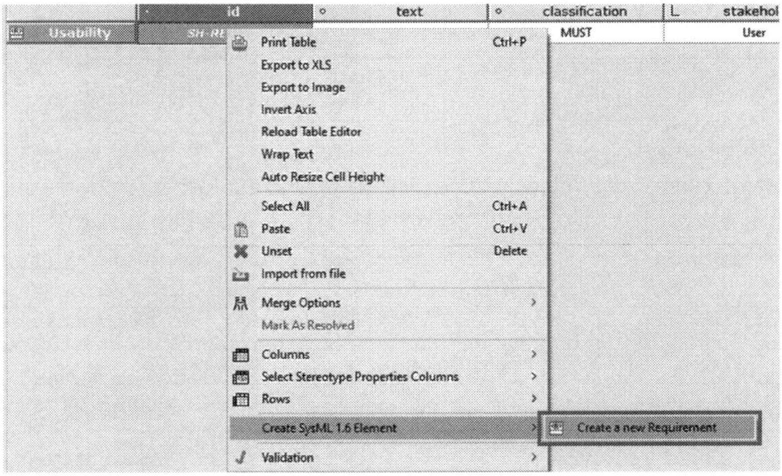

Figure 5.66. Create a new requirement from the table

The new requirement will be added to the same package in which the table from which it was created is stored and can be seen in the Model Explorer.

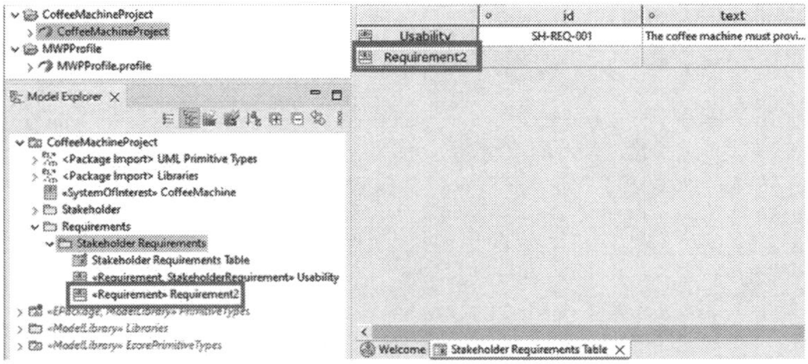

Figure 5.67. The new requirement in the Model Explorer

Now, create all the other stakeholder requirements. Name them and set the values of the "StakeholderRequirement" stereotype properties.

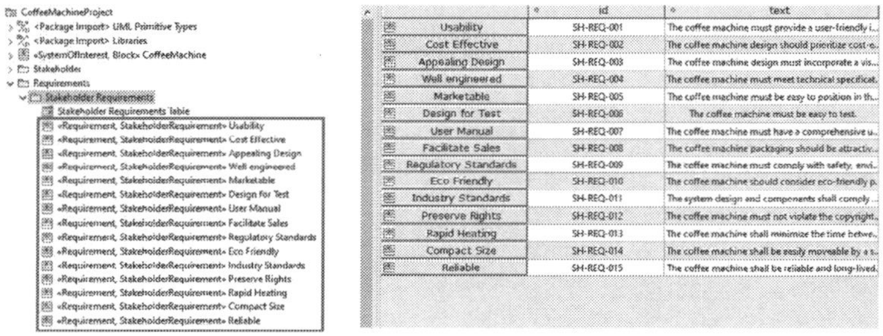

Figure 5.68. All stakeholder requirements

Next, we want to show the relationship between a stakeholder and its requirements on a requirement diagram. To create a new requirement diagram, right-click on the "Stakeholder Requirements" package and select *New Diagram->SysML 1.6 Requirement Diagram.*

Note that this is an optional step to show how a relation that already exists in the model can be shown visually on a diagram. Typically, the methodology will provide some guidance and rules, for which aspects of a system diagrams should be created in the model.

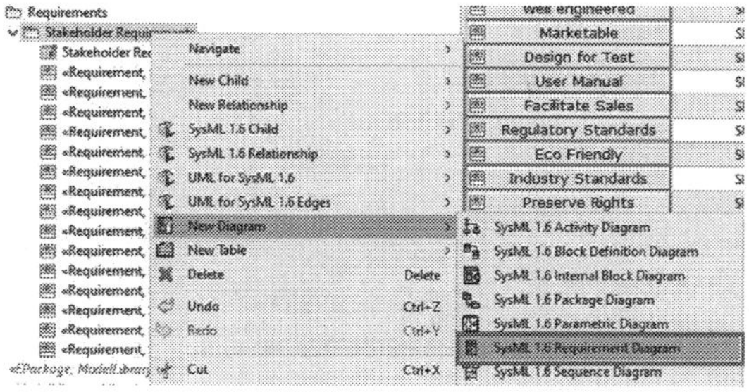
Requirements
Navigate
New Child
New Relationship
SysML 1.6 Child
SysML 1.6 Relationship
UML for SysML 1.6
UML for SysML 1.6 Edges
New Diagram
New Table
Delete
Undo
Redo
Cut
Delete
Ctrl+Z
Ctrl+Y
Ctrl+X
Marketable
Design for Test
User Manual
Facilitate Sales
Regulatory Standards
Eco Friendly
Industry Standards
Preserve Rights
SysML 1.6 Activity Diagram
SysML 1.6 Block Definition Diagram
SysML 1.6 Internal Block Diagram
SysML 1.6 Package Diagram
SysML 1.6 Parametric Diagram
SysML 1.6 Requirement Diagram
SysML 1.6 Sequence Diagram

Figure 5.69. Create a requirement diagram

Name the diagram “Stakeholder Requirements Traceability RD” and add the “User” stakeholder and its four requirements (SH-REQ-001, SH-REQ-013, SH-REQ-014, SH-REQ-015) to the diagram using drag and drop.

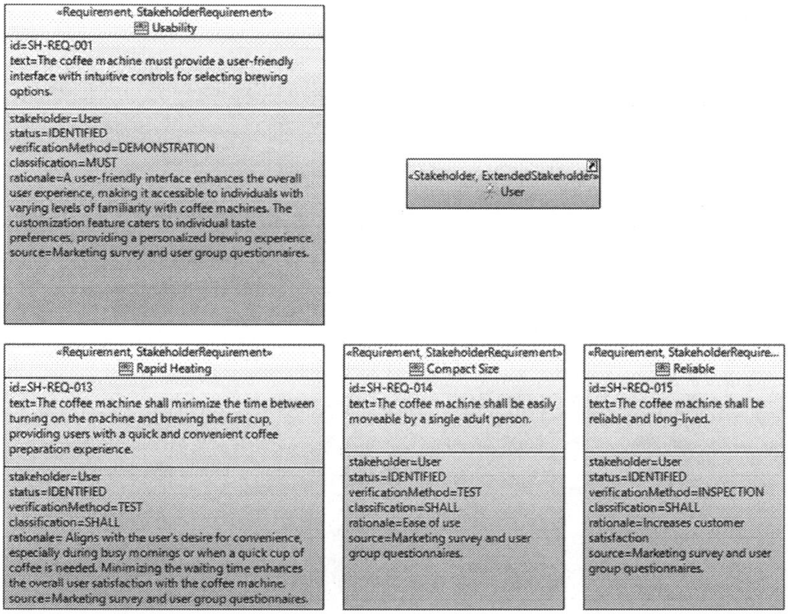
«Requirement, StakeholderRequirement»
Usability
id=SH-REQ-001
text=The coffee machine must provide a user-friendly interface with intuitive controls for selecting brewing options.
stakeholder=User
status=IDENTIFIED
verificationMethod=DEMONSTRATION
classification=MUST
rationale=A user-friendly interface enhances the overall user experience, making it accessible to individuals with varying levels of familiarity with coffee machines. The customization feature caters to individual taste preferences, providing a personalized brewing experience.
source=Marketing survey and user group questionnaires.
«Stakeholder, ExtendedStakeholder»
User
«Requirement, StakeholderRequirement»
Rapid Heating
id=SH-REQ-013
text=The coffee machine shall minimize the time between turning on the machine and brewing the first cup, providing users with a quick and convenient coffee preparation experience.
stakeholder=User
status=IDENTIFIED
verificationMethod=TEST
classification=SHALL
rationale= Aligns with the user's desire for convenience, especially during busy mornings or when a quick cup of coffee is needed. Minimizing the waiting time enhances the overall user satisfaction with the coffee machine.
source=Marketing survey and user group questionnaires.
«Requirement, StakeholderRequirement»
Compact Size
id=SH-REQ-014
text=The coffee machine shall be easily moveable by a single adult person.
stakeholder=User
status=IDENTIFIED
verificationMethod=TEST
classification=SHALL
rationale=Ease of use
source=Marketing survey and user group questionnaires.
«Requirement, StakeholderRequire...
Reliable
id=SH-REQ-015
text=The coffee machine shall be reliable and long-lived.
stakeholder=User
status=IDENTIFIED
verificationMethod=INSPECTION
classification=SHALL
rationale=Increases customer satisfaction
source=Marketing survey and user group questionnaires.

Figure 5.70. Requirement diagram with four requirements and one stakeholder

The little arrow symbol next to the “User” stakeholder indicates that this element resides in a different package than the current diagram. The element is imported from the other package and can then be used in the package context of the diagram.

To show the relation between the requirements and the stakeholder, which already exists in the model, on the diagram, select all five elements and then right-click on one of them and select *Filters->Show/Hide Related Link.*

To select multiple elements, you can either hold shift or control and then add the elements to the selection by left-clicking on them. Alternatively, you can draw a selection box on the canvas by clicking and holding the left mouse button. Since the five elements are currently the only elements on the diagram, you can also select all elements using Control+A.

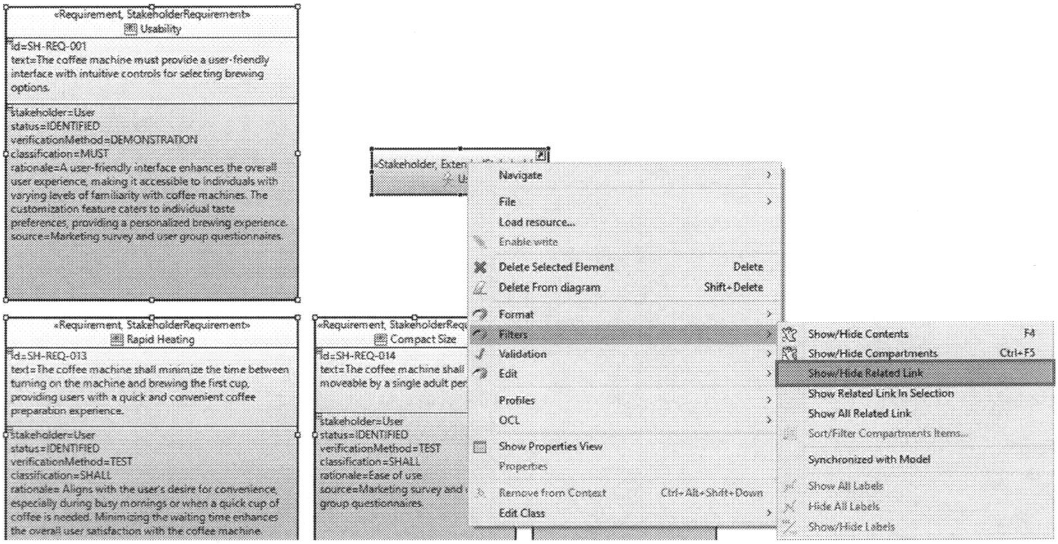
«Requirement, StakeholderRequirement»
Usability
id=SH-REQ-001
text=The coffee machine must provide a user-friendly interface with intuitive controls for selecting brewing options.
stakeholder=User
status=IDENTIFIED
verificationMethod=DEMONSTRATION
classification=MUST
rationale=A user-friendly interface enhances the overall user experience, making it accessible to individuals with varying levels of familiarity with coffee machines. The customization feature caters to individual taste preferences, providing a personalized brewing experience.
source=Marketing survey and user group questionnaires.
«Requirement, StakeholderRequirement»
Rapid Heating
id=SH-REQ-013
text=The coffee machine shall minimize the time between turning on the machine and brewing the first cup, providing users with a quick and convenient coffee preparation experience.
stakeholder=User
status=IDENTIFIED
verificationMethod=TEST
classification=SHALL
rationale= Aligns with the user's desire for convenience, especially during busy mornings or when a quick cup of coffee is needed. Minimizing the waiting time enhances the overall user satisfaction with the coffee machine.
Compact Size
id=SH-REQ-014
stakeholder=User
status=IDENTIFIED
verificationMethod=TEST
classification=SHALL
rationale=Ease of use
Navigate
File
Load resource...
Enable write
Delete Selected Element
Delete
Delete From diagram
Shift+Delete
Format
Filters
Validation
Edit
Profiles
OCL
Show Properties View
Properties
Remove from Context
Ctrl+Alt+Shift+Down
Edit Class
Show/Hide Contents
F4
Show/Hide Compartments
Ctrl+F5
Show/Hide Related Link
Show Related Link In Selection
Show All Related Link
Sort/Filter Compartments Items...
Synchronized with Model
Show All Labels
Hide All Labels
Show/Hide Labels

Figure 5.71. Open the "Show/Hide Links" dialog

In the "Show/Hide Links" dialog select the "stakeholder" properties of the four requirements.

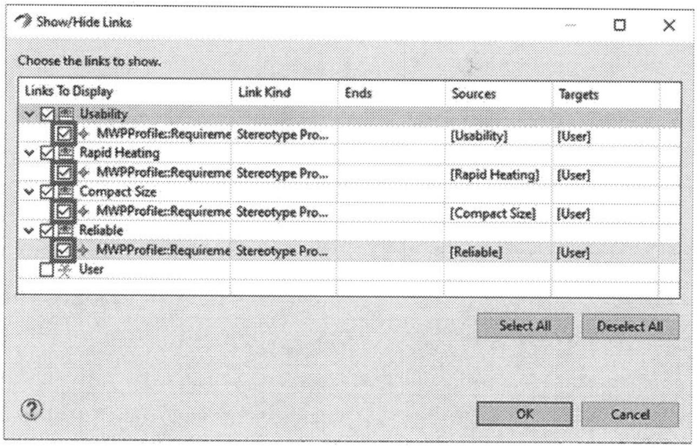
Show/Hide Links
Choose the links to show.
Links To Display
Link Kind
Ends
Sources
Targets
Usability
MWPProfile::Requireme
Stereotype Pro...
[Usability]
[User]
Rapid Heating
MWPProfile::Requireme
Stereotype Pro...
[Rapid Heating]
[User]
Compact Size
MWPProfile::Requireme
Stereotype Pro...
[Compact Size]
[User]
Reliable
MWPProfile::Requireme
Stereotype Pro...
[Reliable]
[User]
User
Select All
Deselect All
OK
Cancel

Figure 5.72. Select links for display

The relation between the requirements and the stakeholder will then be shown on the diagram.

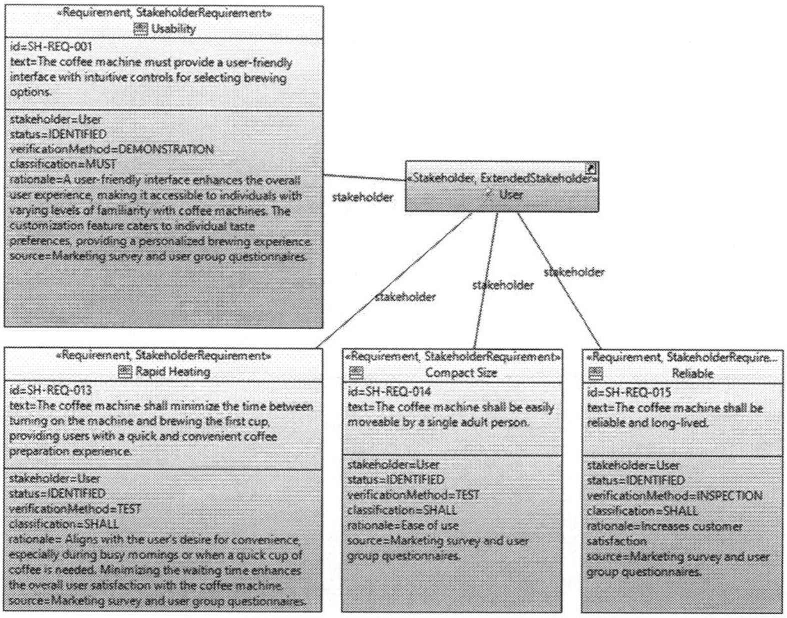
«Requirement, StakeholderRequirement»
Usability
id=SH-REQ-001
text=The coffee machine must provide a user-friendly interface with intuitive controls for selecting brewing options.
stakeholder=User
status=IDENTIFIED
verificationMethod=DEMONSTRATION
classification=MUST
rationale=A user-friendly interface enhances the overall user experience, making it accessible to individuals with varying levels of familiarity with coffee machines. The customization feature caters to individual taste preferences, providing a personalized brewing experience.
source=Marketing survey and user group questionnaires.
«Stakeholder, ExtendedStakeholder»
User
stakeholder
stakeholder
stakeholder
stakeholder
«Requirement, StakeholderRequirement»
Rapid Heating
id=SH-REQ-013
text=The coffee machine shall minimize the time between turning on the machine and brewing the first cup, providing users with a quick and convenient coffee preparation experience.
stakeholder=User
status=IDENTIFIED
verificationMethod=TEST
classification=SHALL
rationale= Aligns with the user's desire for convenience, especially during busy mornings or when a quick cup of coffee is needed. Minimizing the waiting time enhances the overall user satisfaction with the coffee machine.
source=Marketing survey and user group questionnaires.
«Requirement, StakeholderRequirement»
Compact Size
id=SH-REQ-014
text=The coffee machine shall be easily moveable by a single adult person.
stakeholder=User
status=IDENTIFIED
verificationMethod=TEST
classification=SHALL
rationale=Ease of use
source=Marketing survey and user group questionnaires.
«Requirement, StakeholderRequire...
Reliable
id=SH-REQ-015
text=The coffee machine shall be reliable and long-lived.
stakeholder=User
status=IDENTIFIED
verificationMethod=INSPECTION
classification=SHALL
rationale=Increases customer satisfaction
source=Marketing survey and user group questionnaires.

Figure 5.73. Finished stakeholder requirement traceability diagram

5.4 Define System Goals

Setting system goals early in the system development lifecycle is crucial for ensuring that the system meets the needs of its stakeholders and achieves success. Goals provide a clear direction for the development team and help to ensure that all decisions are made in a way that supports the overall objectives of the system.

Measures of Effectiveness (MOEs) are quantifiable criteria used to assess how well a system is achieving its intended objectives and goals. MOEs play a crucial role in evaluating the overall success and performance of a system. These measures help in making informed decisions, managing risks, and ensuring that the system meets the requirements and expectations of stakeholders. Measures of Effectiveness serve as the yardstick by which the success of a system is measured. Defined at the outset of the system development lifecycle (SDLC), MOEs encapsulate the overarching objectives and intended outcomes of the system. These measures encapsulate the "what" of system success, providing a roadmap for development efforts.

MOEs are directly related to stakeholders and their requirements. Stakeholders articulate their needs and expectations and MOEs act as a bridge, translating these requirements into quantifiable metrics. This alignment ensures that system goals are not only defined by technical considerations but are deeply rooted in the aspirations and necessities of those for whom the system is being developed or who otherwise have an influence on the system. A systematic approach to setting system goals thus involves establishing clear linkages between MOEs and stakeholder requirements. Each MOE should have a direct correlation to one or more specific stakeholder needs. For instance, if a stakeholder expresses a need for enhanced system reliability, an associated MOE could measure Mean Time Between Failures (MTBF) or system uptime. MOEs provide a tangible means of quantifying success criteria. This quantification brings precision to the often subjective realm of system goals. Stakeholder requirements, which may initially be expressed in qualitative terms, find a quantifiable counterpart through the formulation of MOEs. This clarity enhances communication, ensuring a shared understanding of success across the project team.

Defining MOEs is a critical aspect of systems engineering as they provide a quantifiable way to assess how well a system meets its objectives and requirements. Defining MOEs serves the goals of systems engineering:

- **Objective evaluation**: Systems engineering aims to design and develop systems that meet specific objectives. MOEs establish clear and measurable criteria for evaluating whether a system is achieving its intended objectives. They provide a basis for objective assessment and help in gauging the success of the system against predefined goals.

- **Performance assessment**: Systems engineering strives to ensure the optimal performance of a system. MOEs define parameters for measuring the performance of a system. Whether it's efficiency, speed, accuracy, or other performance attributes, MOEs provide quantifiable metrics to assess how well the system performs in real-world conditions.
- **User satisfaction**: Systems engineering aims to create systems that satisfy user needs and expectations. MOEs include criteria related to user satisfaction. By defining metrics that capture user experience, usability, and other relevant factors, systems engineers can gauge how well the system aligns with user expectations.
- **Risk management**: Systems engineering seeks to identify and mitigate risks associated with a system. MOEs help in quantifying and assessing risks. By defining measures related to system reliability, safety, and fault tolerance, systems engineers can proactively identify potential risks and implement strategies to mitigate them.
- **Continuous improvement**: Systems engineering seeks continuous improvement in system performance and functionality. MOEs provide a baseline for performance, enabling systems engineers to identify areas for improvement. By regularly assessing and comparing actual performance against defined measures, the system can undergo continuous refinement and enhancement.

5.4.1 System Goals Definition Pitfalls

There are some pitfalls when the system goals and measures of effectiveness are defined:

- **Not involving all stakeholders in the goal-setting process** System goals should be defined with input from all relevant stakeholders, including users, operators, managers, and decision-makers. This ensures that the goals are aligned with the needs and expectations of everyone who will be affected by the system. Failure to involve all stakeholders can lead to goals that are not relevant or achievable.
- **Not making goals SMART**: System goals should be Specific, Measurable, Achievable, Relevant, and Time-bound (SMART). SMART goals are clear, objective, and measurable, which makes them more likely to be achieved. Failure to make goals SMART can lead to confusion and frustration, and can make it more difficult to track progress and make informed decisions.
- **Not considering future needs**: System goals should be flexible enough to accommodate changes in requirements over time. The world is constantly changing, and so will the needs of the users and other stakeholders. Failing to consider future needs can lead to systems that become obsolete or irrelevant before they are even fully deployed.

- **Not using metrics that are meaningful to stakeholders**: Measures of effectiveness should be selected that are relevant to the goals of the system and that are meaningful to the stakeholders. Failure to use meaningful metrics can make it difficult to assess progress and make informed decisions.
- **Not using metrics that are actionable**: Measures of effectiveness should be actionable, meaning that they should provide information that can be used to improve the system. Failure to use actionable metrics can make it difficult to take corrective action when problems arise.
- **Not communicating goals and metrics effectively**: The goals and metrics of the system should be communicated effectively to all stakeholders. This ensures that everyone is on the same page and that they are working towards the same objectives. Failure to communicate goals and metrics effectively can lead to confusion, frustration, and a lack of accountability.

By avoiding these pitfalls, systems engineers can increase the likelihood of success in their projects. By involving all stakeholders, making goals SMART, considering future needs, using meaningful metrics, using actionable metrics, and communicating goals and metrics effectively, systems engineers can ensure that their systems are aligned with the needs of the users and other stakeholders and that they are effective in meeting their goals.

5.4.2 Coffee Machine Measures of Effectiveness

The following list provides some MoEs that can be used for assessing the overall performance, efficiency, and user satisfaction of the coffee machine. Establishing clear traceability between stakeholder requirements and measures of effectiveness is crucial for ensuring that the design and functionality of the coffee machine align with stakeholder expectations.

Brewing Time
Description: The time taken to complete the brewing process.
Measure: Seconds (s)
Objective: Minimize brewing time for user convenience.
Stakeholder requirement: SH-REQ-013

Temperature Control Accuracy
Description: Precision in maintaining the desired brewing temperature.
Measure: Degrees Celsius (°C)
Objective: Ensure consistent and optimal brewing temperature.
Stakeholder requirement: SH-REQ-004

Water Flow Rate
Description: The rate at which water flows through the coffee machine during brewing.
Measure: Liters per minute (L/min)
Objective: Optimize flow rate for efficient extraction.
Stakeholder requirement: SH-REQ-004

Energy Consumption
Description: The amount of energy consumed during the brewing process.
Measure: Kilowatt-hours (kWh)
Objective: Minimize energy consumption for sustainability.
Stakeholder requirement: SH-REQ-010

User Interface Responsiveness
Description: Time taken for the system to respond to user inputs.
Measure: Milliseconds (ms)
Objective: Ensure quick and responsive user interactions.
Stakeholder requirement: SH-REQ-001

Maintenance Interval
Description: Time between required maintenance tasks.
Measure: Days
Objective: Increase the time between maintenance cycles for user convenience.
Stakeholder requirement: SH-REQ-015

Reliability
Description: The system's ability to operate consistently without failures.
Measure: MTBF
Objective: Maximize system reliability for user satisfaction.
Stakeholder requirement: SH-REQ-015

Unit Weight
Description: The weight of the coffee machine.
Measure: Kilograms (kg)
Objective: Minimize weight to make the coffee machine easier to move.
Stakeholder requirement: SH-REQ-014

These measures of effectiveness provide a holistic view of the coffee machine's performance, covering aspects of brewing, user experience, sustainability, and system reliability. Regular monitoring and analysis of these metrics contribute to continuous improvement and the delivery of a high-quality coffee machine.

5.4.3 Value Types, Quantity Kinds and Units

In SysML, value types are data types that represent the values of objects in a system model. They can be used to specify the data that will be stored in blocks, flow through ports, and used in operations and constraints. Value types are used to support unambiguous specifications in SysML by providing a standard way to represent the data that is used in a system model. This makes it easier for stakeholders to understand and communicate about the system, and it helps to ensure that the model is consistent and accurate. Value types support clear specification in SysML:

- **They provide a standard vocabulary for describing data.** This means that stakeholders do not have to guess what the meaning of a data value is, and it helps to ensure that the model is consistent across different stakeholders.
- **They provide a way to constrain the values that can be used.** This helps to ensure that the model is accurate and that it will only represent valid data.
- **They can be used to validate system models.** This helps to identify and correct errors in the model before it is implemented.

SysML supports different value types:

- **Primitive value types** These represent the basic data types integer, complex, and real numbers, as well as Booleans, and strings.
- **Composite value types** These represent complex data types that can contain other value types, such as arrays and structures.
- **Enumeration value types** These represent a set of predefined values, such as the states of a machine or the colors of a traffic light.

Additionally, SysML also allows users to define their own, user-defined value types. This can be useful for modeling specialized data that is not covered by the predefined types. The definition of Value Types can be enhanced using quantity kinds and units. The relation between value type, quantity kind, and unit is crucial for accurately representing and describing quantities within a system model.

- **Value Type** A value type is a classification of values that share common characteristics. It represents the nature or type of a quantity, such as whether it is numerical, textual, or boolean. Example: Real, Integer, Boolean, String.
- **Quantity Kind** A quantity kind represents a general category or type of quantity, often associated with a physical property or abstract concept. It provides a semantic context for values, indicating what aspect of the system is being measured. Example: Length, Mass, Time, Temperature.

- **Unit** A unit is a standardized measure used to express a particular quantity in a specific quantity kind. Units define the scale or magnitude of a quantity, providing a consistent and universally understood way to represent measurements. Example: Meter (m) for length, Kilogram (kg) for mass, Second (s) for time, Celsius (°C) for temperature.

A value type is associated with a quantity kind to specify the nature of the quantity, e.g., a numerical value representing a length. The quantity kind is associated with a unit to provide a standardized measurement scale for values of that quantity kind. The consistent application of value types, quantity kinds, and units ensures clarity and precision in the modeling of a system.

> In 1998, NASA launched the Mars Climate Orbiter (MCO) to study the Martian atmosphere. However, the spacecraft failed to enter Mars' orbit and was lost. The investigation into the cause of the failure revealed that the MCO was using imperial units (pounds-force) for thruster measurements while the ground control team was using metric units (newtons). This resulted in a command error that caused the MCO to burn up in the Martian atmosphere. The root cause of the MCO failure was a lack of communication and coordination between the MCO team and the ground control team. The MCO team was responsible for the conversion of data from imperial to metric units, but they did not communicate this to the ground control team. As a result, the ground control team was unaware of the conversion error and sent commands to the spacecraft in metric units.– Sauser, Reilly and Shenhar, 2009

> Since SysML 1.5, the SysML specification contains in Annex E a non-normative extension that defines a model library of SysML quantity kind and unit definitions for a subset of quantities and units defined by the International System of Quantities (ISQ) and the International System of Units (SI). While some of the available SysML tools implement this library and thereby make the quantity kinds and units available to the user, unfortunately, Papyrus does not. So it is up to the users to create these elements. Of course, once this is done, the quantity kinds and units can be reused; they do not have to be modeled from scratch for every modeling project.

5.4.4 Value Types Relevant to the Coffee Machine

The measures of effectiveness that have been defined for the coffee machine require the following value types:

Time[second]
Quantity Kind: Time
Unit: second
Symbol: s

Time[millisecond]
Quantity Kind: Time
Unit: millisecond
Symbol: ms

Time[day]
Quantity Kind: Time
Unit: day
Symbol: d

Temperature[celsius]
Quantity Kind: Temperature
Unit: degree celsius
Symbol: °C

Mass[kilogram]
Quantity Kind: Mass
Unit: kilogram
Symbol: kg

Length[centimeter]
Quantity Kind: Length
Unit: centimeter
Symbol: cm

Flow Rate[l/min]
Quantity Kind: Volume over time
Unit: liter per minute
Symbol: l/min

Power consumption[kWh]
Quantity Kind: Power over time
Unit: kilowatt hour
Symbol: kWh

MTBF[h]
Quantity Kind: Time
Unit: hour
Symbol: h

5.4.5 Modeling the Coffee Machine Measures of Effectiveness and Value Types

Add a property element to the "CoffeeMachine" block, by right-clicking on the block and selecting *UML for SysML 1.6->Property.*

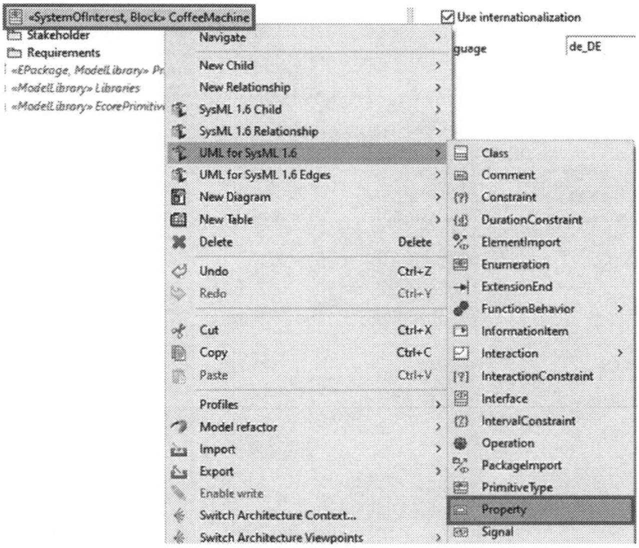

Figure 5.74. Create new property

SysML establishes different kinds of properties belonging to a SysML Block or ValueType, e.g., a part property, is a property typed by a block and a constraint property is typed by a constraint block. A property typed by a SysML ValueType is a value property. The kind of a property is defined by its type, the element itself is always the generic element property.

Name the new property "Brewing Time" and apply the stereotype "MeasureOfEffectiveness" from the MWP profile.

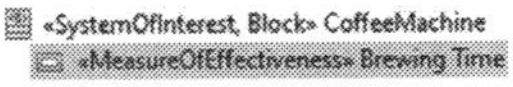

Figure 5.75. New property with name and stereotype

Repeat this process for the other MoEs of the coffee machine and, optionally, add them to a generic table.

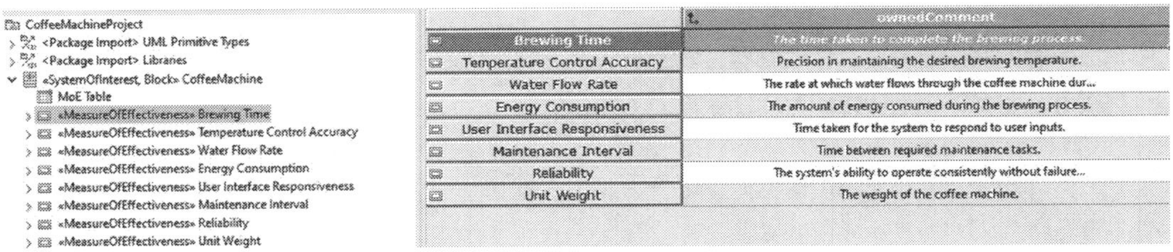

Figure 5.76. Coffee machine MoEs in the Model Explorer and in a table

Create a new package called "Value Types" in the model, then add a new value type to this package by right-clicking on the package and selecting *SysML 1.6 Child->ValueType.*

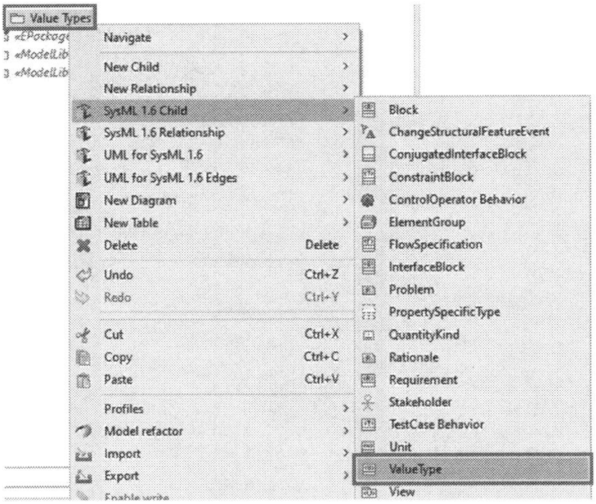

Figure 5.77. Create a new value type

Name the new value type “Time[second]”.

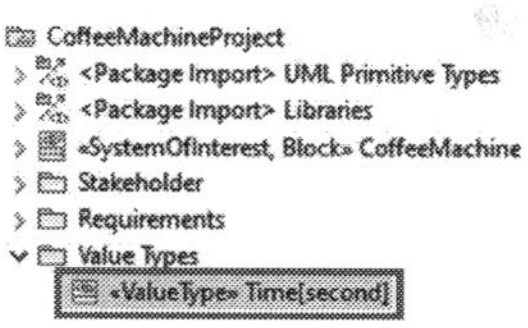

Figure 5.78. New value type “Time[second]”

Add a new quantity kind to the same package by right-clicking on the package and selecting *SysML 1.6 Child->QuantityKind.* Name it “Time”.

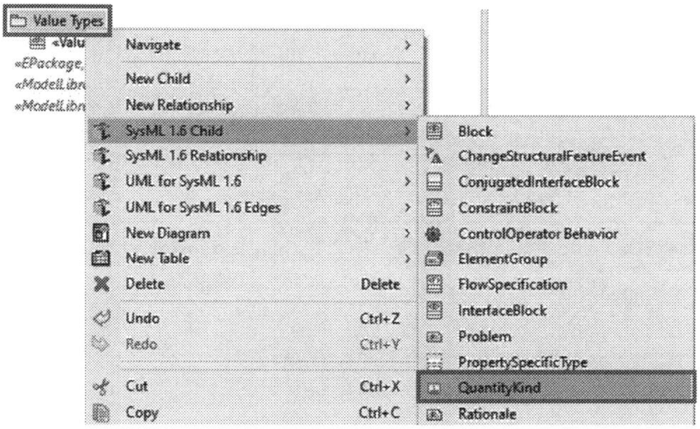

Figure 5.79. Create a new quantity kind

Add a new unit to the same package by right-clicking on the package and selecting *SysML 1.6 Child->Unit.* Name it “second”.

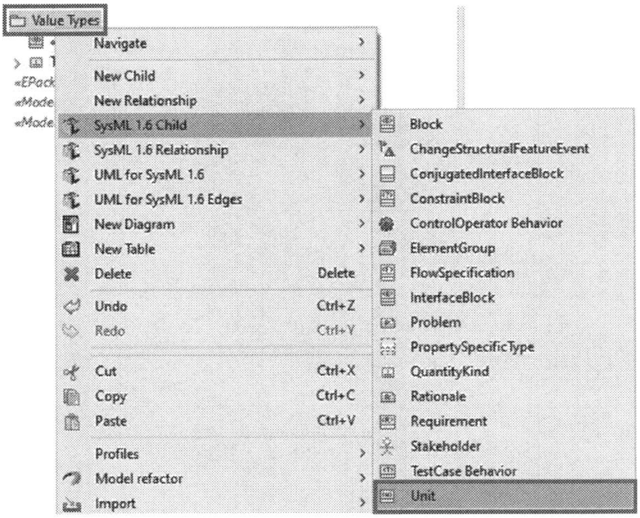

Figure 5.80. Create a new unit

To set the symbol for the "second" unit, expand the properties of the "second" element in the Model Explorer and select the slot "symbol".

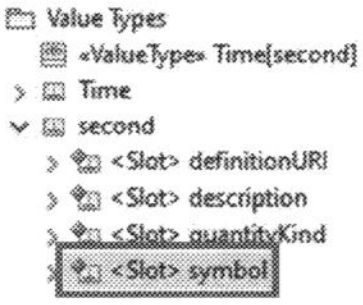

Figure 5.81. Select symbol slot

Go to the Properties View of the symbol slot and click on the button with the pen symbol.

Figure 5.82. Open dialog for setting the symbol of a unit

A new dialog appears that enables setting the symbol for the unit. Enter the value "s" for second.

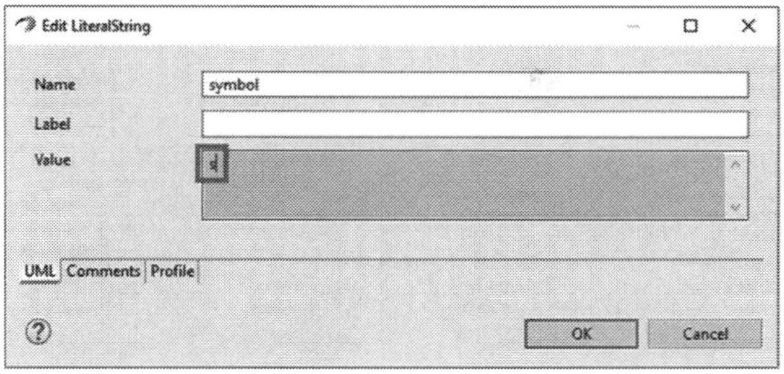

Figure 5.83. Dialog for setting the symbol of a unit

Now, that the quantity kind "Time" and the unit "second" have been modeled, they can be used to specify the value type "Time[second]". Select the value type "Time[second]" in the Model Explorer and click on the button with the three dots next to the Unit property in the SysML 1.6 tab of Properties View.

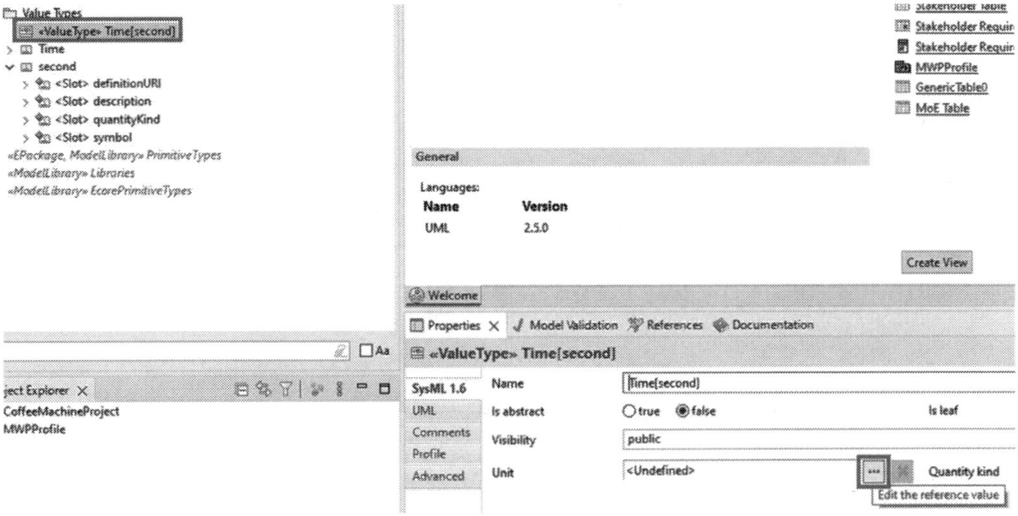

Figure 5.84. Edit value for the unit of a value type

Select the unit "second".

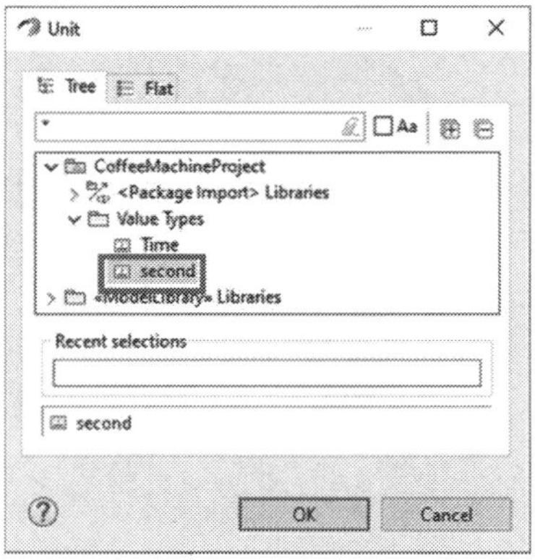

Figure 5.85. "Select unit" dialog

Now, open the dialog for setting the quantity kind by clicking on the button with the three dots next to the Quantity kind property.

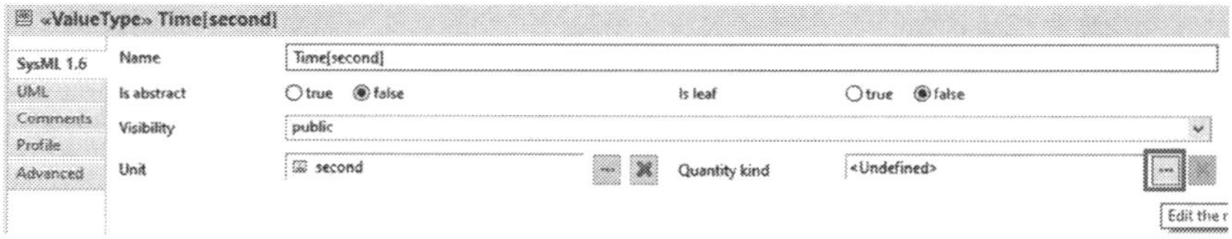

Figure 5.86. Edit value for the quantity kind of a value type

Select the quantity kind "Time".

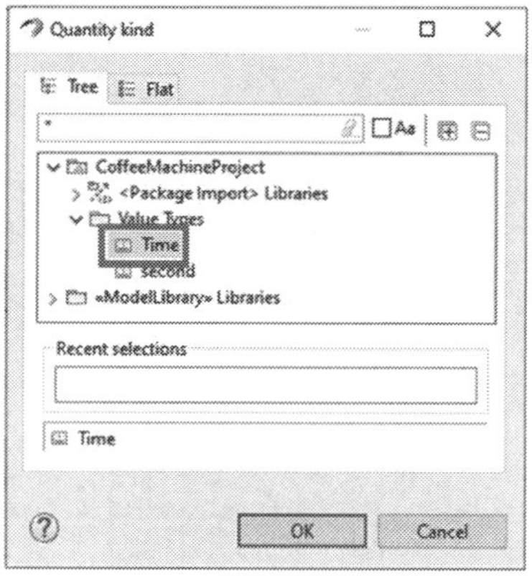

Figure 5.87. "Select quantity kind" dialog

The unit and the quantity kind of the value type are now set.

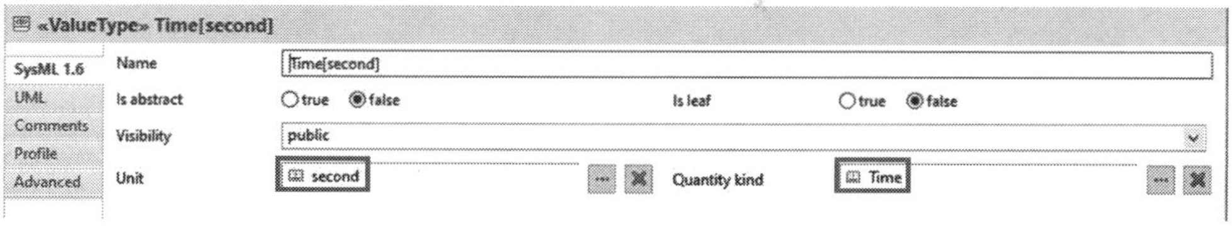

Figure 5.88. Value type with unit and quantity kind

Repeat the process for the ValueTypes "Time[day]", "Time[hour]" and "Time[millisecond]".

Next, we need to specify the primitive value type that is used to set the value of a value property typed by the value type "Time[second]". Since we do not want to use fractions or decimals for setting the time in seconds, we use integers as the primitive value type. To model this, create a new BDD, in the "Value Types" package and name it "Value Types BDD".

Drag and drop all four value types on the newly created BDD.

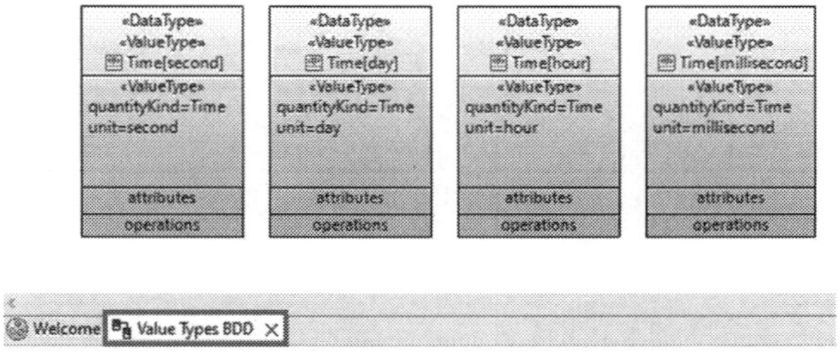

Figure 5.89. "Value Types BDD" with four value types

Open the "Libraries" package in the Model Explorer and add the value type "Integer" from the package "PrimitiveBlocks:ValueTypes" to the diagram.

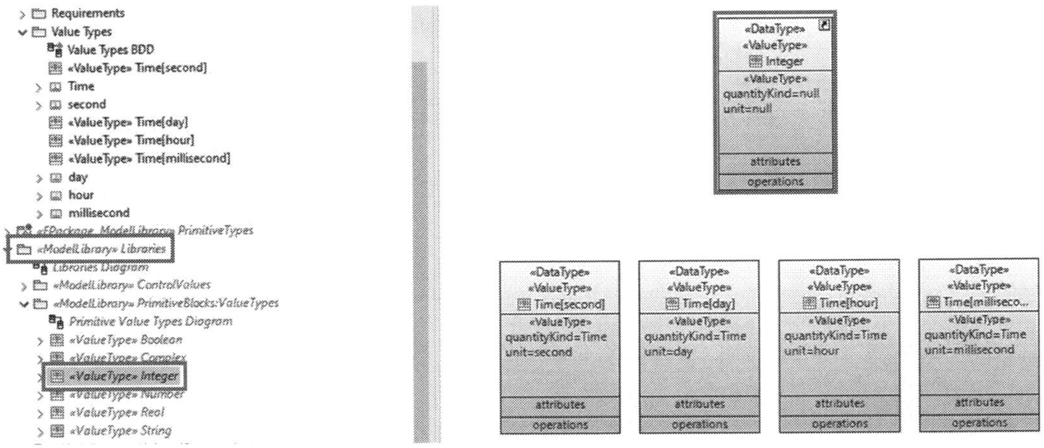

Figure 5.90. Add the "Integer" value type to the diagram

To model the fact, that all values of type "Time[second]" are integers, model a generalization relation between the "Time[second]" value type and the "Integer" value type.

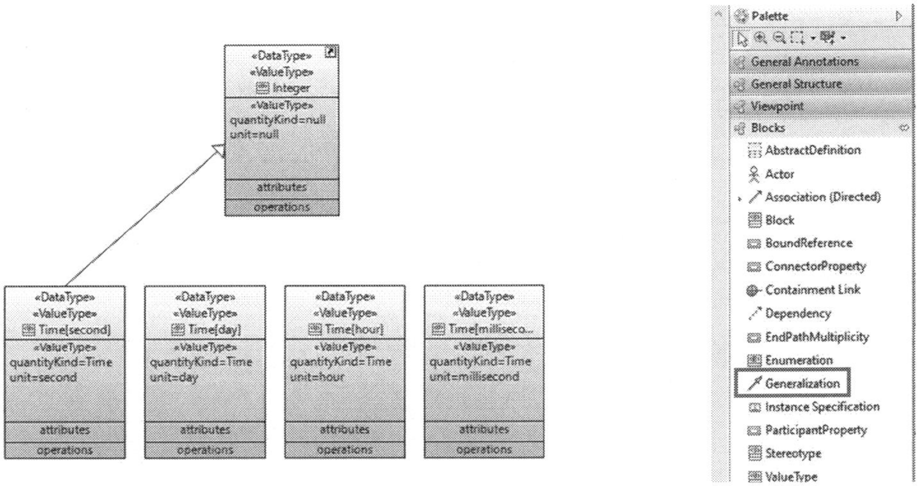

Figure 5.91. Create a generalization

Repeat this process for the other value types.

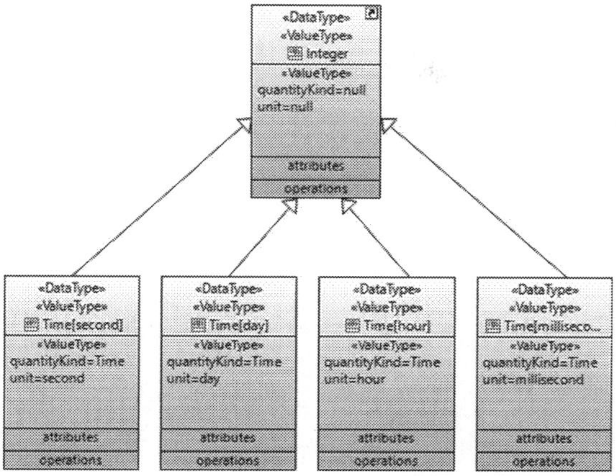

Figure 5.92. "Value Types BDD" with five value types and their relations

Repeat the process for the value type "Temperature[celsius]". Since we want to use decimal numbers to set a value for the temperature, instead of modeling a generalization to the "Integer" value type, model a generalization from the "Temperature[celsius]" value type to the "Real" value type.

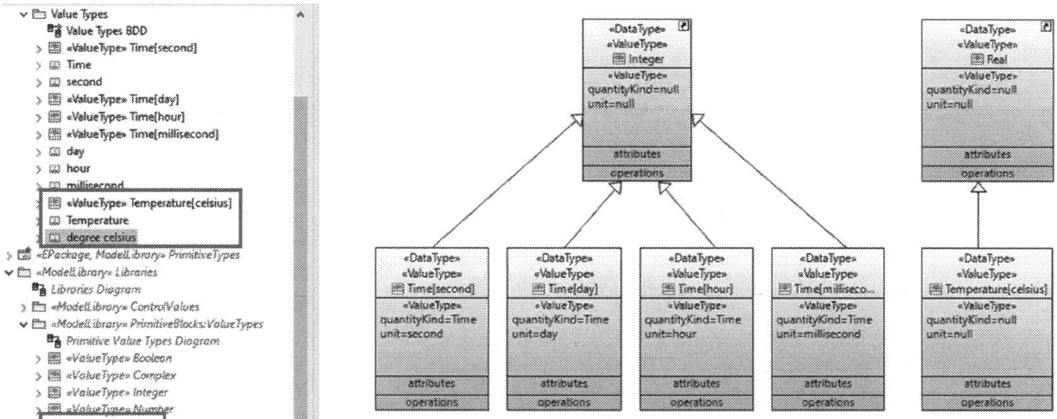

Figure 5.93. New value type, quantity kind and unit for temperature in degrees Celsius

Repeat the process for all the remaining value types.

The default view of a value type on a BDD adds compartments for attributes and operations that only clutter the diagram. To remove them, select all the value types on the diagram, right-click on one of them and select *Filters->Show/Hide Compartments.*

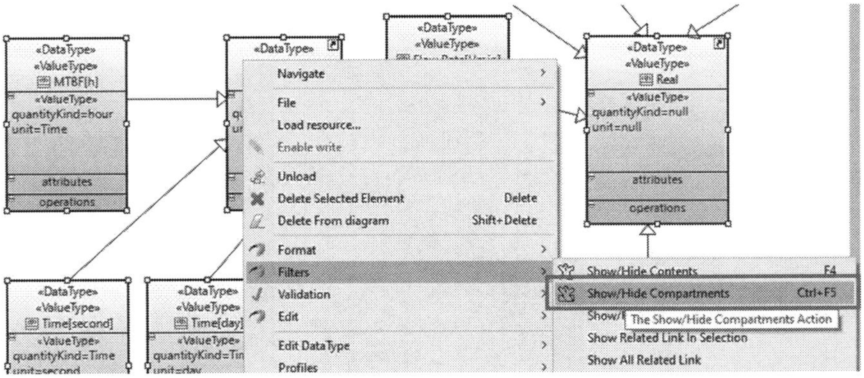

Figure 5.94. Open "Show/Hide Compartments" dialog

Remove all compartments except for the value type compartment from the selection and make this applicable for all the selected value types by clicking the button "Propagate selection to elements of same type".

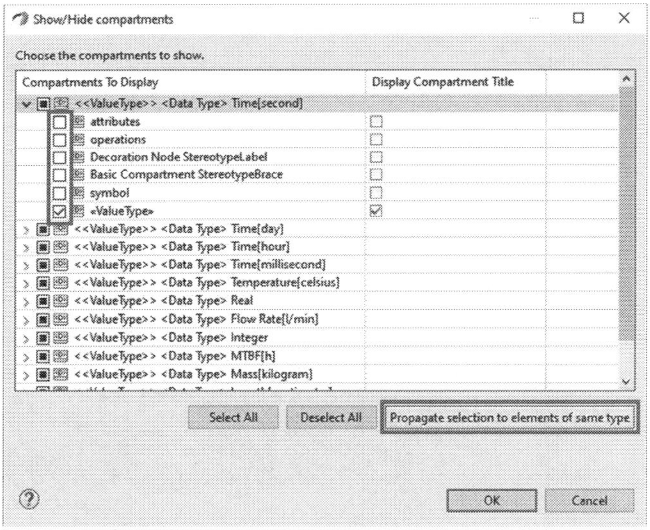

Figure 5.95. Select visible compartments

Move the elements around to create a tidy diagram.

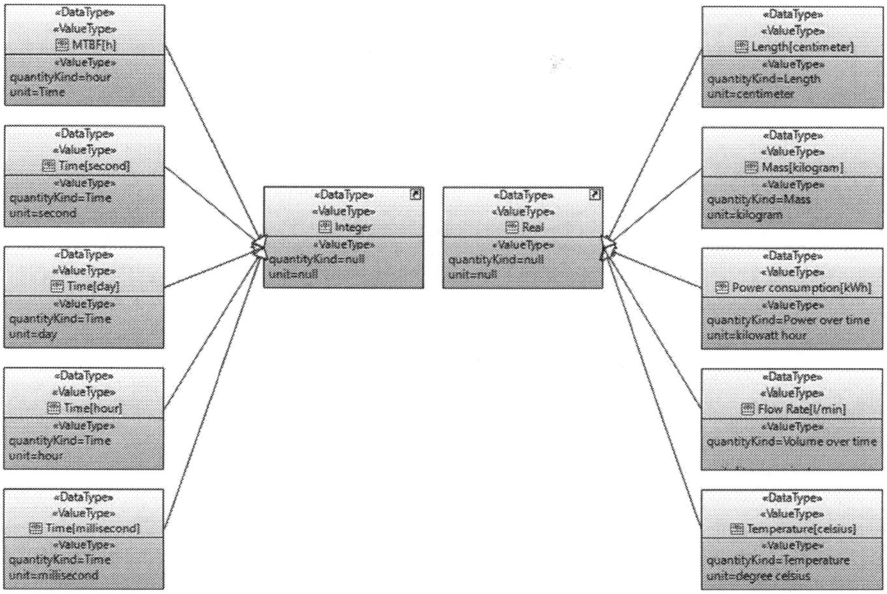

Figure 5.96. Finished ValueTypesBDD

Optionally, create a table for all the value types.

	quantityKind	unit	general
Time[second]	Time	second	Integer
Time[day]	Time	day	Integer
Time[hour]	Time	hour	Integer
Time[millisecond]	Time	millisecond	Integer
Temperature[celsius]	Temperature	degree celsius	Real
Mass[kilogram]	Mass	kilogram	Real
Length[centimeter]	Length	centimeter	Real
Flow Rate[l/min]	Volume over time	liter per minute	Real
Power consumption[kWh]	Power over time	kilowatt hour	Real
MTBF[h]	hour	Time	Integer

Figure 5.97. ValueTypes Table

Now, we want to specify the types of the MoEs that we have previously created using the new value types. Select the “Brewing Time” MoE of the “CoffeeMachine” block in the Model Explorer. Go to the Properties View and click on the button with the three dots next to the Type property.

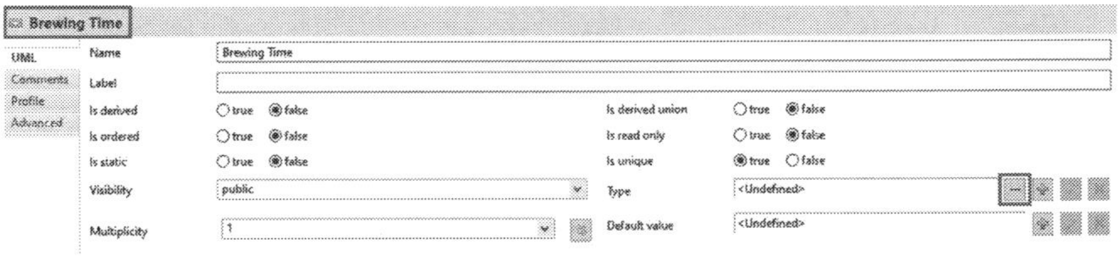

Figure 5.98. Edit value for the type of a MoE

Select the “Time[second]” value type.

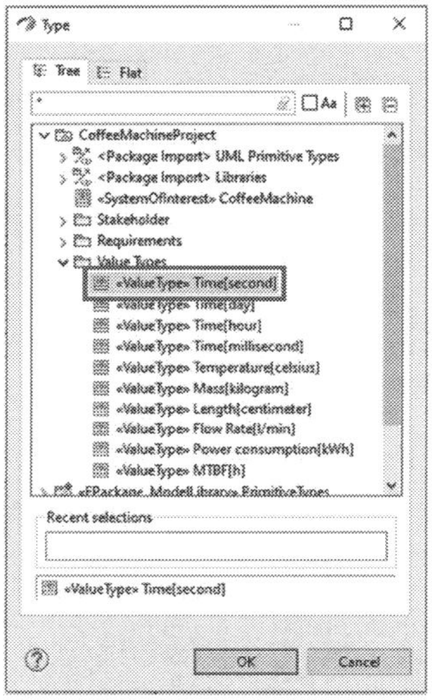

Figure 5.99. “Select Type” dialog

The type of the MoE will also be displayed in the Model Explorer.

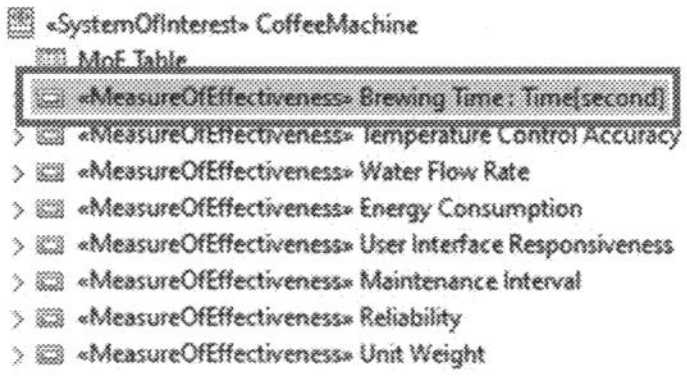

Figure 5.100. MoE with type in the Model Explorer

Repeat the process for all the other MoEs. If there is a table for the MoEs, you can add the type of the MoEs as an additional column.

	ownedComment	type
Brewing Time : Time[second]	The time taken to complete the brewing process.	Time[second]
Temperature Control Accuracy : ...	Precision in maintaining the desired brewing temperature.	Temperature[celsius]
Water Flow Rate : Flow Rate[l/min]	The rate at which water flows through the coffee machine dur...	Flow Rate[l/min]
Energy Consumption : Power con...	The amount of energy consumed during the brewing process.	Power consumption[kWh]
User Interface Responsiveness : ...	Time taken for the system to respond to user inputs.	Time[millisecond]
Maintenance Interval : Time[day]	Time between required maintenance tasks.	Time[day]
Reliability : MTBF[h]	The system's ability to operate consistently without failure...	MTBF[h]
Unit Weight : Mass[kilogram]	The weight of the coffee machine.	Mass[kilogram]

Figure 5.101. MoE table with types

Knowing which stakeholder requirement is related to a measure of effectiveness is important information. To add this information to the model, we need to establish a traceability link from the MoEs to their relevant stakeholder requirements. Since we also want to make these traceability links visible, we create them on a diagram. First, create a new requirement diagram in the "Stakeholder Requirements" package and name it "Stakeholder MoE Traceability RD". Add all the MoEs to the diagram as well as the stakeholder requirements to which they are related (see table above).

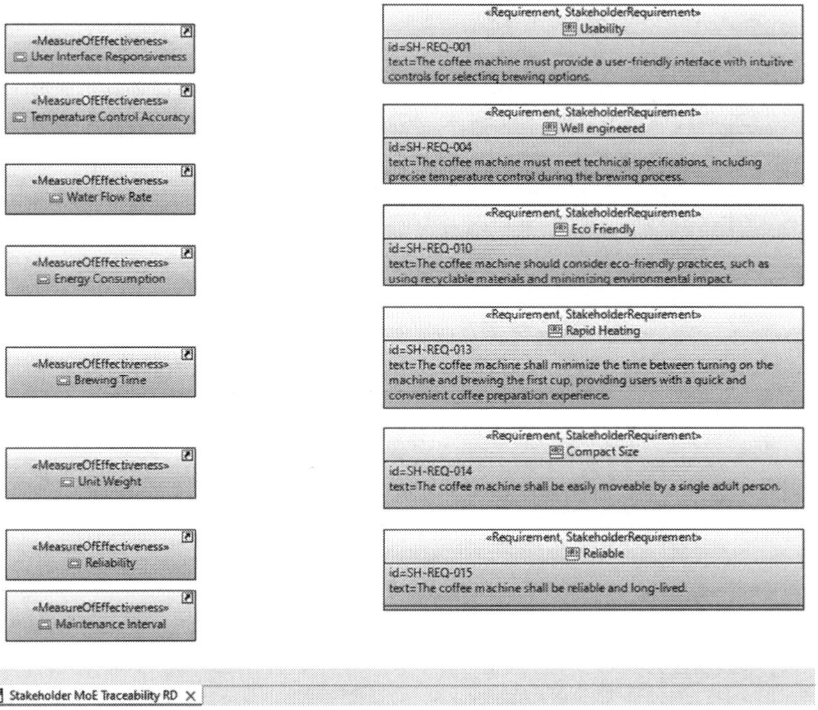
«MeasureOfEffectiveness»
User Interface Responsiveness
«MeasureOfEffectiveness»
Temperature Control Accuracy
«MeasureOfEffectiveness»
Water Flow Rate
«MeasureOfEffectiveness»
Energy Consumption
«MeasureOfEffectiveness»
Brewing Time
«MeasureOfEffectiveness»
Unit Weight
«MeasureOfEffectiveness»
Reliability
«MeasureOfEffectiveness»
Maintenance Interval
«Requirement, StakeholderRequirement»
Usability
id=SH-REQ-001
text=The coffee machine must provide a user-friendly interface with intuitive controls for selecting brewing options.
«Requirement, StakeholderRequirement»
Well engineered
id=SH-REQ-004
text=The coffee machine must meet technical specifications, including precise temperature control during the brewing process.
«Requirement, StakeholderRequirement»
Eco Friendly
id=SH-REQ-010
text=The coffee machine should consider eco-friendly practices, such as using recyclable materials and minimizing environmental impact.
«Requirement, StakeholderRequirement»
Rapid Heating
id=SH-REQ-013
text=The coffee machine shall minimize the time between turning on the machine and brewing the first cup, providing users with a quick and convenient coffee preparation experience.
«Requirement, StakeholderRequirement»
Compact Size
id=SH-REQ-014
text=The coffee machine shall be easily moveable by a single adult person.
«Requirement, StakeholderRequirement»
Reliable
id=SH-REQ-015
text=The coffee machine shall be reliable and long-lived.
Stakeholder MoE Traceability RD

Figure 5.102. Requirement diagram with MoEs and stakeholder requirements

Select the "Trace" element from the Diagram Palette and create a trace relation from the "User Interface Responsiveness" MoE to the "Usability stakeholder" requirement.

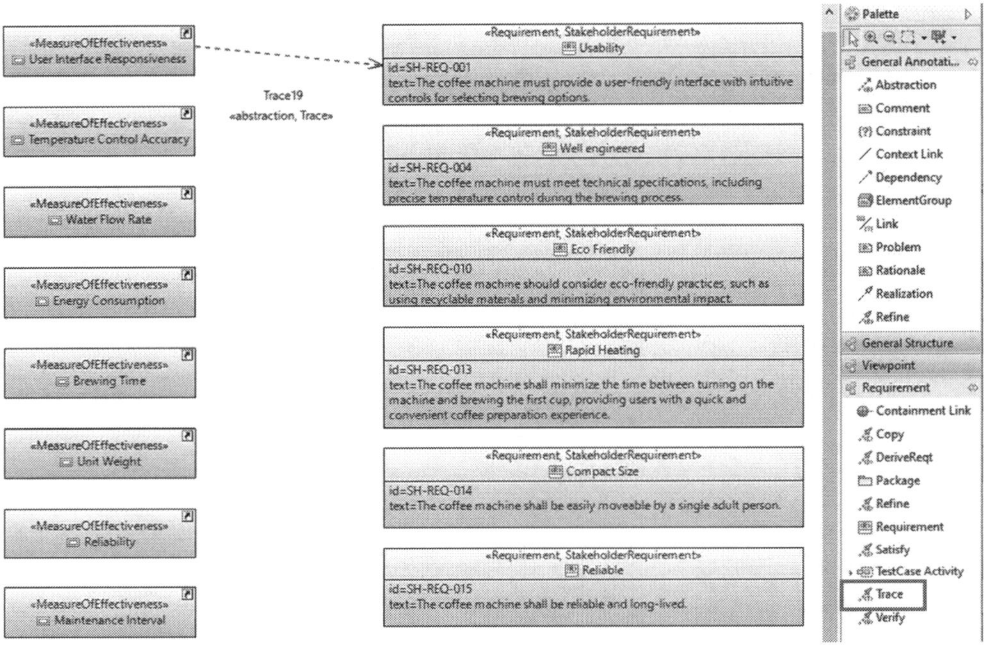

Figure 5.103. Create trace link

The SysML trace relation is a specialization of the UML trace relation, which is in turn based on the UML abstraction metaclass. Papyrus displays the stereotype abstraction on the trace relation due to these relations in the metamodel.

Repeat the same process for all the other MoEs.

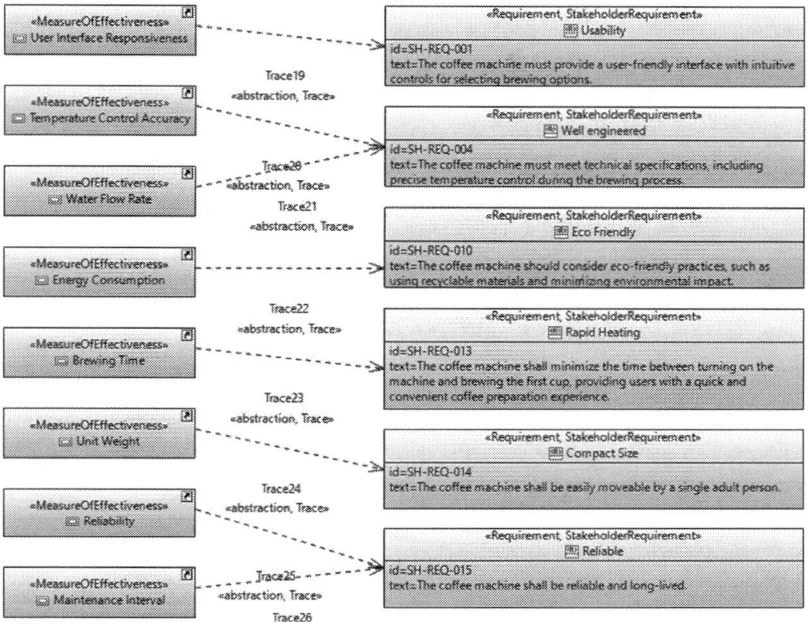

Figure 5.104. MoEs traced to stakeholder requirements

Trace links are model elements in their own right that have a name, which is, by default, displayed on the requirement diagram. Since the names of the trace links do not add information to the diagram, we can remove them from the diagram. To do this, select all the trace links on the diagram by clicking on the *Select Actions* button and then selecting *Select All Connectors.* Alternatively, you could achieve the same thing by holding Shift or Control and selecting all the trace links individually.

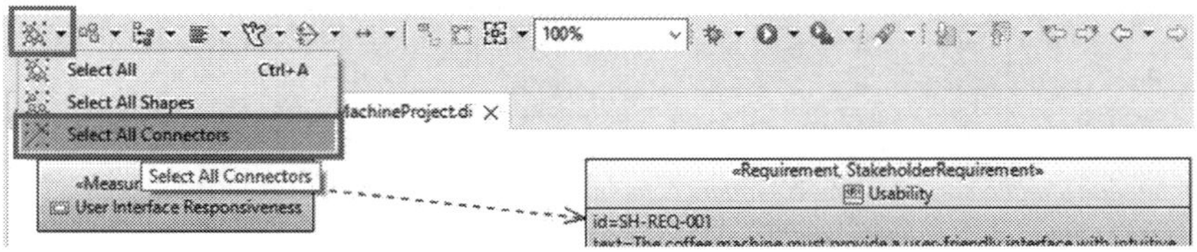

Figure 5.105. Select all connectors

While all the trace links are selected, right-click on one of them and select *Filters->Show/Hide Labels.*

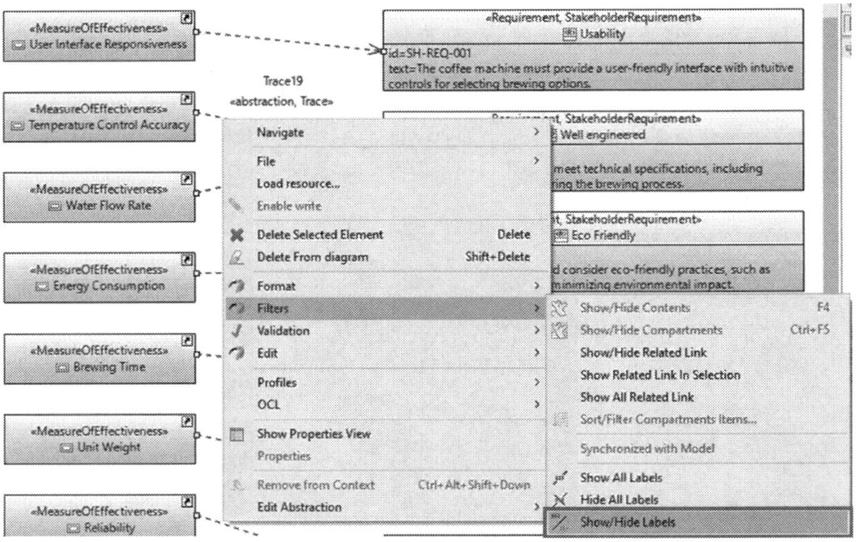

Figure 5.106. Open the "Show/Hide Labels" dialog

Remove the selection name for all of the trace links.

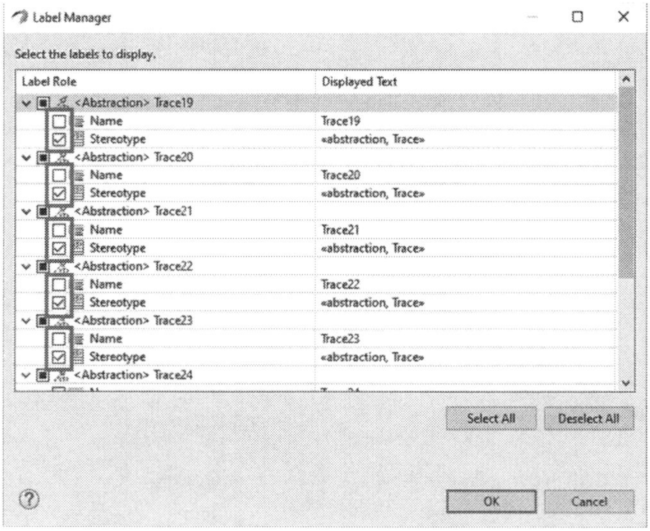

Figure 5.107. "Show/Hide Labels" dialog

The finished "Stakeholder MoE Traceability RD" diagram should look like this.

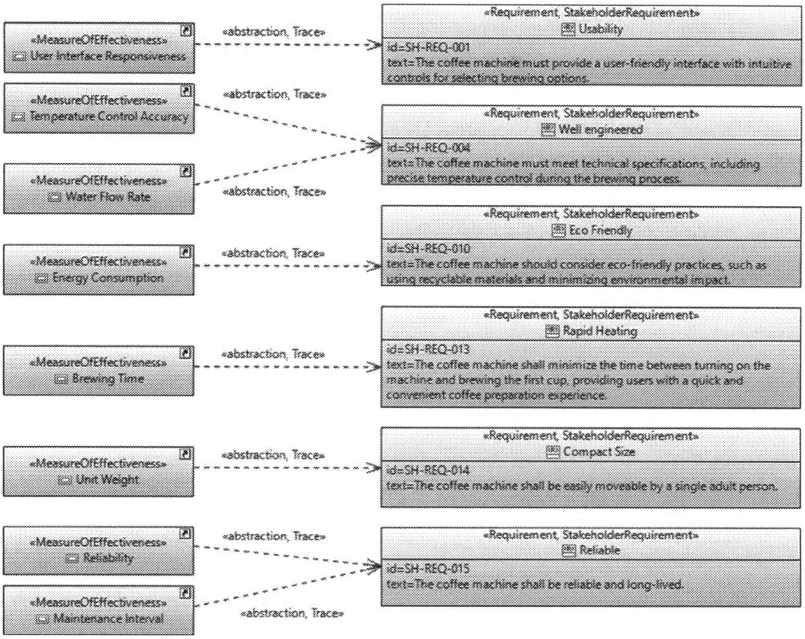

Figure 5.108. Finished "Stakeholder MoE Traceability RD"

Now that we have modeled several elements and established relations between them, we can use the navigate feature of Papyrus to jump between related elements or jump from an element to its usage in a diagram. Right-click on a MoE and select *Navigate*. The first entry in the Navigate menu, Go to element, allows jumping to diagrams on which the element appears.

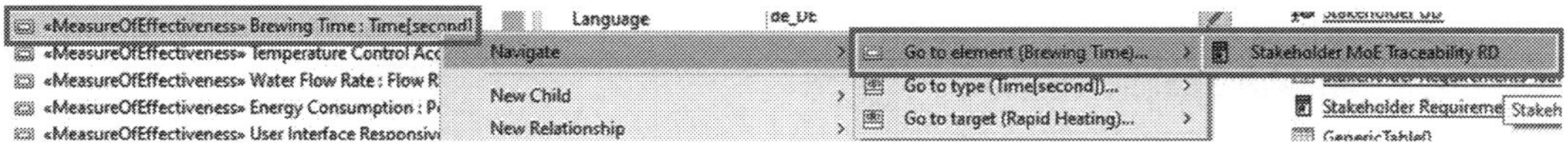

Figure 5.109. Navigate->Go to element

The second entry in the Navigate menu, Go to type, allows navigating to the type of the MoE. The additional sub-menu that opens allows jumping to the type in the Model Explorer or to a diagram on which the type element appears.

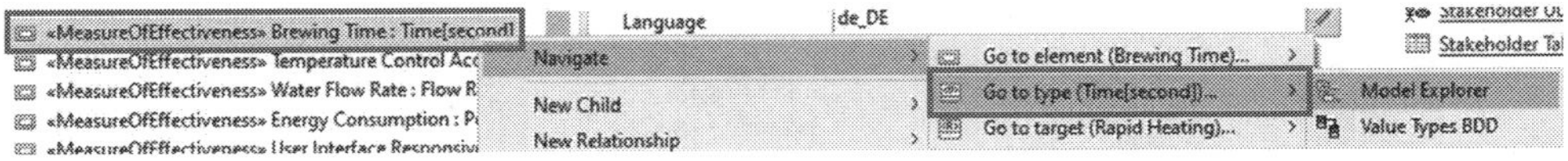

Figure 5.110. Navigate->Go to type

The third entry in the Navigate menu, Go to the target, allows navigating to any other elements to which a trace link had been modeled. In the case of the MoE, we can jump to the stakeholder requirement to which the MoE is related. And, again, we can either jump to the stakeholder requirement in the Model Explorer or to any of the diagrams on which it appears.

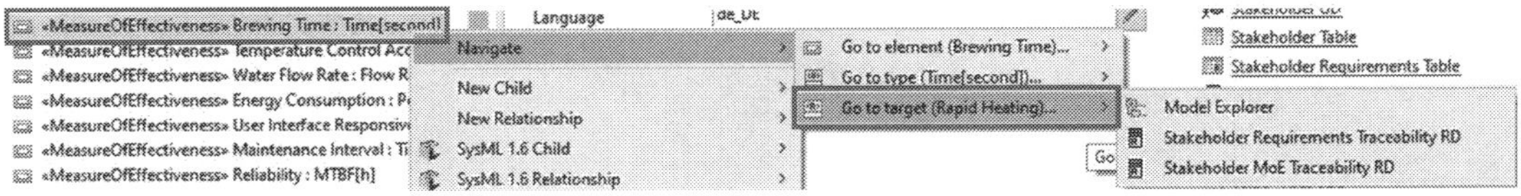

Figure 5.111. Navigate->Go to target

The navigation between elements works in both directions. If you click on the stakeholder requirement "Rapid Heating," it is possible to navigate to the related MoE "Brewing Time" by right-clicking and selecting *Navigate->Go to source.*

Figure 5.112. Navigate->Go to source

Navigation in SysML models is a valuable feature and one of the main selling points of MBSE in the first place. It enables:

- **Enhanced understanding**: Navigating through related elements in a SysML model can help to improve understanding of the system's overall structure and behavior. By seeing how different elements are connected, it becomes easier to see the big picture and identify potential problems or areas for improvement.
- **Improved traceability**: Navigation can also help to improve traceability, which is the ability to follow the flow of information or requirements through the system development lifecycle. By being able to easily navigate between related elements, it becomes easier to ensure that changes are made in a consistent and controlled manner.
- **Reduced rework**: Navigation can also help to reduce rework in the system development process. By being able to quickly identify related elements, it becomes easier

to find and fix problems early in the development process, before they have the opportunity to cause significant delays or costs.

- **Collaboration**: Navigation can also facilitate collaboration between different stakeholders in the system development process. By being able to easily share and view models, it becomes easier for engineers, managers, and other stakeholders to work together effectively.
- **Reusability**: Navigation can also promote the reuse of SysML models. By being able to easily identify and access related elements, it becomes easier to reuse existing models in new projects. This can save time and effort, and it can also help to ensure that the system development process is consistent and efficient.

5.5 Define Context and High-Level Interfaces

Context analysis in systems engineering serves the crucial purpose of understanding the broader environment in which a system operates. It provides a comprehensive view of the external factors, stakeholders, and influences that can impact the system's design, development, and operation. The primary goals of context analysis include:

- **System boundaries definition**: Identify and define the boundaries of the system of interest, distinguishing it from its external environment.
- **Scope definition**: Clearly define the scope of the system, ensuring that the system's objectives align with the needs and expectations of stakeholders.
- **Environmental considerations**: Examine the external environment to understand the physical, organizational, regulatory, and cultural factors that may impact the system.
- **Requirements elicitation**: Gather and document initial requirements and constraints based on the system's context, ensuring that the system is designed to meet external expectations.
- **Risk assessment**: Identify potential risks and challenges associated with the system's context, enabling proactive risk management strategies.

The main output of the context analysis is a context diagram. A context diagram visually represents the system, its boundaries, and the external entities with which it interacts. It provides a high-level overview of the system's context and the interfaces it has with external entities.

By conducting context analysis early in the systems engineering process, teams can establish a solid foundation for subsequent activities, ensuring that the system is designed and developed with a clear understanding of its external environment and stakeholder needs.

5.5.1 Context Analysis Pitfalls

System context analysis is a critical phase in systems engineering that involves understanding the broader environment in which a system operates. While conducting system context analysis, it's essential to be aware of potential pitfalls that can affect the accuracy and completeness of the analysis. Here are some common pitfalls to avoid:

- **Limited Consideration of Lifecycle Phases**: Focusing only on the current phase of the system lifecycle can result in overlooking future changes and requirements. Take a lifecycle approach, considering the entire lifespan of the system, from concept through design, implementation, operation, and retirement.
- **Static Analysis Without Evolution Consideration**: Treating the system context as static without considering how it might evolve over time can lead to inadequate long-term planning. Incorporate a dynamic perspective by forecasting how the system context might change and identifying potential future challenges.
- **Lack of Regular Updates**: Failing to update the system context analysis regularly can result in outdated information and an inaccurate representation of the current environment. Establish a process for regular reviews and updates to ensure that the system context analysis remains relevant and reflects changes in the external environment.
- **Failure to Validate with Stakeholders**: Conducting the context analysis in isolation without validation from stakeholders can lead to misinterpretations and inaccuracies. Regularly engage with stakeholders to validate assumptions, gather feedback, and ensure the accuracy of the system context analysis.

By avoiding these common pitfalls, system engineers can conduct a more thorough and accurate system context analysis, laying a solid foundation for subsequent phases of the systems engineering process.

5.5.2 Context Elements and High-Level Interfaces of the Coffee Machine

The elements of the context of the coffee machine:

Name	Description
Power Network	The electrical power source that provides energy to the coffee machine.
Water Network	The water source supplying water for the coffee brewing process.
User	The user of the coffee machine.
Coffee Machine	The system of interest itself is also part of its context.

The High-level interface of the coffee machine:

Name	From	To	Exchanged Items
Power Network	Power Network	Coffee Machine	Power
Water Network	Water Network	Coffee Machine	Water
User to Coffee Machine	User	Coffee Machine	Coffee Pad, User Request
Coffee Machine to User	Coffee Machine	User	Status Information, Coffee

5.5.3 Modeling the Coffee Machine Context

Create a new package for the context analysis of the coffee machine and name it "Context". In the new package, create a new BDD called "Coffee Machine Context BDD".

> Note that a system can have different contexts based on the lifecycle or operating phase that it is currently in, e.g., the operational context, the maintenance context, or the decommissioning context. Since we are only looking at the operational context of the coffee machine here, it is just called context.

In previous sections, we created model elements in the Model Explorer and put them on a diagram afterward. However, we can also create model elements directly in a diagram. To do this, select the "Block" element in the Diagram Palette and left-click on the diagram canvas.

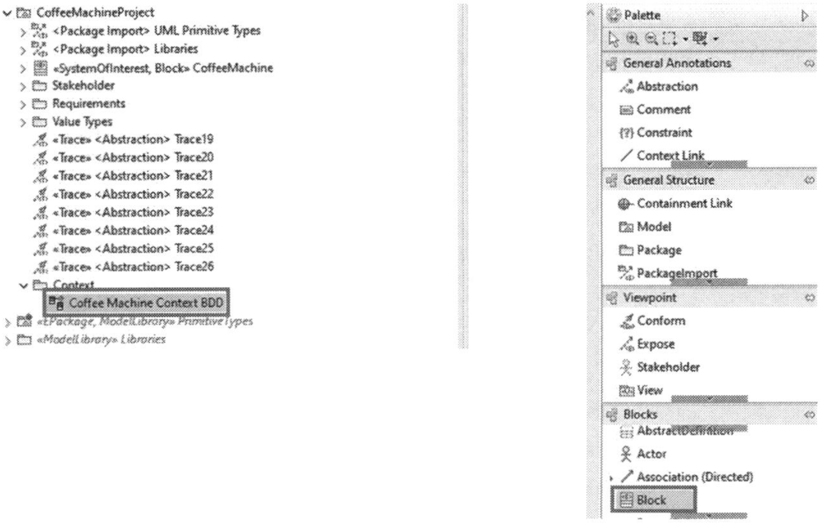

Figure 5.113. Create a new block

Name the newly created block “Coffee Machine Context”.

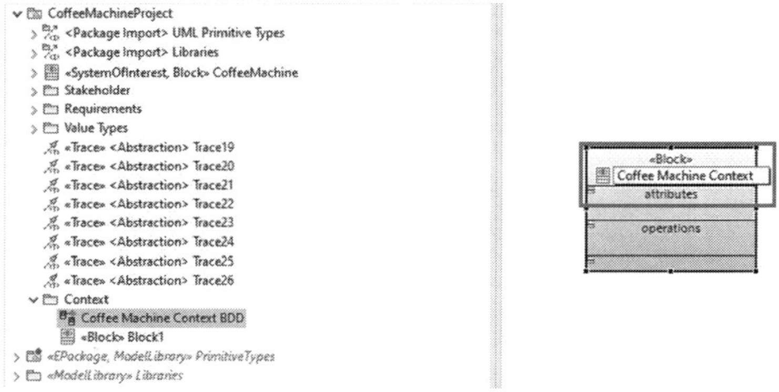

Figure 5.114. Name the new block

Repeat the same process for the context elements “Power Network” and “Water Network”. Note that the new blocks are also shown in the Model Explorer in the same package in which the diagram resides.

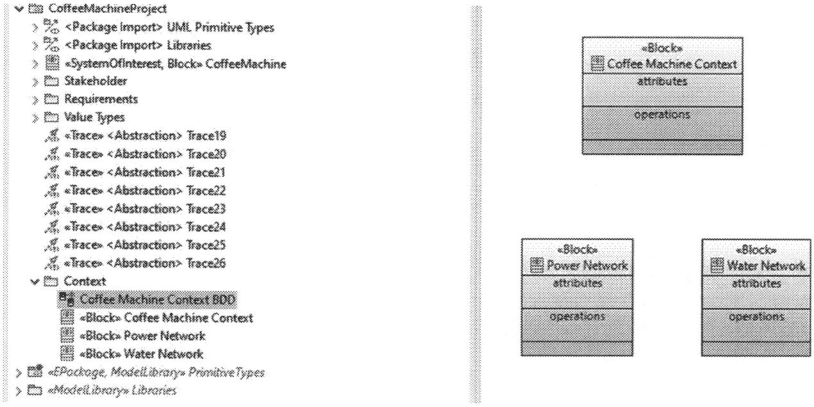

Figure 5.115. "Power Network" and "Water Network" added

Do not create a new block for the coffee machine. Remember, that creating the block representing our system of interest was one of the very first modeling steps. So, drag and drop the "CoffeeMachine" block that was already created on the diagram.

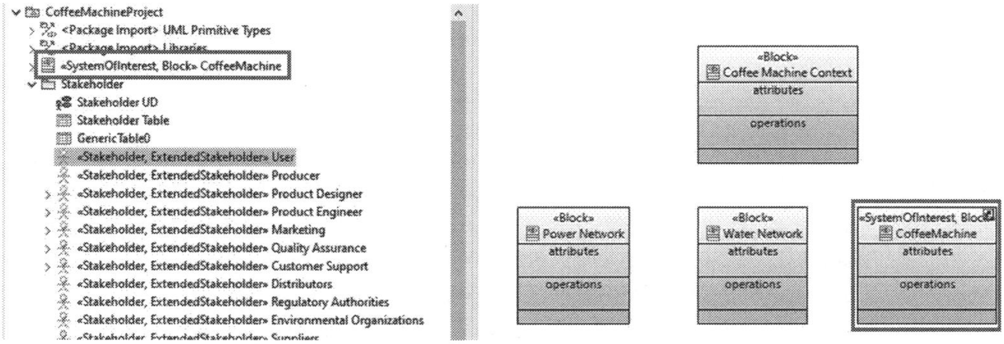

Figure 5.116. "CoffeeMachine" added to context BDD

The user is also part of the (operational) context and was also created before during the stakeholder analysis. Add the "User" stakeholder to the context BDD.

> **i** Unfortunately, Papyrus is not able to display the names of actor elements on Block Definition Diagrams. This is a known issue that might be fixed in future versions. If there are several actors on a diagram, a possible workaround is to create a note containing the name of an actor and add it to the diagram.

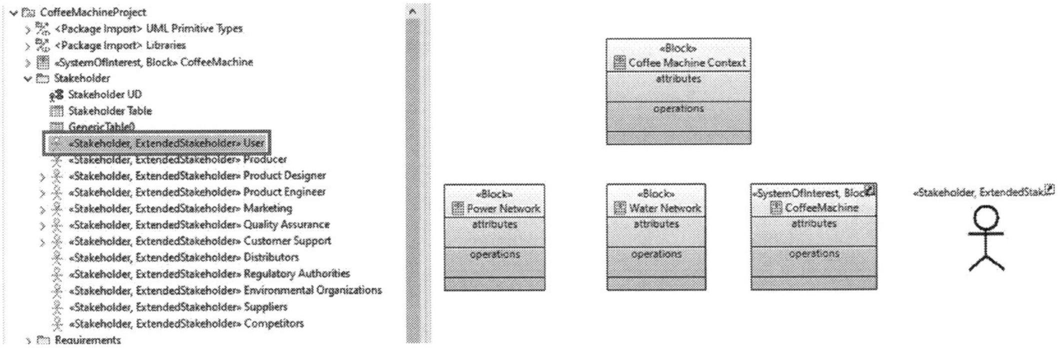

Figure 5.117. "User" added to context BDD

So far, all the blocks that have been created exist next to each other. Now, it is time to model their relationship, i.e., the "User", the "Power Network", the "Water Network," and the "CoffeeMachine" are parts of the context.

A part property is another kind of property belonging to a SysML Block. It is a property that is not typed by a value type but by another block and represents the usage of one block inside another block. Part properties are used to model the hierarchical structure of a system.

To model the composition of the context block, select the "Composite Association (Directed)" relation in the Diagram Palette. To find the "Composite Association (Directed)" relation, expand the "Association (Directed)" relation in the Diagram Palette by clicking on the small downward-facing triangle next to it.

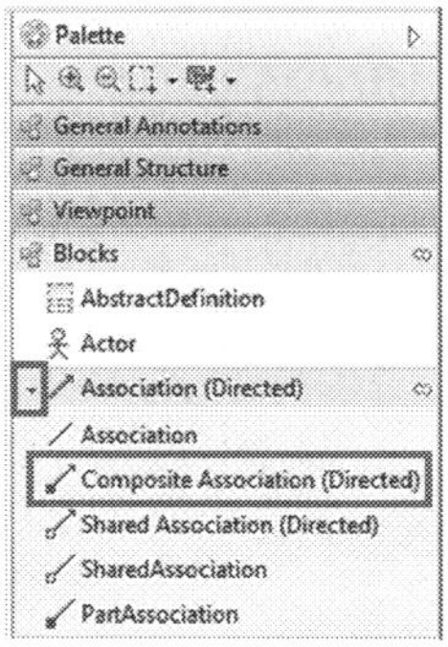

Figure 5.118. Select the "Composite Association (Directed)" relation

After selecting the "Composite Association (Directed)" relation, left-click first on the "Coffee Machine Context" block and then on the "Power Network" block. The composition relation is displayed as an arrow pointing from the owner, the "Coffee Machine Context", to the block which types the part, the "Power Network" block. The arrow has a filled diamond at one end and an open arrowhead at the other end.

The additional black dot at the part end of the composite association signifies that any properties belonging to the association are owned by the classifier, i.e., the block. If the properties are owned by the association itself, then the black dot is not displayed. Typically, in other tools, the properties are, by default, owned by the association, and, thus, the black dot is not shown. You can remove the black dot by setting the owner property of the association from "Classifier" to "Association". It does not have a significant bearing on the model and is very rarely used in practice.

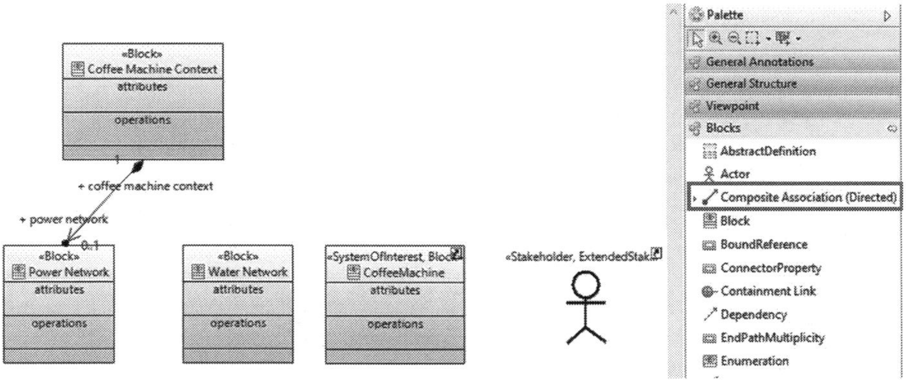

Figure 5.119. Newly created composition relation between two blocks

Repeat the same process for the other blocks and the "User" stakeholder.

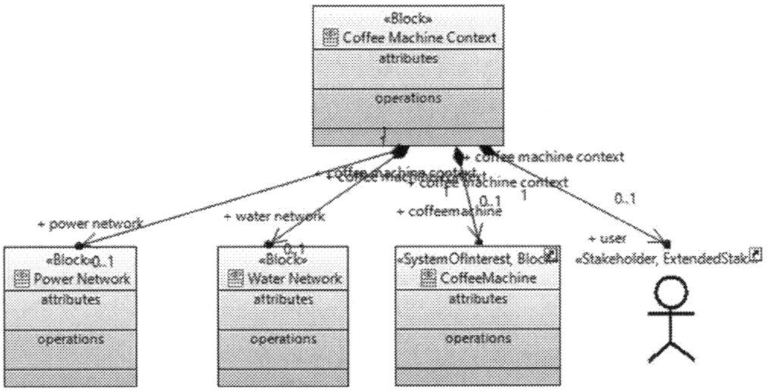

Figure 5.120. Decomposition of the coffee machine context

Note that modeling the decomposition relation in the diagram led to the addition of the part properties in the "Coffee Machine Context" block in the Model Explorer. The part properties have, by default, the same name as the blocks by which they are typed but start with a lowercase letter.

Renaming the part properties is not necessary for the simple model that we are building here; since every block is only used once as a part in the context. If a block would be used several times with a different role then it would be appropriate to name the part properties with the name of the role. If a "Car" block has four different part properties that are typed by a "Wheel" block, it would be a good idea to name the part properties "front left wheel", "front right wheel", "back left wheel", and "back right wheel" respectively.

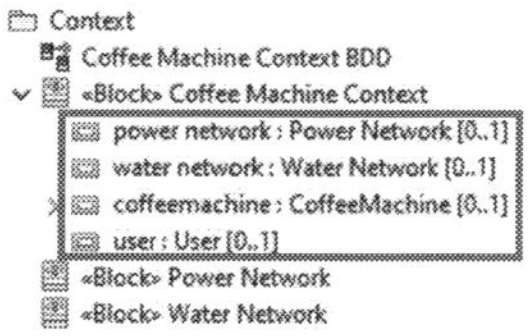

Figure 5.121. Part properties in the Model Explorer

The diagram is now complete but looks a bit cluttered. The first step is to remove the attributes and operations compartments of the blocks from being displayed.

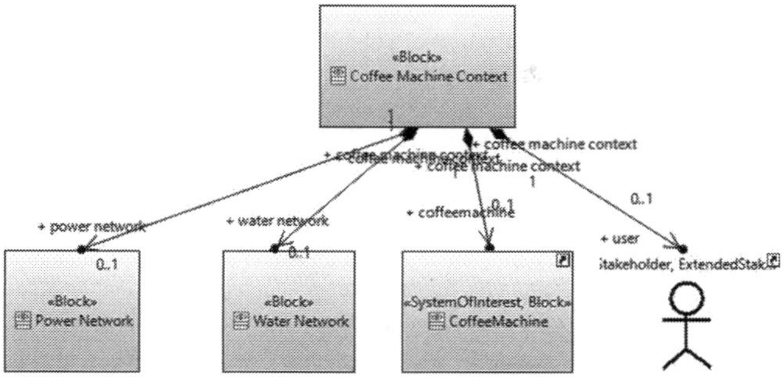

Figure 5.122. Remove attributes and operations compartments

Then, remove all the labels on the composition relation arrows except for the target multiplicity.

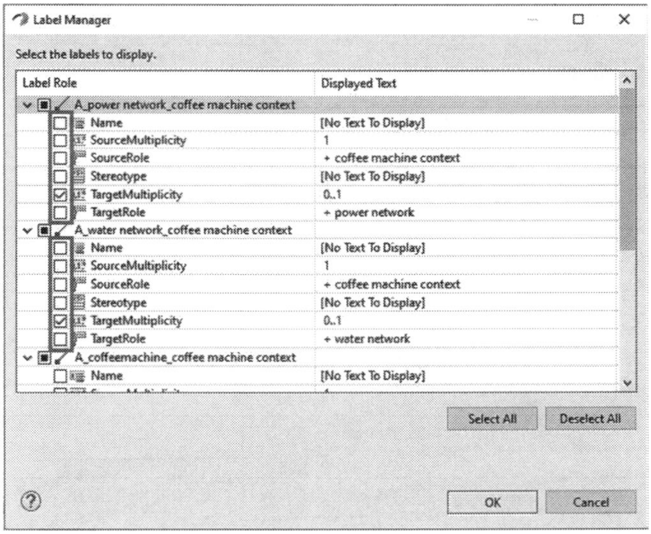

Figure 5.123. Remove labels from relations

Part properties can have multiplicity, which is a property that is used to specify the number of instances of a part or relationship that can exist. In a "Car" Block, the "wheel" part property may have a multiplicity of 4, indicating that there are four wheels on the car.

For our analyses, we do not need multiples of the context parts, so we can set the multiplicity of the part properties to 1. Select the composition relation, and in the UML tab of the Properties View, set the multiplicity to 1.

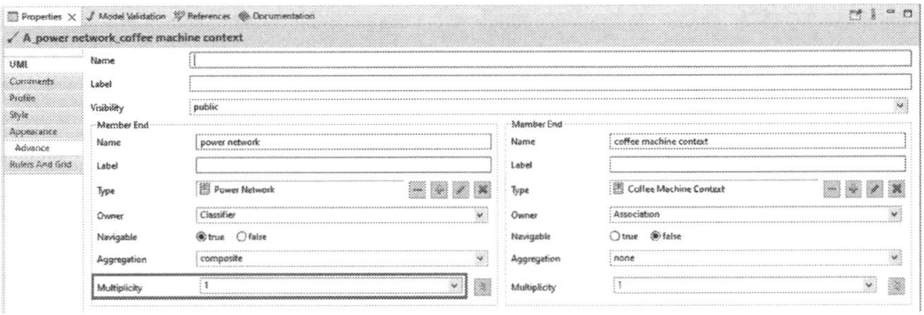

Figure 5.124. Set the multiplicity of a part

Papyrus has a feature for automatic layouting, i.e., to arrange elements on a diagram automatically. Click on the Papyrus menu and select *Diagram->Arrange->All* to activate it and see how the diagram is changed.

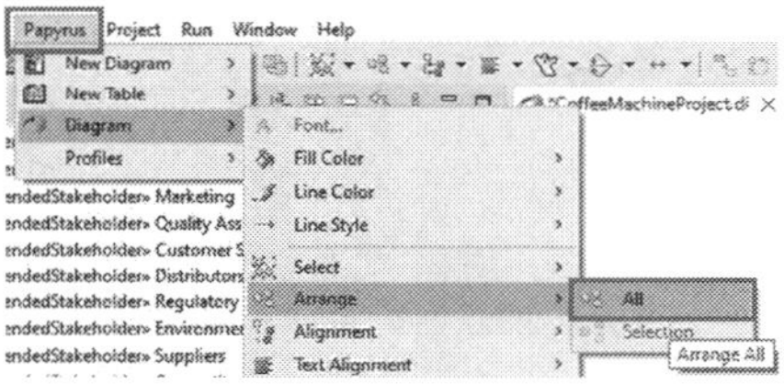

Figure 5.125. Arrange all elements

Papyrus tries to organize the elements on the diagram.

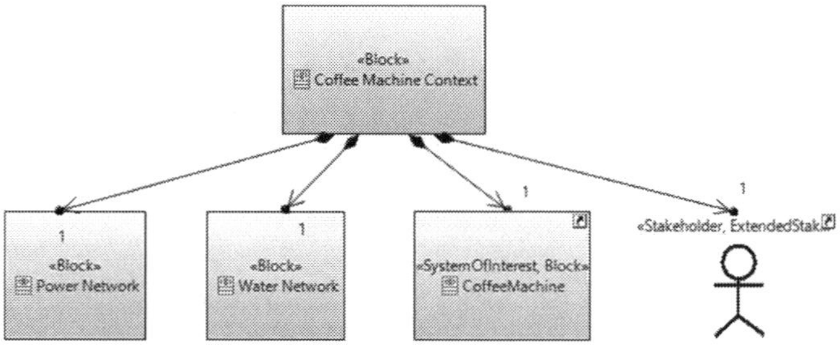

Figure 5.126. Context BDD after auto-layout

Relations between blocks on a BDD are typically not displayed as straight arrows. Instead, a tree-like structure with rectilinear lines is normally used. To change the line style, select all the composite relations on the diagram, right-click on one of them, and select *Format->Line Style->Tree Style Routing.*

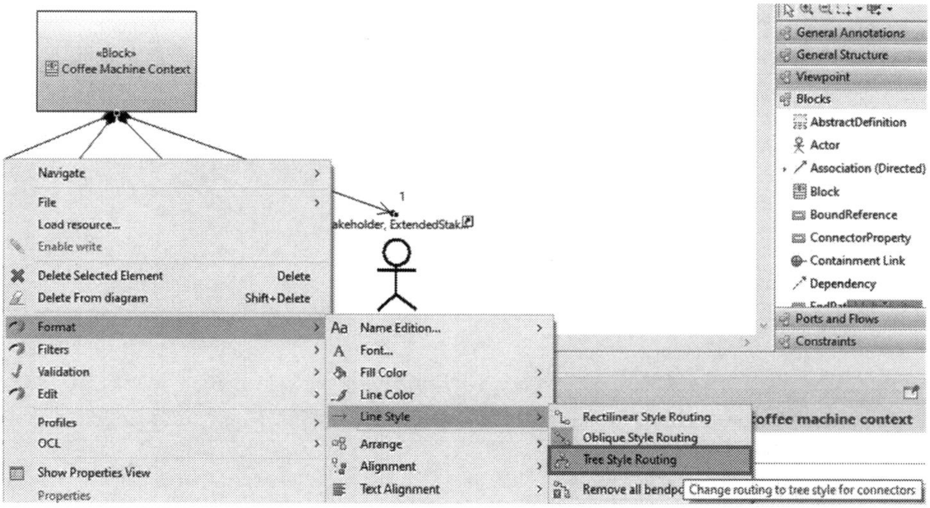

Figure 5.127. Change the line style

If necessary, move the lines around until the diagram is finished.

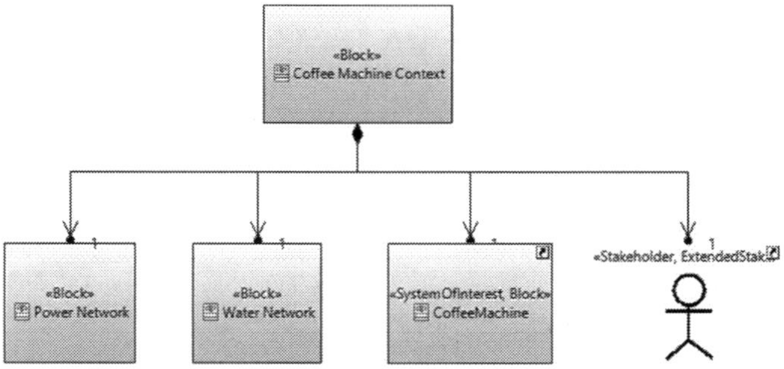

Figure 5.128. Finished "Coffee Machine Context" BDD

Next, we want to model the relations between the parts inside the context block. For this, we need an Internal Block Diagram (IBD). Note that all of the diagrams, that were created so far belonged to a package. The IBD is different as it belongs to the block, whose internal structure it shows. Right-click on the "Coffee Machine Context" block and select *New Diagram->SysML 1.6 Internal Block Diagram.*

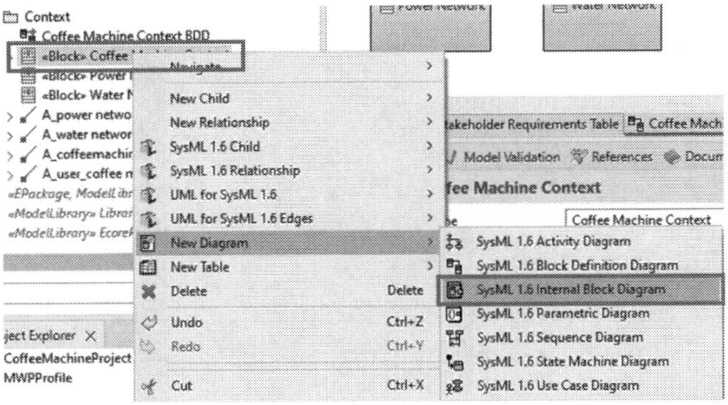

Figure 5.129. Create a new IBD

Name the new IBD "Coffee Machine Context IBD". Note how it is displayed as a part of the Coffee Machine Context block in the Model Explorer. The "Coffee Machine Context" block is automatically added to the new IBD.

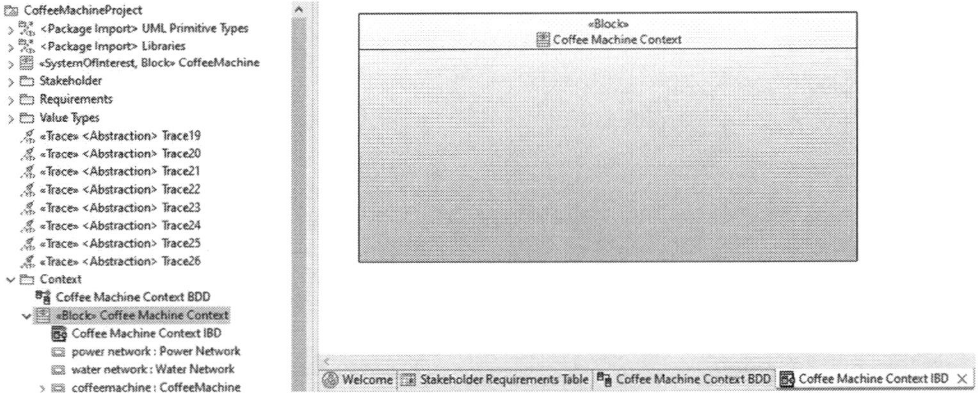

Figure 5.130. New "Coffee Machine Context IBD"

Drag and drop the part properties of the "Coffee Machine Context" block onto the diagram.

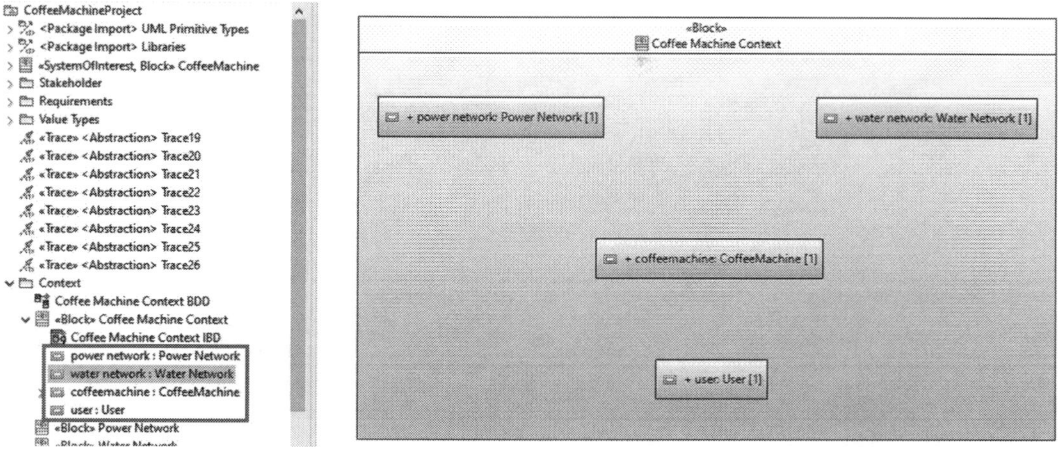

Figure 5.131. Part properties added to the IBD

To model the high-level interfaces between the context elements and the coffee machine, we use item flows. To create a new item flow, select the "Item Flow" element in the Diagram Palette and click and draw from the "power network" part to the "coffeemachine" part.

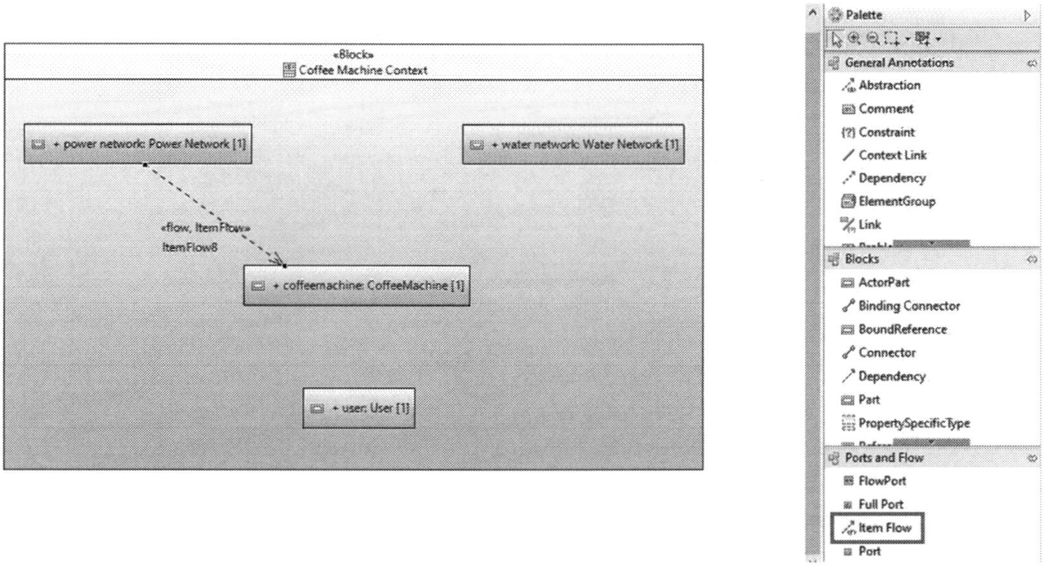

Figure 5.132. Create an item flow

Repeat the same process for the other high-level interfaces. Do not worry if the labels on some of the created elements overlap. It is good practice to make the diagram look nice after all the elements are created.

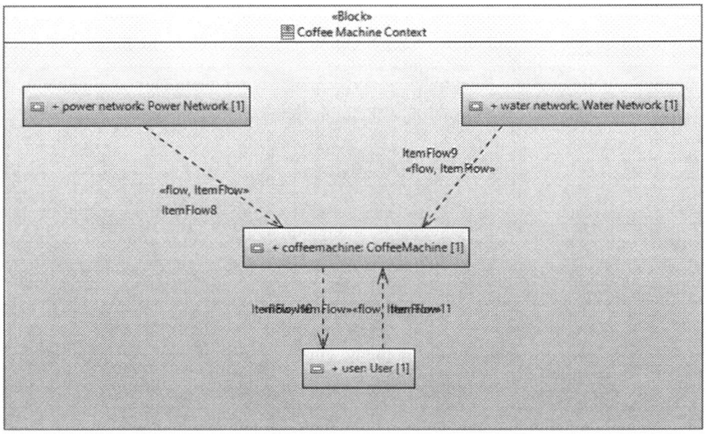

Figure 5.133. Several item flows on the IBD

Rename the item flows to reflect the name of the identified high-level interfaces.

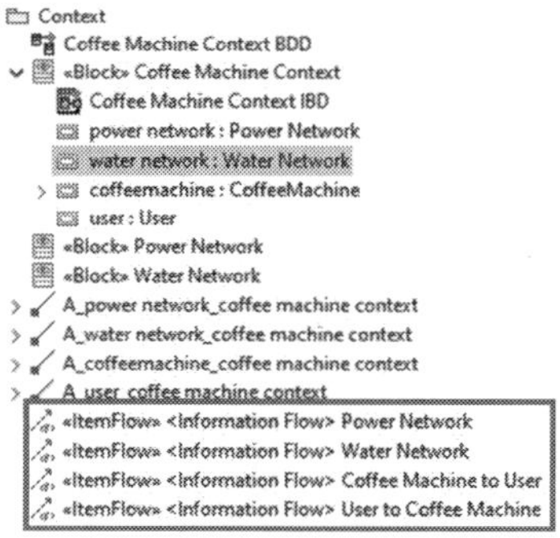

Figure 5.134. Renamed item flows

In the next step, we want to model, what is exchanged between the context elements. Create new value types in the “Value Types” package for each of the items that are exchanged between the parts.

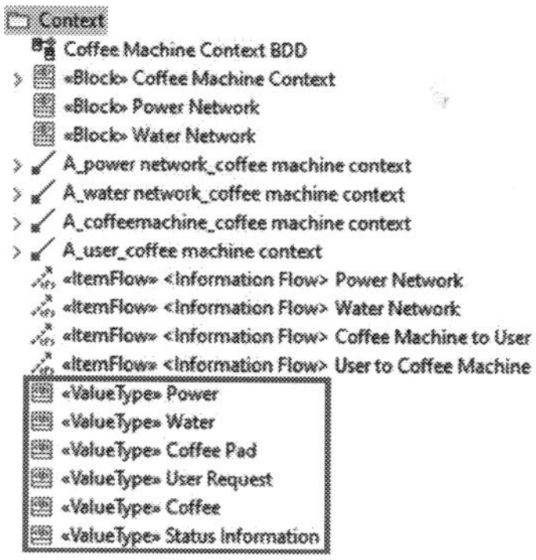

Figure 5.135. Value types for high-level interfaces

To add the "Power" value type to the "Power Network" item flow, select the item flow and click on the button with the plus symbol in the conveyed section of the UML tab of the Properties View of the item flow.

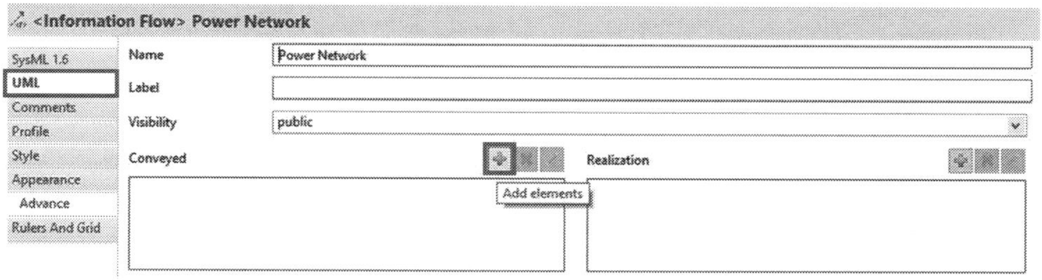

Figure 5.136. Add value type to item flow

In the "Conveyed" dialog, select the value type "Power" and move it to the right side by clicking on the button with the single arrow from left to right.

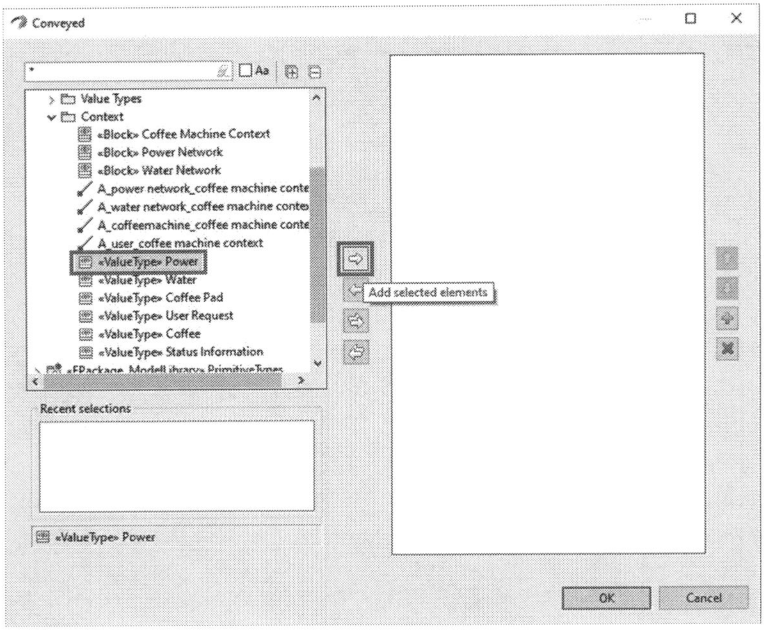

Figure 5.137. Add the “Power” value type

The diagram is now complete but looks a bit cluttered.

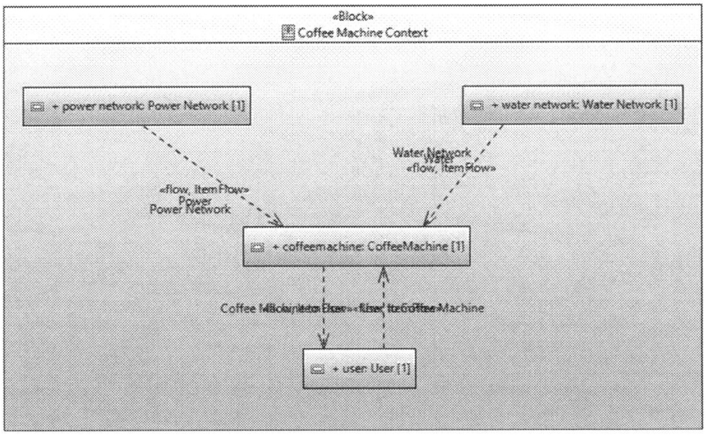

Figure 5.138. Complete but unreadable context IBD

To make it more readable, again, we need to remove some of the labels. Select all the item flows on the diagram and open the “Label Manager” dialog. Remove all of the labels except for the conveyed elements.

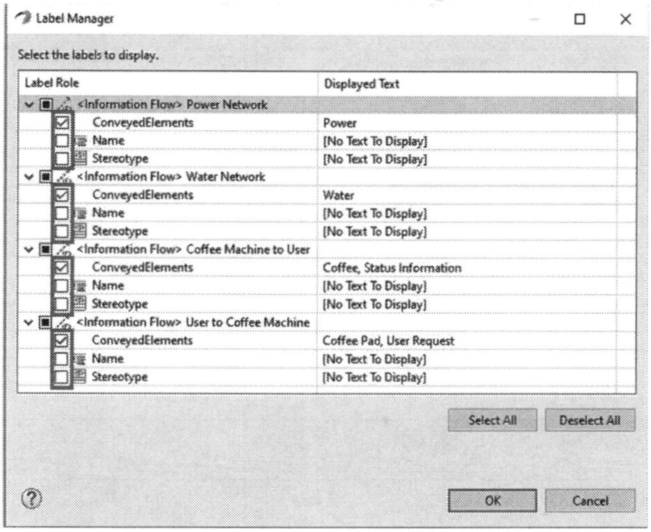

Figure 5.139. "Label Manager" dialog

If necessary, arrange the elements until the diagram is finished.

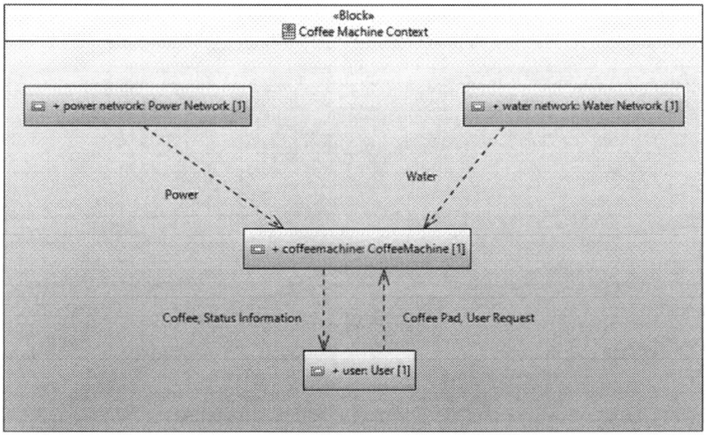

Figure 5.140. Finished context IBD

Optionally, a table to summarize the high-level interfaces of the coffee machine can be created. Create a new generic table, name it "High Level Interfaces Table", and add the item flows of the context to the table. Open the "Select Columns" dialog for the table and add “source”, “target”, and “conveyed” properties to the displayed columns. If you do not find one of the properties in the tree, you can enter the name you are looking for in the text field at the top of the “Select Columns” dialog.

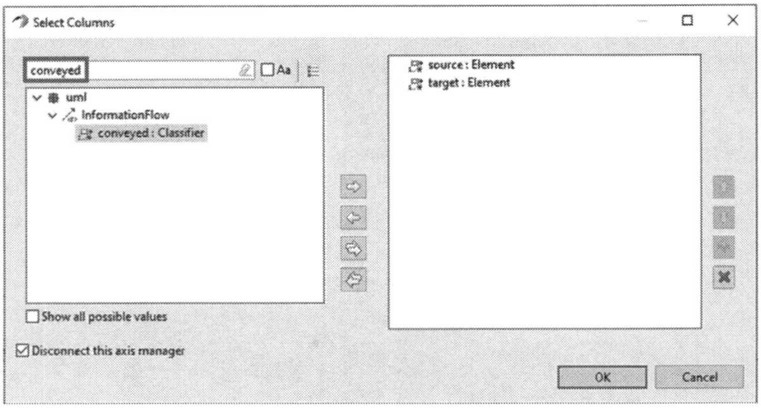

Figure 5.141. Select properties for the table

It is possible to change the labels of the columns in a table. To do this, right-click on the source column and select *Rename Header*.

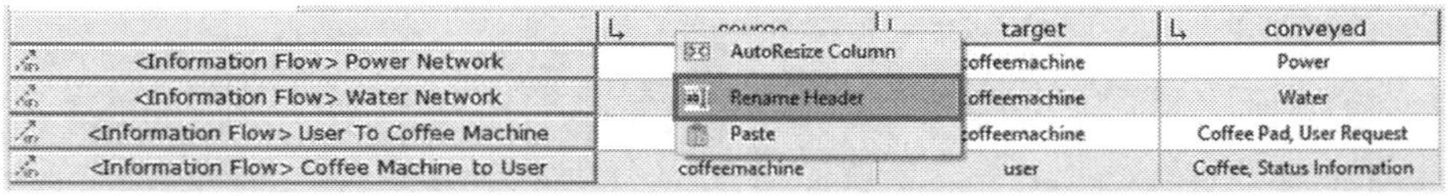

Figure 5.142. Rename the header in the table

Set the new name of the column to “From”.

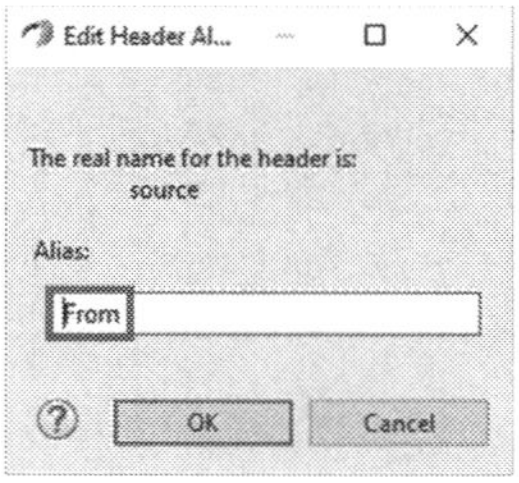

Figure 5.143. Set the displayed name of a column

Repeat the process for the target column and rename it to “To”.

	↳ From	↳ To	↳ conveyed
<Information Flow> Power Network	power network	coffeemachine	Power
<Information Flow> Water Network	water network	coffeemachine	Water
<Information Flow> Coffee Machine to User	coffeemachine	user	Coffee, Status Information
<Information Flow> User to Coffee Machine	user	coffeemachine	Coffee Pad, User Request

Figure 5.144. Finished "High Level Interfaces Table"

5.6 Identify Use Cases

Use cases are a modeling technique commonly used in systems engineering to capture and describe the interactions between users or other external entities and a system. They provide a detailed and concrete representation of how users interact with a system to accomplish specific tasks or achieve particular goals. Use cases are essential for understanding system functionality from an end-user perspective. Use Cases support the systems engineering process in various aspects:

- **Requirements elicitation**: Use cases are employed to identify and document functional requirements by capturing various scenarios in which users interact with the system to achieve specific objectives.
- **Enhanced user experience** Use case analysis focuses on understanding the user's perspective, ensuring that the system is designed to meet their needs and expectations. By identifying and prioritizing user tasks and scenarios, systems engineers can create a system that is intuitive, user-friendly, and effective in supporting user goals.
- **Communication and collaboration**: Use cases serve as a common language for stakeholders to discuss and understand system requirements. They provide a shared framework for communicating requirements across different teams, including developers, users, and managers. This shared understanding fosters collaboration and ensures that everyone is working towards a common goal of creating a system that meets user expectations.
- **System design**: Use cases inform system design by detailing the expected interactions and behaviors between the system and its users. They guide the development of user interfaces and the overall system architecture.
- **Verification and validation**: Use cases contribute to the V&V processes by providing test scenarios. Test cases can be derived from use cases to ensure that the system meets specified requirements.
- **Scope definition**: Use cases help define the scope of the system by identifying the major functionalities that need to be supported. They provide a structured way to prioritize and organize system features.

- **User training**: Use cases are valuable for user training purposes. They help in creating user documentation and training materials by illustrating how the system is intended to be used.
- **Early identification of problems** Use case analysis can identify potential problems and gaps in requirements early in the development process, before significant effort is invested in implementation. By analyzing use cases and their interdependencies, systems engineers can identify inconsistencies, missing requirements, and potential usability issues. This early detection allows for timely mitigation of problems, reducing the risk of rework and delays.

Use cases play a crucial role in systems engineering by providing a user-centric perspective on system functionality. They contribute to requirements analysis, system design, and validation efforts, ensuring that the system is developed to meet the needs and expectations of its users.

5.6.1 Use Case Identification Pitfalls

Identifying use cases accurately is essential for capturing user requirements and informing system design. Some common pitfalls in use case identification and how to mitigate them:

- **Focusing solely on functionalities**: Narrowing the focus exclusively to functional aspects and neglecting non-functional requirements can lead to an incomplete set of use cases. Consider both functional and non-functional aspects of user interactions, including performance, security, and usability requirements.
- **Ignoring exceptional cases**: Overlooking exceptional or error scenarios can result in use cases that do not address all possible user interactions. Explicitly identify and document alternative flows, error-handling procedures, and exceptional scenarios to provide a comprehensive view of user interactions.
- **Being too detailed too soon**: Going into excessive detail in the early stages of use case identification can lead to analysis paralysis and hinder progress. Focus initially on high-level use cases and gradually refine the details as the analysis progresses. Start with a broad view and iteratively add more granularity.
- **Assuming a single correct set of use cases**: Assuming there is only one correct set of use cases can limit creativity and result in a narrow view of user interactions. Encourage brainstorming sessions and diverse perspectives to explore various ways users may interact with the system. Recognize that multiple valid sets of use cases may exist.

- **Failure to validate with users**: Not validating identified use cases with end-users can result in requirements that do not accurately represent user needs and expectations. Conduct regular validation sessions with end-users to ensure that use cases accurately reflect their workflows and requirements.
- **Using use cases for functional decomposition**: Use cases are a valuable tool for capturing and describing the interactions between external actors and a system's functionalities in a user-friendly and comprehensible manner. However, they are not suitable for functional decomposition, which involves breaking down a system into smaller, more manageable parts or functions.

By being aware of these common pitfalls and proactively addressing them, practitioners can enhance the accuracy and effectiveness of the use case identification process, laying a solid foundation for subsequent stages in systems engineering.

5.6.2 Coffee Machine Use Cases

A comprehensive use case analysis using a combination of the methods listed above resulted in the following table of relevant use cases for the coffee machine:

Name	Description
Make Coffee	The user initiates the brewing process for a standard cup of coffee using prefabricated coffee pads.
Clean and Descale Coffee Machine	The user initiates a cleaning and descaling cycle of the coffee machine to ensure its proper functioning.
Power Coffee Machine On/Off	The user powers the coffee machine on or off.

5.6.3 SysML Use Case Diagrams

SysML Use Case diagrams are fairly simple diagrams. The main elements of a SysML use case diagram are:

- **System Boundary**: The system boundary defines the scope of the system under consideration, separating it from its external environment.
- **Actors**: Actors are external entities that interact with the system to achieve a specific goal.

- **Use cases**: Use cases are sequences of actions that the system performs to achieve a specific goal from the perspective of an actor. Use cases are related to a block representing a system via the subject relation.
- **Relationships**: Relationships can be used to describe the interactions between actors and use cases. The three types of relationships are:
  - **Association**: Associations establish relationships between actors and use cases, indicating that an actor is involved in a particular use case.
  - **Extend**: An extend relationship indicates that one use case can extend the behavior of another use case. The extending use case can be triggered by a particular condition in the base use case.
  - **Include**: An include relationship indicates that one use case can include the behavior of another use case. The included use case is embedded within the including use case.
  - **Generalization**: Represents a generalization or inheritance relationship between use cases, where one use case is a more specific version of another.

5.6.4 Modeling the Coffee Machine Use Cases

Create a new package called "Use Cases" and add a new use case diagram by right-clicking on the new package and selecting *New Diagram->SysML 1.6 Use Case Diagram.* Name the new diagram "Coffee Machine Use Cases UD".

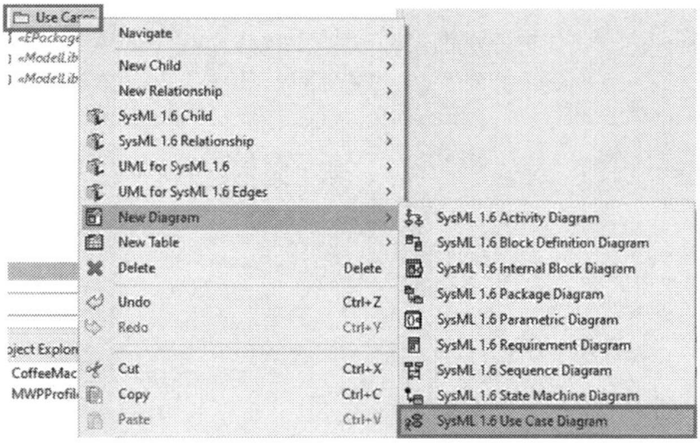

Figure 5.145. Create a new use case diagram

Add the "CoffeeMachine" Block and the "User" actor to the diagram using drag and drop.

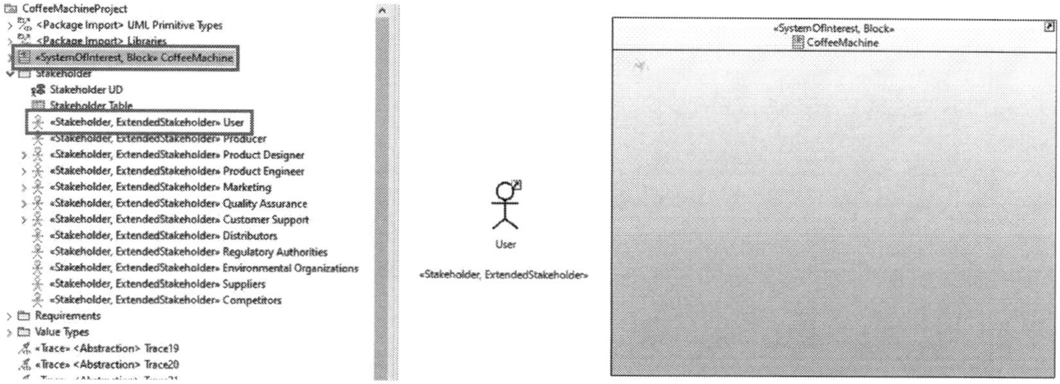

Figure 5.146. Use case diagram with actor and block

Note that, unlike other tools that have a distinct element to model the system boundary, Papyrus just uses the block representing the system to which the use cases belong directly.

Select the “Use Case” element from the Diagram Palette, and left-click inside the “CoffeeMachine” Block to create a new use case. Name it “Make Coffee”.

Figure 5.147. Create a new use case

To associate the “User” stakeholder with the “Make Coffee” use case, select the “Communication Path” relation from the Diagram Palette and left-click first on the “User” and then on the “Make Coffee” use case.

Figure 5.148. Associate the actor with the use case

Repeat the same process for the "Clean and Descale Coffee Machine" use case. Next, create a new use case element for the "Power Coffee Machine On/Off" use case. But, instead of relating it to the "User" directly, we want to indicate that the "Power Coffee Machine On/Off" use case is not a use case in its own right but is included in the other two use cases. Select the "Include" relation from the Diagram Palette and left-click first on the "Make Coffee" use case and then on the "Power Coffee Machine On/Off" use case to model the include relation between the two.

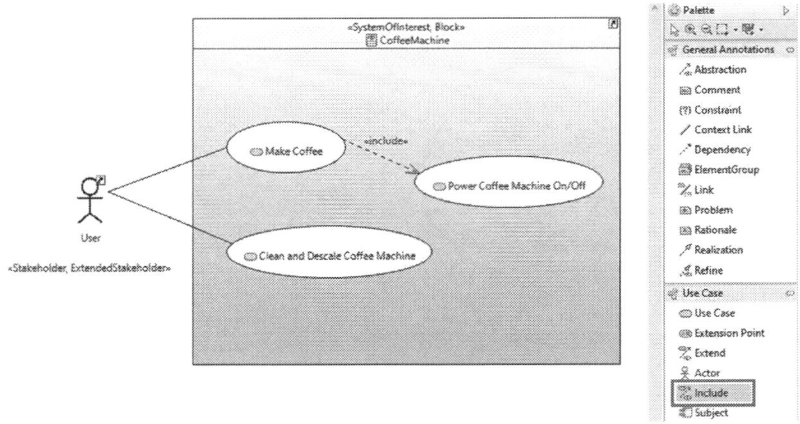

Figure 5.149. Create relations between use cases

Repeat the process to create another include relation between the "Clean and Descale Coffee

Machine" use case and the "Power Coffee Machine On/Off" use case.

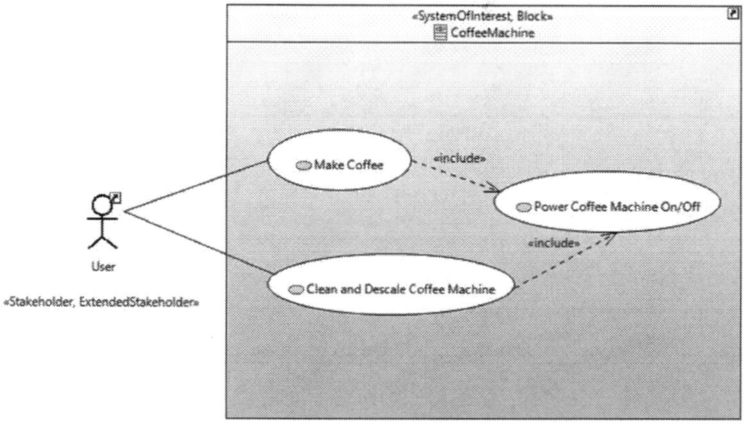

Figure 5.150. Finished use case diagram

Note that all the use cases have been created under the "CoffeeMachine" block by Papyrus. To support model organization, all the use cases are supposed to be stored in the "Use Cases" package. To move the use cases, simply select all three use cases, by holding Shift or Control and drag and drop them to their destination.

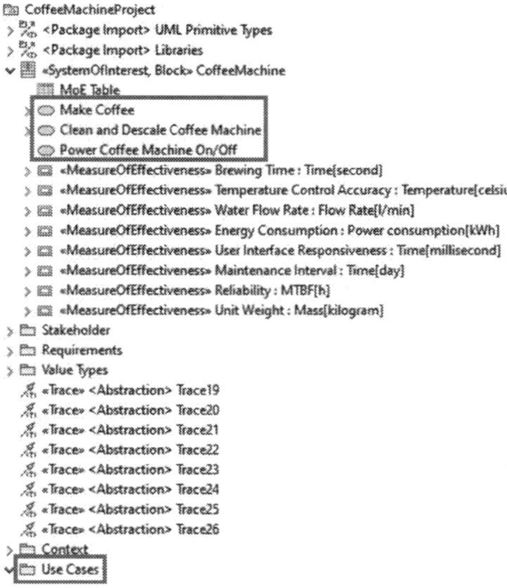

Figure 5.151. Moving elements using drag and drop

The use cases will now be shown in the "Use Cases" package in the Model Explorer.

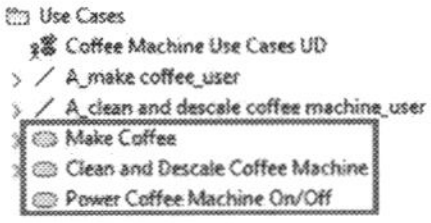

Figure 5.152. Moved use cases

To indicate that the use cases are related to stakeholder requirements, we want to establish more traceability links in the model. Create a new requirement diagram in the "Use Cases" package and call it "Use Cases Requirements Traceability RD". Then, add all three use cases to the diagram.

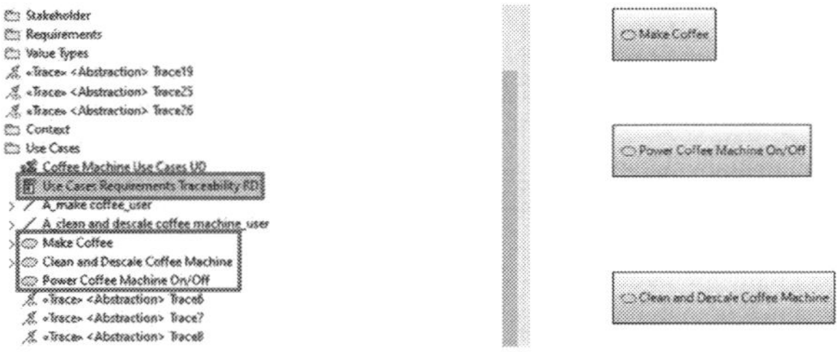

Figure 5.153. New diagram for use case traceability

Add the relevant stakeholder requirements to the diagram as well and create traceability links between the use cases and the requirements.

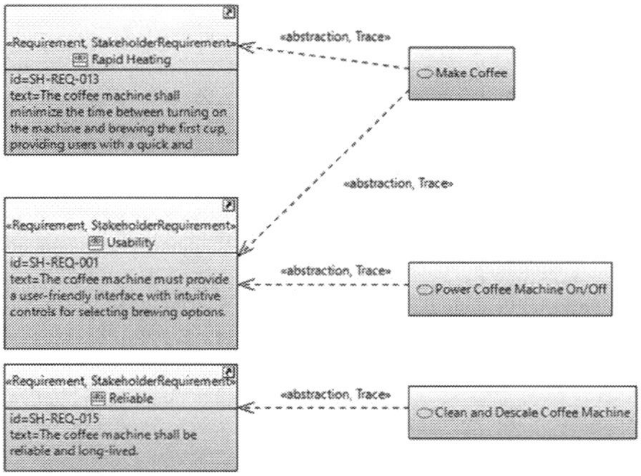

Figure 5.154. Use case to requirements traceability

5.7 Analyze Use Cases

Analyzing a use case involves a systematic examination of the detailed interactions and behaviors that occur between users (actors), the system, and potentially other external systems to achieve a specific goal. The goal of use case analysis is to gain a comprehensive understanding of the steps involved in the use case, potential variations, and the conditions under which the interactions occur. The general workflow for analyzing a use case is the following:

1. **Identify and select the use case**: Start by identifying and selecting the specific use case you want to analyze. A use case typically represents a specific functionality or task that a user wants to perform within the system.
2. **Understand the context**: Review any relevant system documentation, including the use case diagram, to understand the context of the use case. Identify the actors involved and their roles in the use case.
3. **Define preconditions**: Determine any preconditions that must be satisfied before the use case can be executed. Preconditions are conditions or states that must be true for the use case to start.
4. **Identify main flow**: Define the main flow of events for the use case. This is the primary sequence of steps that occur when the use case is executed under normal conditions.
5. **Identify alternate flows**: Identify and document any alternate flows or variations that can occur within the use case. These may represent different paths or conditions that deviate from the main flow.
6. **Define postconditions**: Specify the postconditions or the state of the system after the successful completion of the use case. Postconditions describe the changes or results that should be observable.
7. **Identify exceptional conditions**: Consider exceptional conditions, such as error scenarios or situations where the use case cannot be completed as expected. Define how the system should handle these exceptional conditions.
8. **Identify inputs and outputs**: Document the inputs required for the use case to start and the outputs produced as a result. This includes user inputs, system responses, and any other relevant data.
9. **Specify performance requirements**: If applicable, specify any performance requirements associated with the use case, such as response times, throughput, or other measures of system performance.

10. **Validate and iterate**: Validate the use case analysis with stakeholders, including end-users and system developers. Incorporate feedback and iterate on the analysis to ensure accuracy and completeness.
11. **Document the analysis**: Document the results of the use case analysis clearly and concisely. This documentation serves as a reference for system developers, testers, and other stakeholders.

Following these steps, leads to a thorough analysis of a use case, ensuring that the interactions and behaviors within the system are well-understood and accurately represented. This analysis forms a critical foundation for subsequent stages in the systems engineering process, including design, implementation, and testing.

5.7.1 Use Case Analysis Pitfalls

Avoiding common pitfalls in use case analysis is essential to ensure that the identified use cases accurately represent user requirements and contribute effectively to system development. Here are some common pitfalls to be aware of:

- **Ignoring alternative flows**: Focusing solely on the main flow of a use case and neglecting alternative flows can result in incomplete or unrealistic scenarios. Identify and document alternative flows, exceptions, and variations to provide a comprehensive view of possible interactions.
- **Incomplete use case descriptions**: Providing incomplete or ambiguous descriptions for use cases can lead to misunderstandings and misinterpretations. Ensure that each use case is thoroughly documented, including preconditions, postconditions, and detailed steps for the main and alternative flows.
- **Overcomplicating use case models**: Creating overly complex use case diagrams and models can hinder communication and understanding among stakeholders. Keep use case models clear and concise, avoiding unnecessary details and ensuring that they effectively communicate the essential interactions.
- **Lack of validation with users**: Failing to validate use cases with end-users can result in requirements that do not align with user expectations. Conduct regular validation sessions with end-users to ensure that use cases accurately represent their needs and workflows.

By being mindful of these common pitfalls, practitioners can conduct more effective use case analysis, leading to a better understanding of user requirements and ultimately contributing to the success of system development.

5.7.2 Use Case Analysis with SysML

The SysML Activity Diagram (AD) is a powerful visual tool that can be employed to model the dynamic aspects of a system, including the flow of activities and actions within a use case. When applied to use case analysis, activity diagrams provide a graphical representation of the workflow, interactions, and behavior of the system during the execution of a specific use case. The SysML Sequence Diagram (SD) provides a dynamic view of a system, emphasizing the interactions between various components or actors over time. In the context of use case analysis, sequence diagrams play a valuable role in depicting the dynamic behavior and communication flows associated with specific use cases.

Activity Diagrams and Sequence Diagrams both support use case analysis by:

- **Visualization of use case execution**: The diagrams allow for the visual representation of the steps and activities involved in the execution of a use case. This visual clarity aids stakeholders, including system analysts and developers, in understanding the dynamic aspects of the use case.
- **Specification of activity flows**: The diagrams provide a means to specify the flow of activities within a use case. This includes the sequence of actions, decision points, loops, and parallel flows, offering a detailed view of how the system behaves during the execution of the use case.
- **Identification of actors and roles**: Actors, represented as swimlanes in ADs or Lifelines in SDs, and their roles in the execution of activities can be visually identified on the diagrams. This helps in understanding the interaction between actors and the system during the use case.
- **Integration with use case descriptions**: The diagrams can complement textual use case descriptions by offering a visual representation of the described interactions. This integration enhances the overall understanding of the use case and facilitates communication among stakeholders.
- **Handling decision points and conditions**: Both diagrams support the modeling of alternative flows and variations within a use case, including different paths based on conditional logic. Conditions associated with decisions can be explicitly modeled, providing a detailed view of how the system responds to various scenarios.
- **Modeling parallel activities**: The diagrams support the modeling of parallel activities, allowing for the representation of concurrent actions within the use case. This is particularly useful when multiple activities can be performed simultaneously.
- **Specification of timing and duration**: The Timing and duration of activities can be annotated on the diagrams, providing insights into the temporal aspects of use case

execution. This is essential for understanding the timing constraints and performance considerations associated with the use case.

During use case analysis workshops or meetings, both diagram types can serve as visual aids to facilitate discussions among stakeholders. They provide a shared understanding of the system's behavior and support collaborative analysis. Activity and sequence diagrams offer a valuable means to model the dynamic aspects of use cases, providing a visual and structured representation of how the system behaves during the execution of specific functionalities. They enhance the overall understanding of use cases, promote effective communication among stakeholders, and contribute to a more comprehensive use case analysis.

The choice between a sequence diagram and an activity diagram in use case analysis depends on the specific requirements of the analysis and the type of information you want to emphasize. Both diagram types serve different purposes and provide distinct perspectives on the system's behavior. There are no fixed rules for considering one diagram or the other, but rather some considerations for choosing between a sequence diagram and an activity diagram in use case analysis.

Choose a Sequence Diagram when:

- **Emphasizing interaction details**: Use sequence diagrams when you want to focus on the detailed interactions and communication between actors, objects, or components during the execution of a specific use case. Sequence diagrams provide a chronological sequence of messages and responses.
- **Modeling message passing**: Sequence diagrams are particularly effective for modeling the flow of messages between different elements of the system. If the emphasis is on how objects or components communicate and collaborate to achieve the use case goals, a sequence diagram is suitable.
- **Showing time sequences**: When temporal relationships and the order of events are critical to understanding the behavior of the system during the execution of a use case, sequence diagrams are more appropriate. They provide a clear representation of time sequences.
- **Illustrating object lifelines**: Sequence diagrams are well-suited for illustrating the lifelines of objects or actors participating in the use case. They show the duration of an object's existence and the sequence of messages it sends or receives.

Choose an Activity Diagram when:

- **Capturing high-level workflow**: Activity diagrams are suitable for capturing the high-level workflow of a use case, providing a broader overview of the activities and actions involved. They focus on the flow of control rather than detailed interactions.

- **Modeling concurrent behavior**: When the use case involves concurrent activities or parallel processing, activity diagrams are more appropriate. They provide a visual representation of how multiple activities can occur simultaneously.
- **Emphasizing workflow steps**: Use activity diagrams when you want to emphasize the flow of activities and the sequence of steps involved in the execution of a use case. They are effective for illustrating the procedural aspects of the use case.
- **Showing decision points and branching**: If decision points, conditional flows, and branching are central to the understanding of the use case, activity diagrams allow for a clear representation of decision nodes and alternative paths.

The choice between a sequence diagram and an activity diagram in use case analysis depends on the level of detail you want to capture, the emphasis on interaction vs. workflow, and whether temporal sequences or concurrent behavior are more critical to understanding the system behavior. It's often beneficial to use both diagram types in conjunction to provide a comprehensive view of the system from the user's perspective.

The SysML State Machine Diagram (STM) is a third type of diagram that could be used for modeling a use case. However, STMs are more suited for modeling the behavior of a single object instead of the interaction between multiple actors and they are already covered in the section about system states and mode analysis, so they are not used for modeling the use cases of the coffee machine.

5.7.3 The "Make Coffee" Use Case

What follows is a detailed textual description of the "Make Coffee" use case, covering preconditions, postconditions, and inputs.

Use Case: Make Coffee

Preconditions:

- The coffee machine is powered on.
- The coffee pad compartment is empty.

Inputs:

- The user selects the standard cup size for brewing.

Flow of Events:

- User initiates brewing process:
  - The user places a new prefabricated coffee pad in the designated compartment.
  - The user navigates the coffee machine interface and selects the standard cup size.
  - The user initiates the brewing process.
- Coffee machine processing:
  - The coffee machine reads the user's selection and activates the brewing mechanism.
  - The machine extracts coffee from the pad and dispenses it into the cup.
- Completion of brewing:
  - Once the brewing process is complete, the coffee machine signals the user, typically with a sound or visual notification.
  - The user can retrieve the brewed coffee.

Postconditions:

- The standard cup of coffee is brewed and ready for consumption.
- The used coffee pad is stored in the designated compartment for disposal.
- The coffee machine returns to a standby or ready state, awaiting the next user interaction.

Alternative Scenario - Insufficient Water:

Preconditions:

- The coffee machine is powered on.
- The water reservoir is empty or contains an insufficient amount of water.

Inputs:

- User's attempt to brew coffee.

Flow of Events:

- The user attempts to brew coffee without refilling the water reservoir.
- The coffee machine detects insufficient water and displays an alert on the interface.
- The user is prompted to refill the water reservoir.
- If the user refills the water reservoir, then the use case can begin anew.

5.7.4 Modeling the "Make Coffee" Use Case

We model the "Make Coffee" use case using a SysML Activity Diagram. Add a new AD to the "Make Coffee" use case by right-clicking on the "Make Coffee" use case in the Model Explorer and selecting "New Diagram->SysML 1.6 Activity Diagram". Name the new diagram "Make Coffee AD".

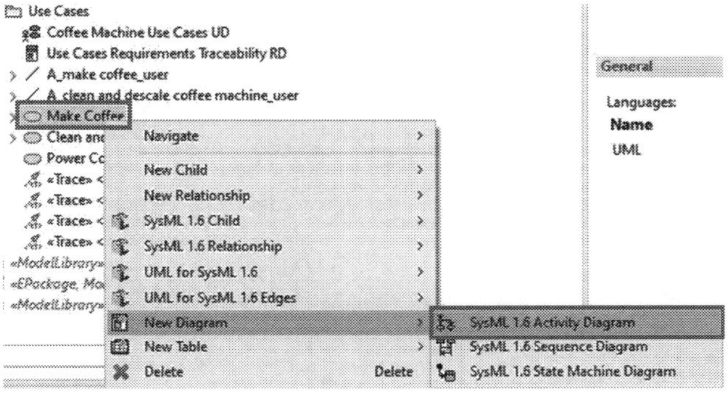

Figure 5.155. Create a new activity diagram

Since ADs require an activity, Papyrus automatically creates a new activity with a default name when the AD is created and adds it to the AD. Rename the activity to "Make Coffee".

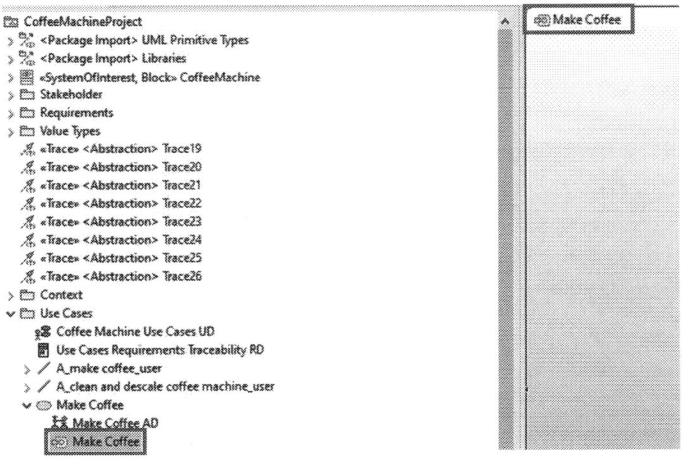

Figure 5.156. Renamed activity "Make Coffee"

An activity represents a specific behavior of a system, something that it can do, or a function that a system can perform. An activity diagram is a visual representation of the flow of activities and can show the sequence of steps involved in completing a task, such as a use case. An activity diagram cannot exist on its own. It belongs to an activity.

We want to use swimlanes, or more formally allocate activity partitions, on the AD to show, what activities are performed by which participant in the use case. Create a new allocate activity partition by right-clicking on the "Make Coffee" activity and selecting *SysML 1.6 Child->AllocateActivityPartition.*

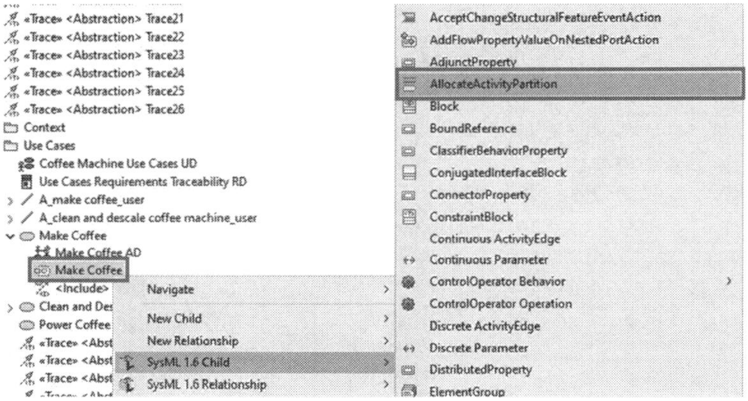

Figure 5.157. Create a new allocate activity partition

To relate the newly created allocate activity partition with a responsible block or actor, go to the UML tab of the Properties View of the element and click on the button with the three dots next to the Represents property to open the "Represents" dialog.

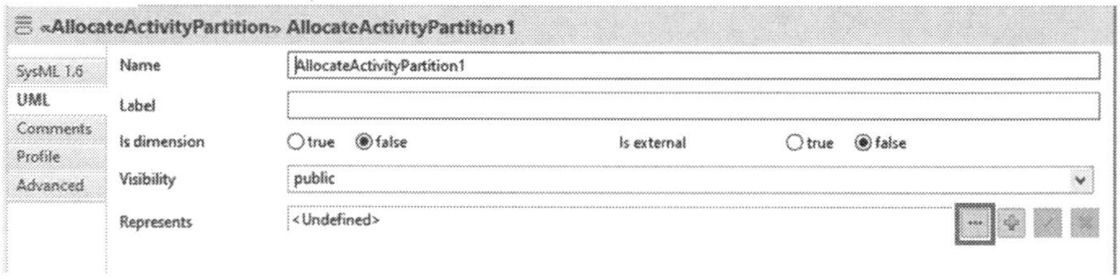

Figure 5.158. Open the "Represents" dialog

Select the "User" stakeholder from the "Stakeholder" package.

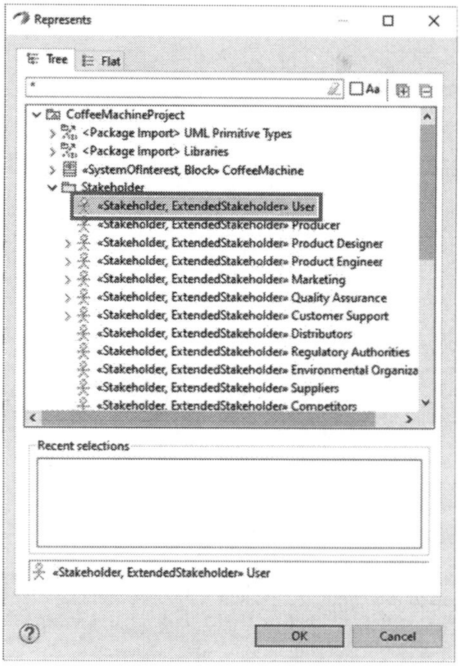

Figure 5.159. Select the "User" stakeholder

Additionally, name the allocate activity partition "User".

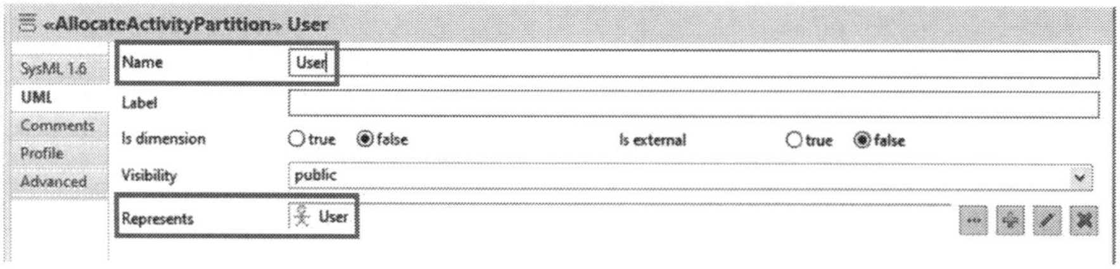

Figure 5.160. Name the allocate activity partition

Repeat the same process for the "CoffeeMachine" block.

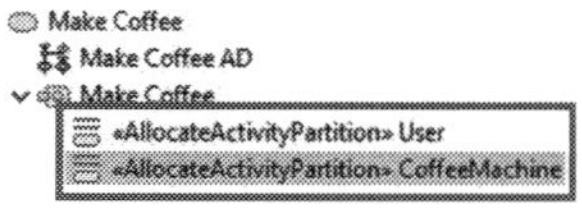

Figure 5.161. Allocate activity partitions belonging to the "Make Coffee" activity

Now, drag and drop both allocate activity partitions onto the AD and make them take the whole space of the activity diagram.

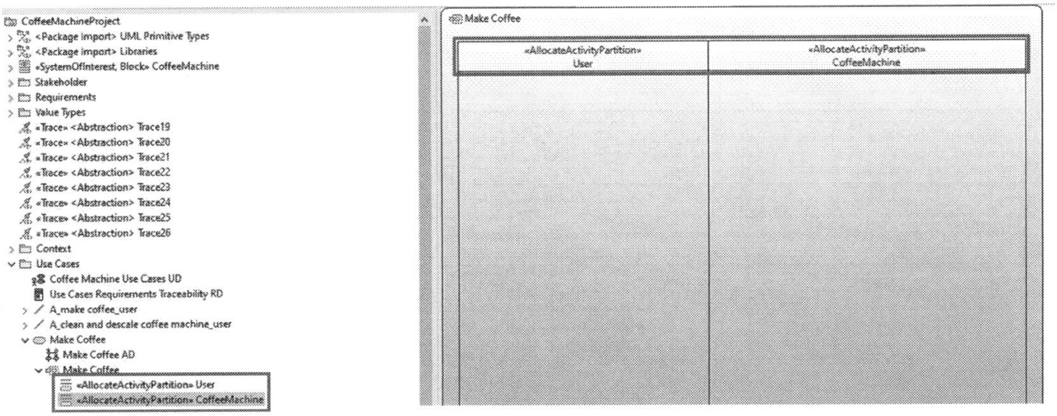

Figure 5.162. Allocate activity partitions on the activity diagram

Select the "Call Behavior Action" element from the Diagram Palette, and left-click inside the activity diagram to create a new call behavior action for every step of the use case.

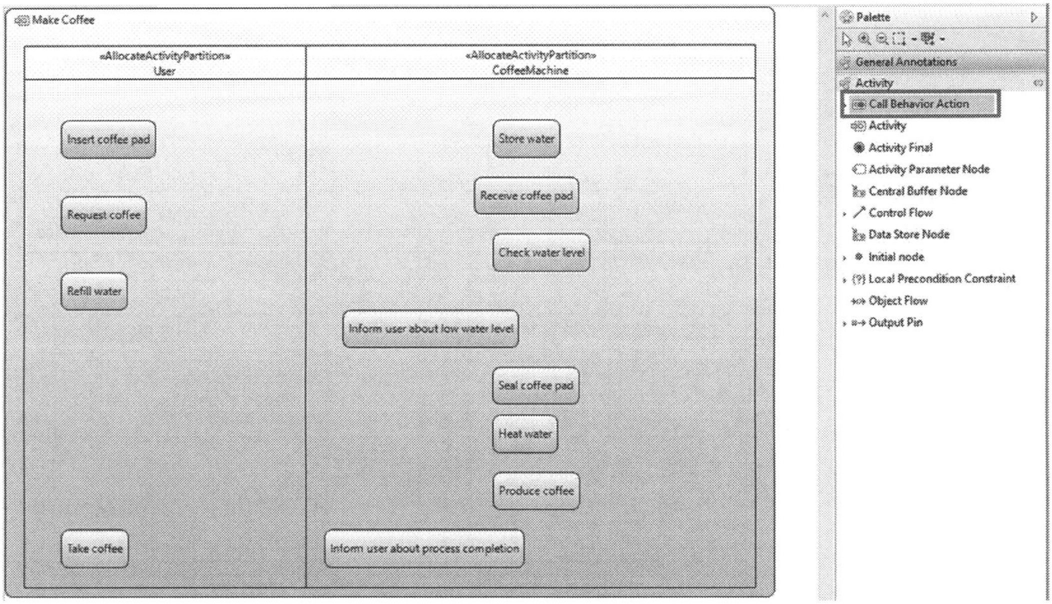

Figure 5.163. Activity Diagram with call behavior actions

Even if the grid for aligning elements on a diagram has been turned off, Papyrus, by default, nevertheless aligns the elements with the invisible grid points. This is helpful for diagrams with larger elements like block diagrams but can make arranging the elements on an activity diagram hard. To turn off the "Snap to grid" feature temporarily, hold the ALT key while moving an element. To toggle the "Snap to grid" feature on a per-diagram basis, open the "Papyrus" menu and select or deselect *Diagram->View->Snap to Grid.* To toggle the "Snap to grid" feature for all newly created diagrams, open the Papyrus preferences and select or deselect the "Snap to grid for new diagrams" option in the Diagrams settings.

Additionally, create an initial node to represent the start of the use case, an activity final node to represent the end of the use case, a merge node, and a decision node.

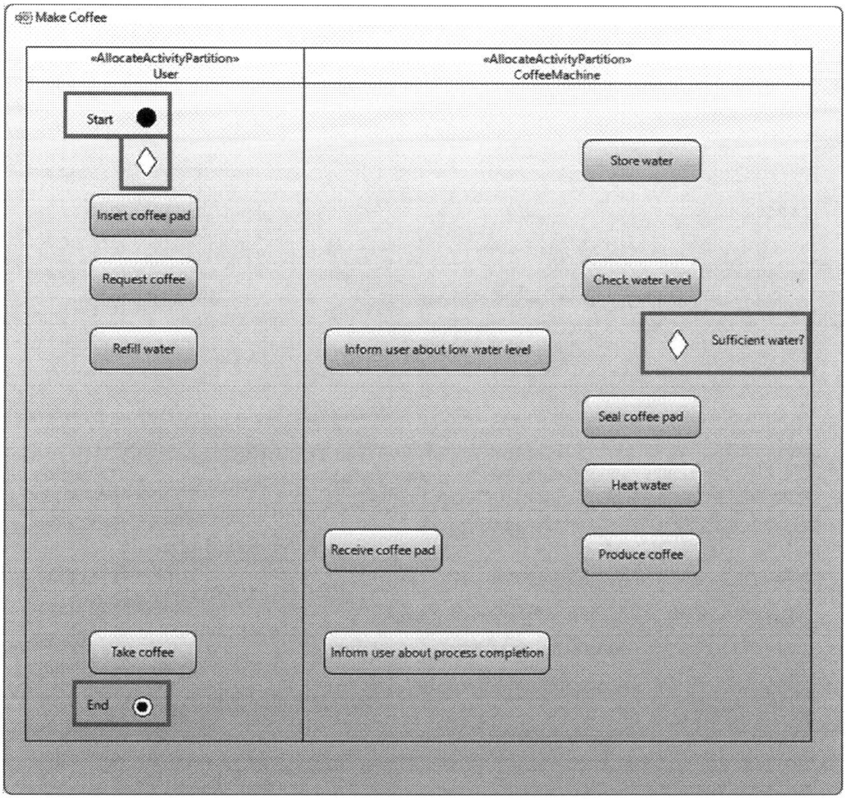

Figure 5.164. Activity Diagram with more elements

The merge node and the decision node can be found by expanding the initial node in the Diagram Palette.

The merge node does not need a name, as it does not add information to the model, so you can just delete the default name that Papyrus assigns to the element.

Select the "Insert Coffee Pad" call behavior action and open its Properties View. To create a new activity that will be called by this call behavior action, right-click on the button with the plus next to the Behavior property and select Activity.

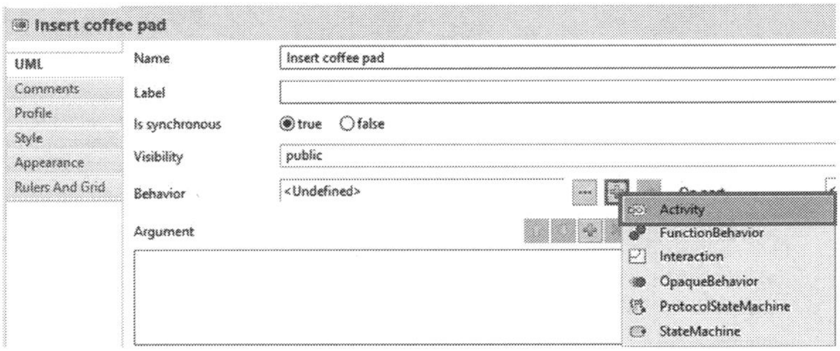

Figure 5.165. Create a new activity

The first step in creating this new activity is to select its owner. Click on the buttons with the three dots and select the "User" stakeholder from the dialog.

Figure 5.166. Select the owner of the new activity

Name the activity "Insert coffee pad". To model the output of the activity, click on the button with the plus sign in the Owned Parameter section.

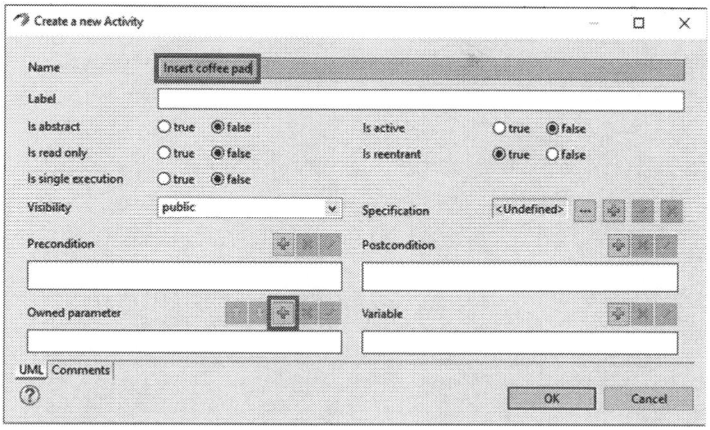

Figure 5.167. **Name the new activity and add a parameter**

Name the new parameter "coffee pad" and set its Direction to "output".

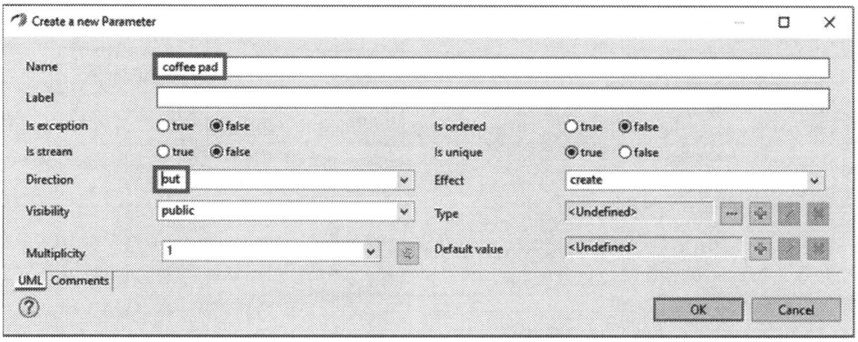

Figure 5.168. **Name the parameter and set its direction**

The parameters of an activity and the input and output pins of call behavior action calling that activity have to match. Papyrus can synchronize the elements and add new parameters of an activity to a call behavior action as pins. Unfortunately, sometimes, this does not happen automatically. To trigger the synchronization process, click on the button with the three dots next to the Behavior property of the "Insert Coffee Pad" call behavior action and immediately click on the "OK" button to re-set the already selected activity and trigger the synchronization. Then the call behavior action "Insert coffee pad" on the diagram gets a new output pin called "coffee pad".

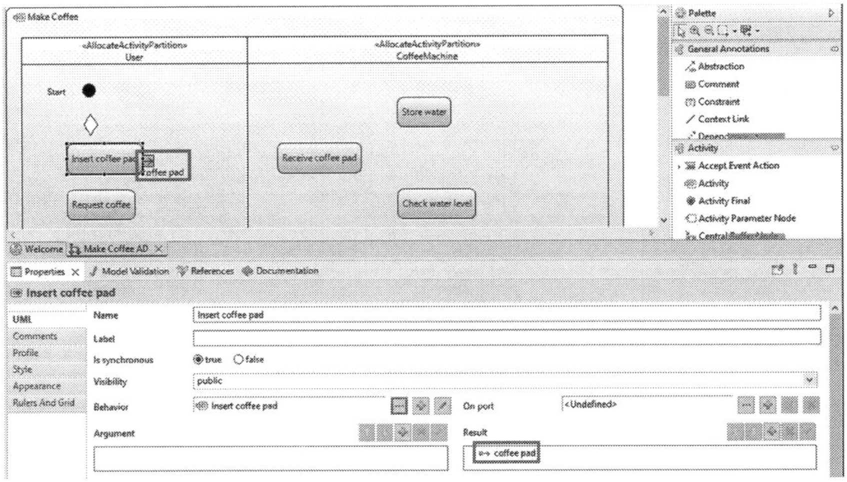

Figure 5.169. Call behavior action with new output pin

Select the "Produce Coffee" call behavior action and open its Properties View. To create a new activity that will be called by this call behavior action, click on the button with the plus next to the Behavior property and select Activity. Since the "Produce Coffee" activity is performed by the coffee machine, select the "CoffeeMachine" block as the container in the dialog. Then click on the button with the three dots next to the Reference property.

Actors have only one property, "ownedBehavior", that can be set to an activity so it is automatically chosen by Papyrus when a new activity is created for an actor. A block has two properties, "ownedBehavior" and "nestedClassifier", that can reference an activity, and Papyrus can not automatically choose the appropriate property. This is why the reference must be set by the user in one case but not the other.

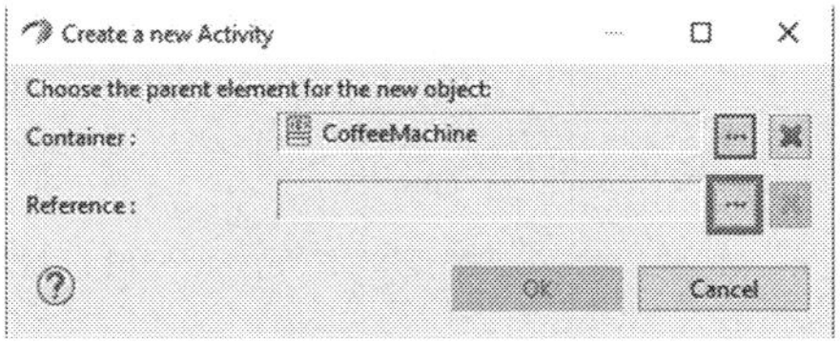

Figure 5.170. Open the dialog for setting the reference property

Choose the "ownedBehavior" property.

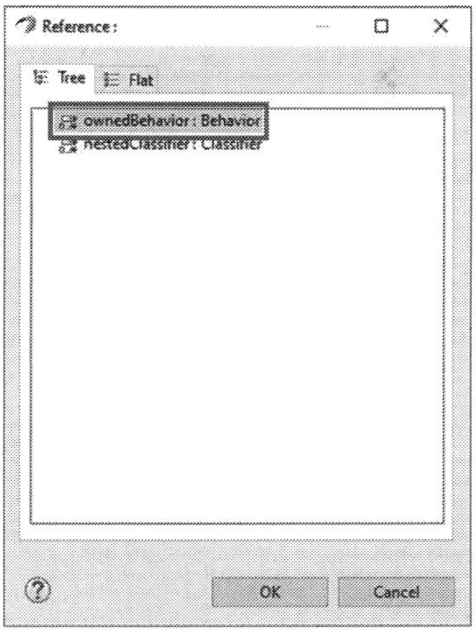

Figure 5.171. Set the reference property

Add three parameters to the "Produce Coffee" activity: two inputs ("water" and "coffee pad") and one output ("coffee").

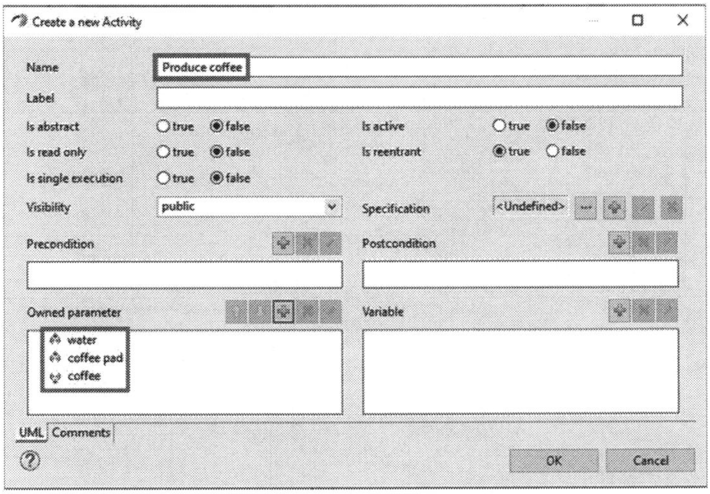

Figure 5.172. Parameters of the "Produce Coffee" activity

Repeat the same process for all the other call behavior actions of the use case.

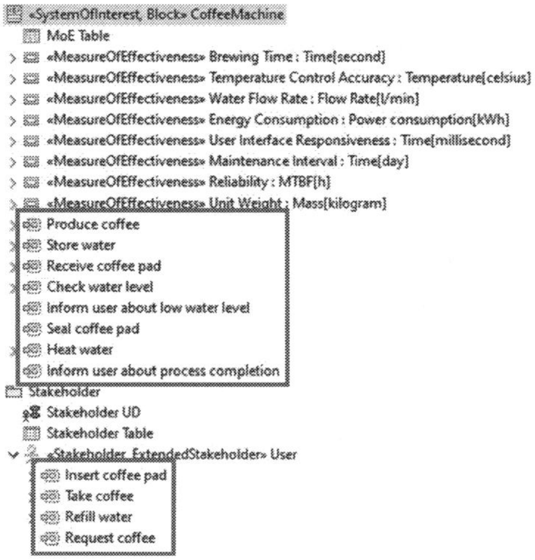

Figure 5.173. All new activities in the Model Explorer

To model the sequence of actions in the activity diagram, select the "Control Flow" element from the Diagram Palette. Then, left-click first on the "Start" node and then on the merge node.

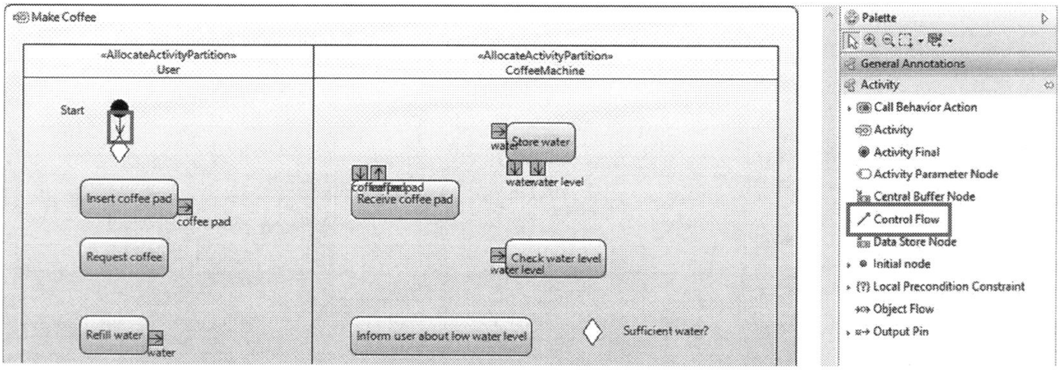

Figure 5.174. Create a control flow

Repeat the same process to add additional control flows to the diagram to complete the sequence of operations.

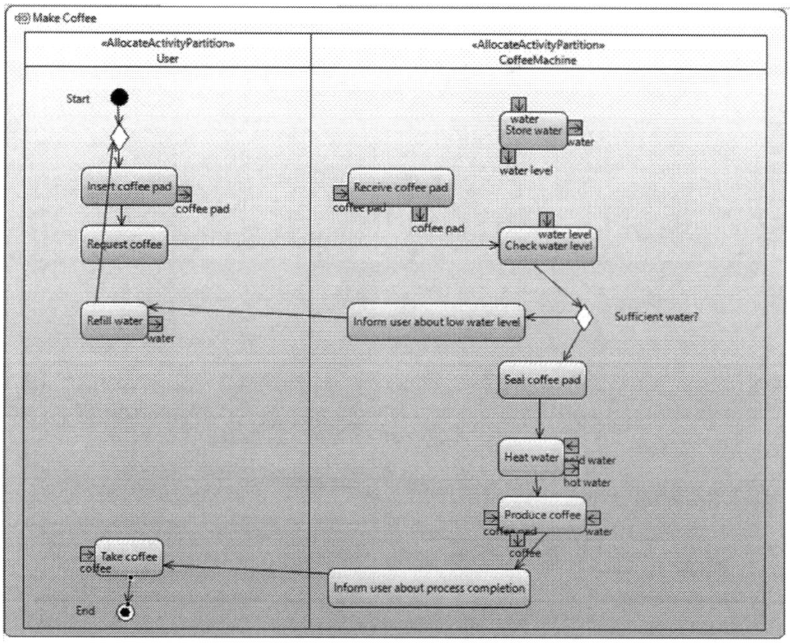

Figure 5.175. Activity diagram with control flows

To model the flow of objects or information in the activity diagram, select the "Object Flow" element from the Diagram Palette. Then, left-click first on the "coffee pad" output pin of the "Request Coffee" call behavior action and then on the "coffee pad" input pin of the "Receive coffee pad" call behavior action.

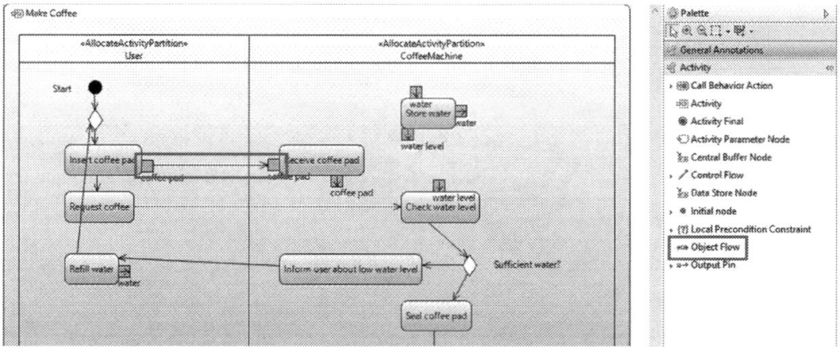

Figure 5.176. Create an object flow

Repeat the same process to add additional object flows to the diagram to complete the flow of objects.

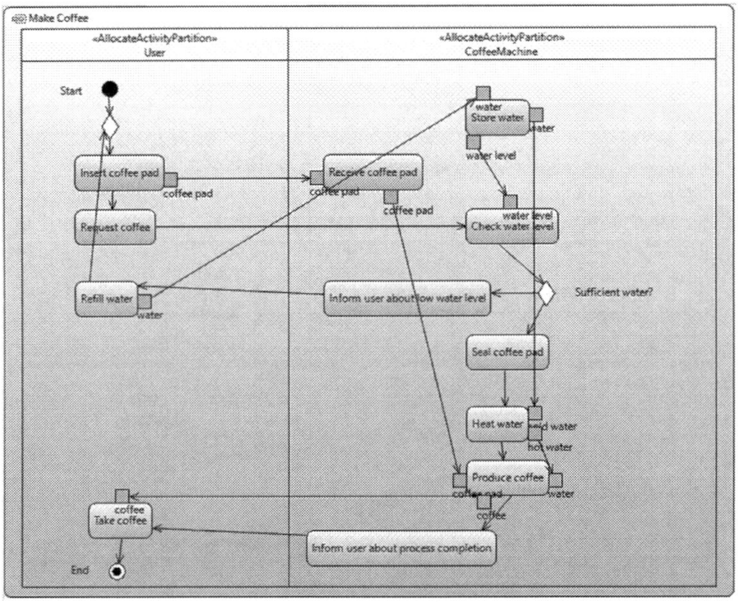

Figure 5.177. The complete activity diagram for the "Make Coffee" use case

The activity diagram is now complete but not very nice to look at. To enable a better routing of the flows, change their line style. Select all of the flows on the diagram, right-click on one of them, and select *Format->Line Style->Rectilinear Style Routing.*

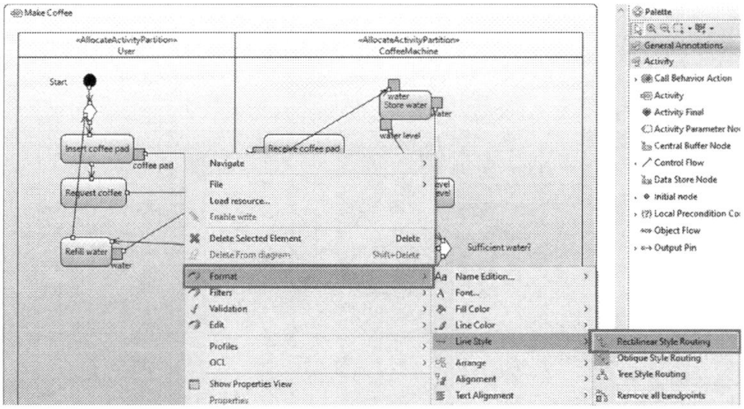

Figure 5.178. Change the line style of the flows

This already improves the readability of the AD.

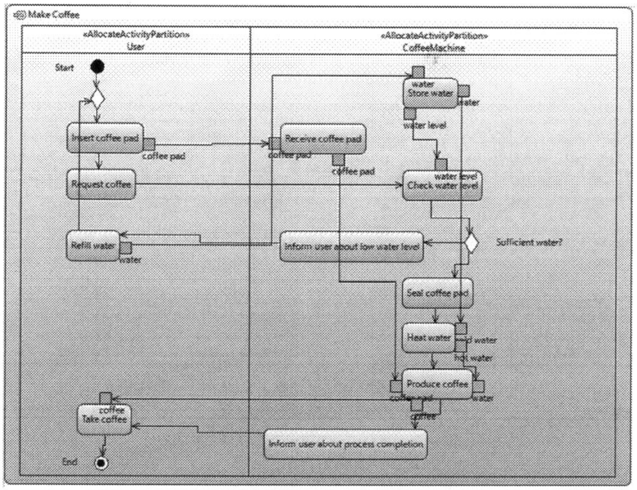

Figure 5.179. Activity Diagram with rectilinear line style

Left-click on the flows to add additional bend points where necessary and ensure that the flows do not cross each other unless necessary.

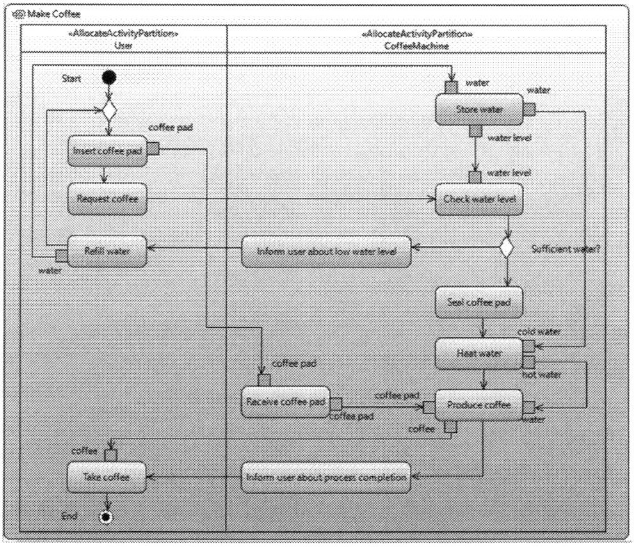

Figure 5.180. Activity diagram with improved layout

The “Sufficient water?” decision node has two exits. To model, which transition from the decision node is selected, guards on the outgoing edges need to be added. Select the control flow, which leads to the "Seal coffee pad" call behavior action and open its Properties View. Click on the button with the plus next to the guard property and select "OpaqueExpression".

UML has a concept of opaque expression, which refers to code or maths expressed in a language other than UML represented as a text string. Such expressions can be used, for example, in an opaque behavior, on guards on state machine diagram transitions, and in constraints on a constraint block in SysML parametric diagrams.

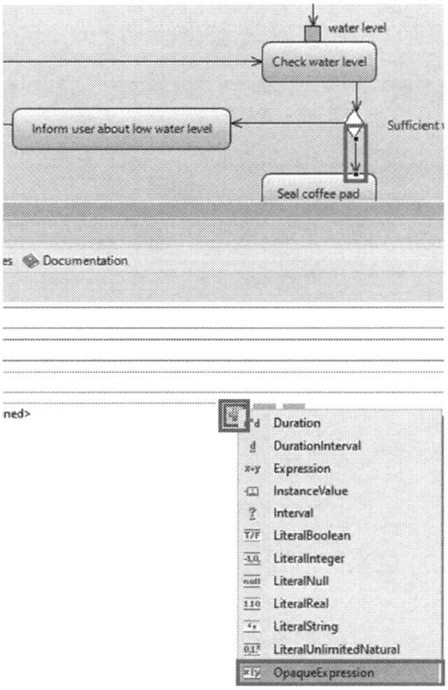

Figure 5.181. Add a guard

This opens the dialog for creating a new opaque expression. To specify the guard, first set the language of the opaque expression. To do this, click on the button with the plus next to the language section.

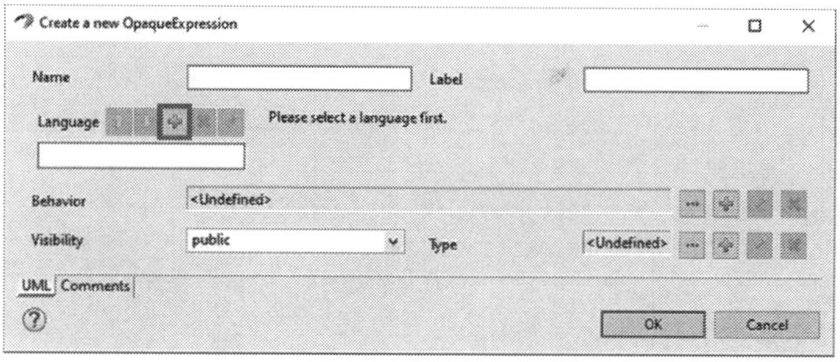

Figure 5.182. Dialog for creating a new opaque expression

Select "Natural Language" and move it to the right using the button with the single arrow pointing to the right.

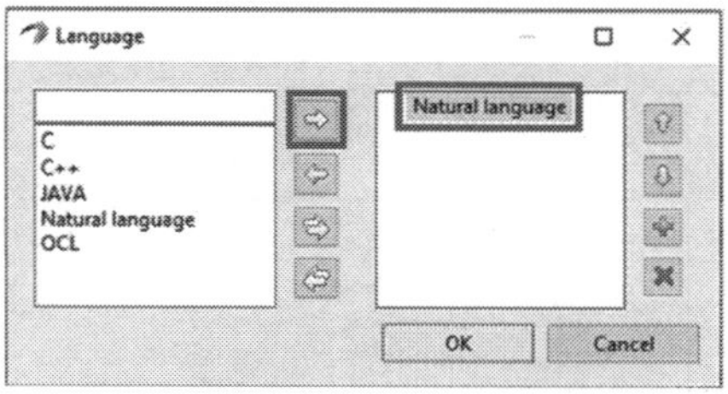

Figure 5.183. Select the language for the opaque expression

Then, specify the statement of the opaque expression. In this case, it is simply "Sufficient water".

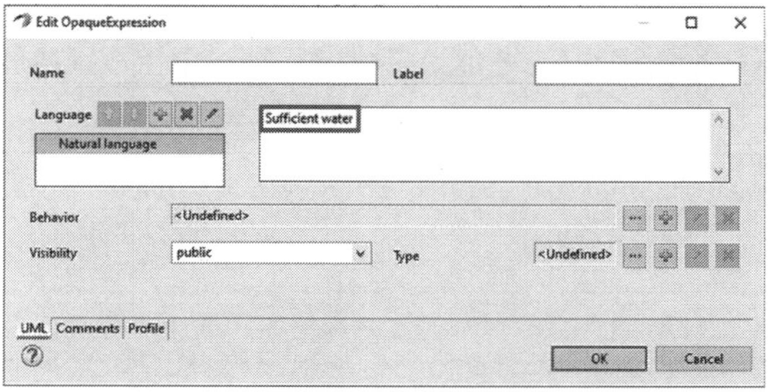

Figure 5.184. Specify the statement of the opaque expression

Repeat the same process for the other control flow, leaving the "Sufficient water?" decision node to add an "else" guard condition.

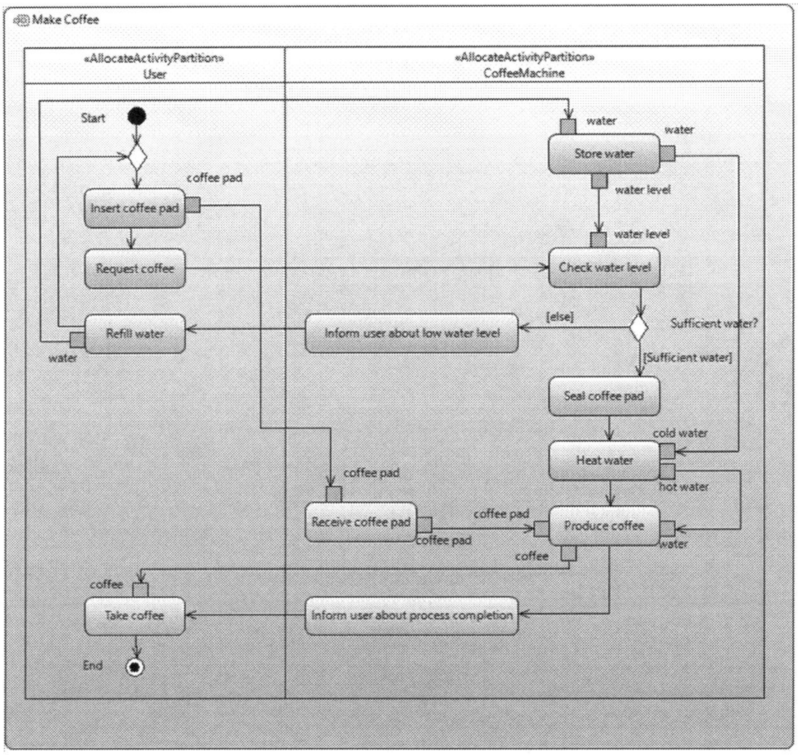

Figure 5.185. Finished activity diagram for the "Make Coffee" use case

5.7.5 The "Clean and Descale Coffee Machine" Use Case

What follows is a detailed textual description of the "Clean and Descale Coffee Machine" use case, covering preconditions, postconditions, inputs, and outputs.

Use Case: Clean and Descale Coffee Machine

Preconditions:

- The coffee machine is powered on.
- The coffee machine requires cleaning and descaling.

Inputs:

- The coffee machine indicates to the user that cleaning and descaling are required.

- The user initiates the cleaning and maintenance process through the coffee machine interface.

Flow of Events:

- The user initiates cleaning:
  - The user navigates the coffee machine interface to access the cleaning and maintenance options.
  - The user selects the option to start the cleaning and descaling process.
- Cleaning cycle:
  - The coffee machine begins a cleaning cycle.
  - The machine may dispense water to flush the internal components.
  - At the end of the cleaning cycle, the coffee machine indicates to the user that the cleaning is complete.
- Descaling cycle:
  - The coffee machine indicates to the user that insertion of the descaling agent into the water tank is required.
  - The coffee machine begins a descaling cycle.
  - The machine flushes the internal components with the water containing the descaling agent.
- Completion of cleaning and descaling cycle:
  - The coffee machine signals the completion of the cleaning and descaling process cycle through visual indicators or messages on the display.
  - The coffee machine performs any finalization steps, such as a final rinse cycle or a confirmation message on the display.

Postconditions:

- The coffee machine is cleaned and descaled.
- The coffee machine is ready for regular use.

Outputs:

- The coffee machine provides feedback to the user, indicating the successful completion of the cleaning and maintenance process.

Alternative Scenario - Insufficient User Action:

Preconditions:

- The coffee machine is powered on.
- The coffee machine requires cleaning and descaling.

Inputs:

- User's lack of initiation for the cleaning process.

Flow of Events:

- The coffee machine prompts the user to initiate the cleaning and descaling process.
- The user neglects to start the cleaning process within a specified time.

Postconditions:

- The coffee machine may display a reminder or alert indicating that cleaning is overdue.
- The machine may enter a standby state until the cleaning process is initiated.

Outputs:

- Alert or reminder message on the machine's display.

5.7.6 Modeling the "Clean and Descale Coffee Machine" Use Case

We model the "Clean and Descale Coffee Machine" use case using a SysML Sequence Diagram. Add a new sequence diagram to the "Clean and Descale Coffee Machine" use case by right-clicking on the "Clean and Descale Coffee Machine" use case in the Model Explorer and selecting *New Diagram->SysML 1.6 Sequence Diagram.* Name the new diagram "Clean and Descale Coffee Machine SD".

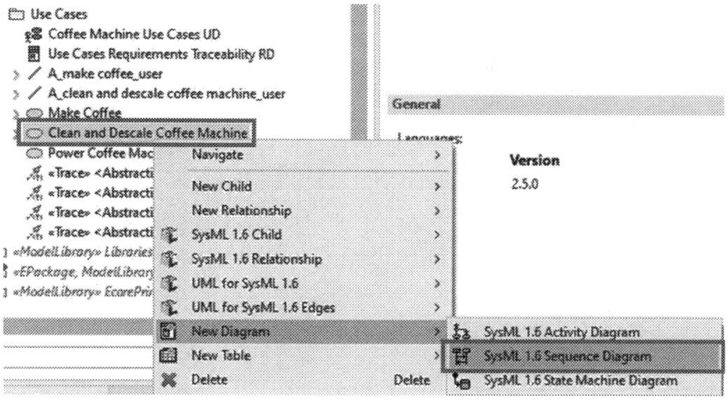

Figure 5.186. Create a new sequence diagram

Since sequence diagrams require an interaction in Papyrus, it automatically creates a new interaction with a default name when the SD is created and adds it to the SD. Rename the interaction to "Clean and Descale Coffee Machine".

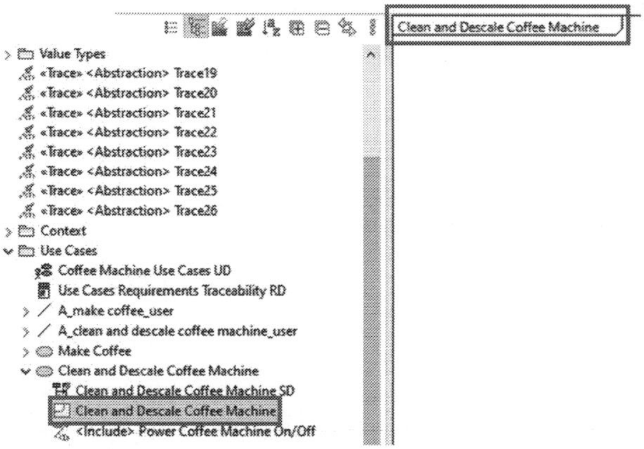

Figure 5.187. Renamed interaction "Clean and Descale Coffee Machine"

Create two lifelines on the newly created sequence diagram: select the "Lifeline" element from the Diagram Palette and click on the diagram.

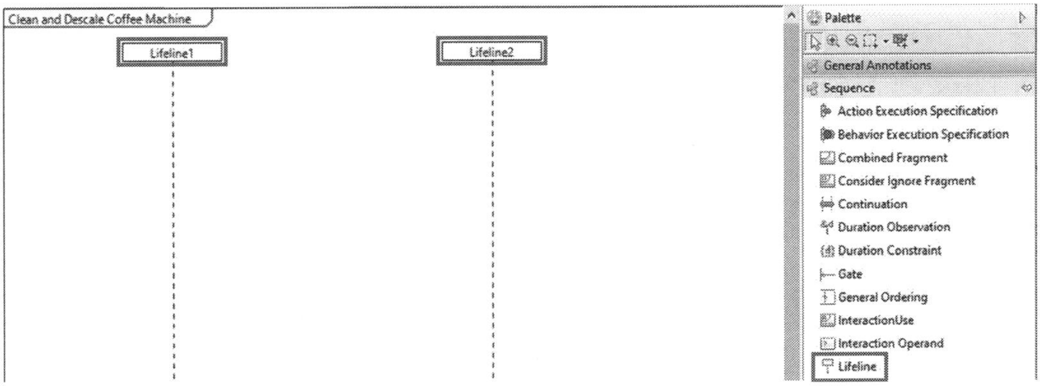

Figure 5.188. Sequence diagram with two lifelines

To relate the lifelines to the participants of the use case, drag and drop the properties "user" and "coffeemachine" of the "Coffee Machine Context" block onto the newly created lifelines.

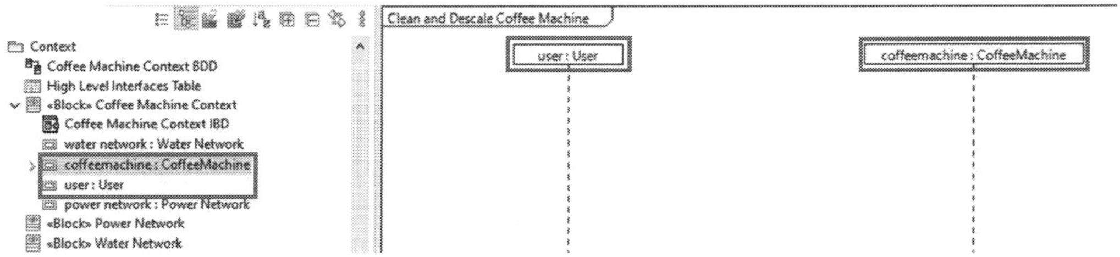

Figure 5.189. Relate lifelines to use case participants

The relation between the lifeline and the use case participant will also be shown in the Property View of the lifeline.

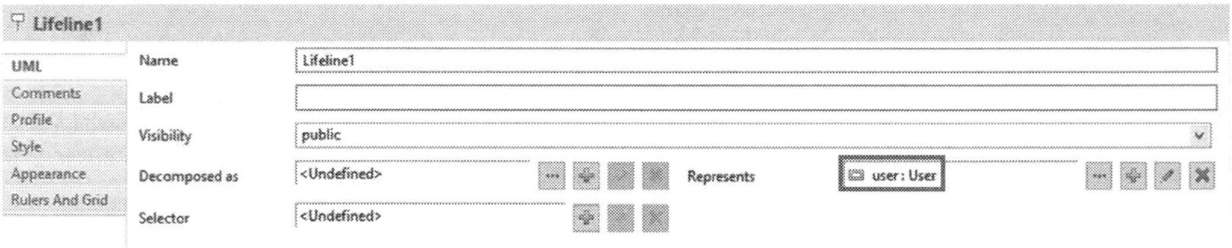

Figure 5.190. Lifeline related to use case participant

To model the prerequisite of the use case, select the "State Invariant" element in the Diagram Palette and click on the lifeline representing the coffee machine. Name the state invariant "Cleaning and descaling required and coffee machine is on".

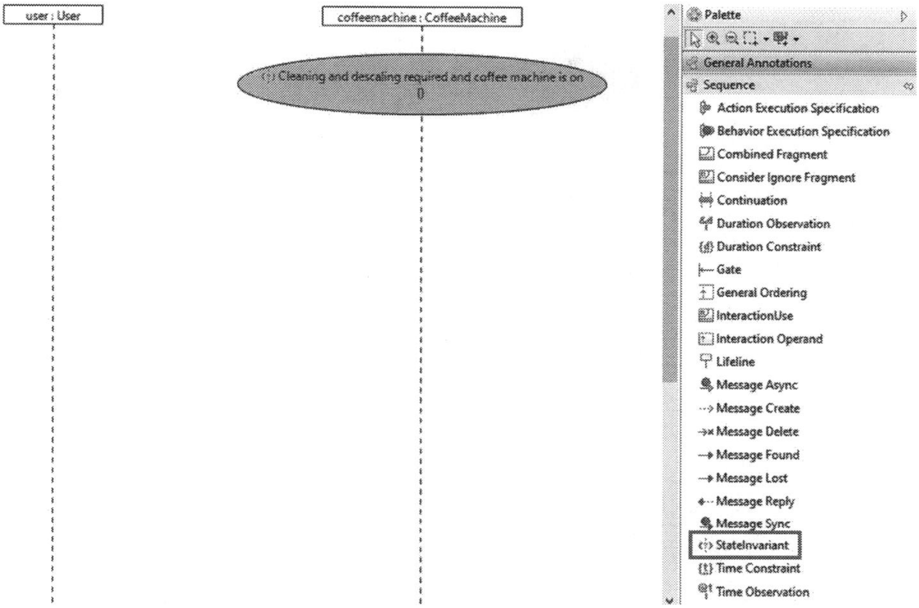

Figure 5.191. State invariant on the sequence diagram

The state invariant should not be confused with a use case element although they look similar. Papyrus uses an ellipse to represent the state invariant element on the sequence diagram instead of the rectangle with rounded corners prescribed by the SysML standard.

Select the "Message Async" element in the Diagram Palette and model the interaction between the coffee machine and the user by first clicking on the lifeline representing the sender of the message and then clicking on the lifeline representing the receiver of the message.

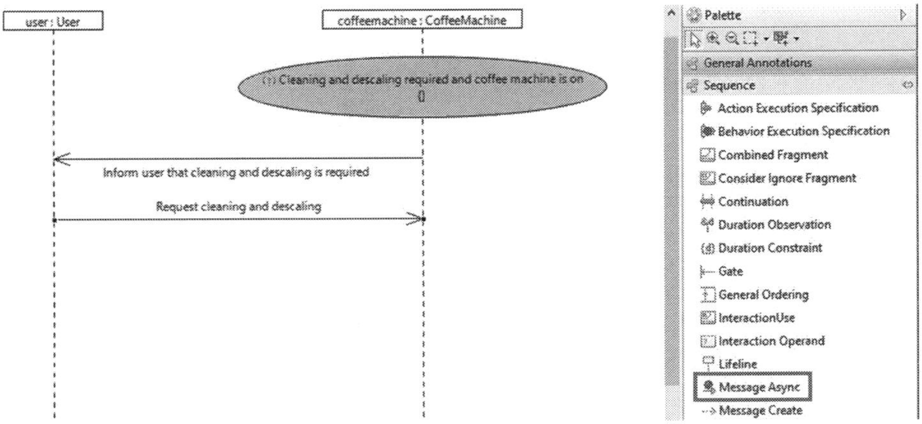

Figure 5.192. Messages on the sequence diagram

The next step is to add a function that needs to be performed by the coffee machine. Select the "Behavior Execution Specification" in the Diagram Palette and left-click on the "CoffeeMachine" lifeline.

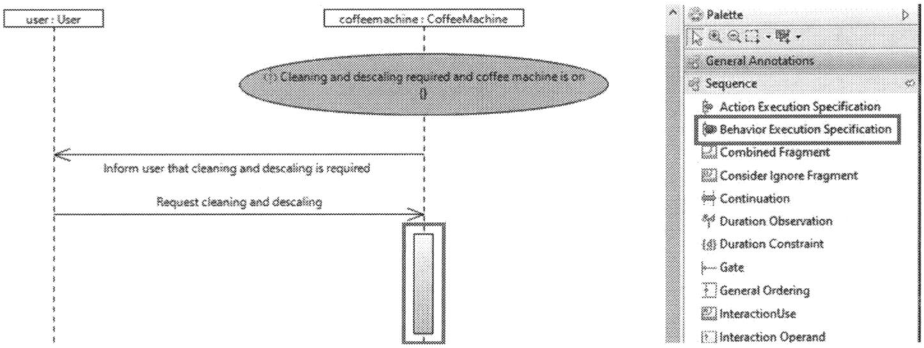

Figure 5.193. Behavior execution specification on the sequence diagram

To create a new activity that will be called by this behavior execution specification click on the button with the plus next to the Behavior property and select "Activity".

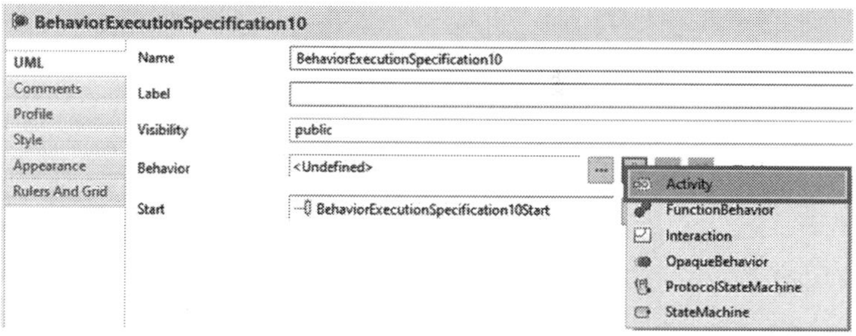

Figure 5.194. Create a new activity

Set the "CoffeeMachine" block as the container in the dialog and "ownedBehavior" as the reference.

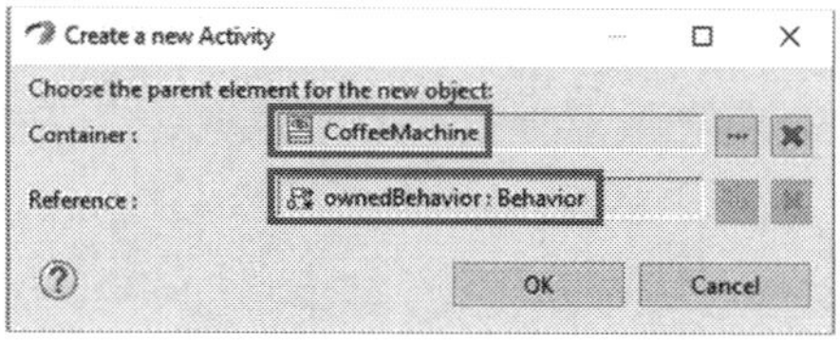

Figure 5.195. Set owner and reference of new activity

Name the new activity "Perform cleaning cycle".

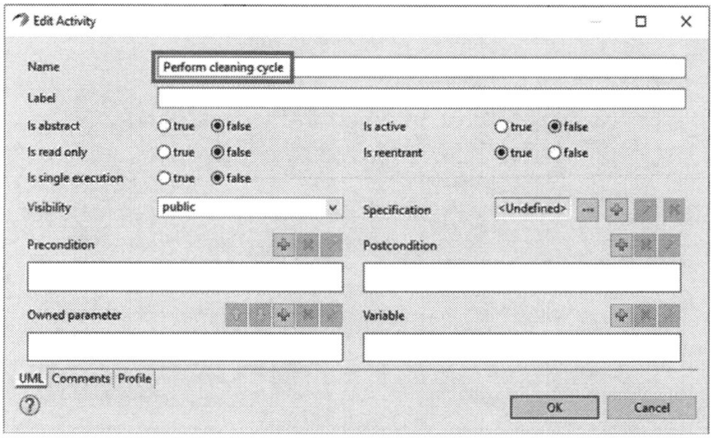

Figure 5.196. Name the new activity

The next step is to specify the signals that are sent during the interaction between the coffee machine and the user. Before you proceed, create a new package called "Interfaces" in the model.

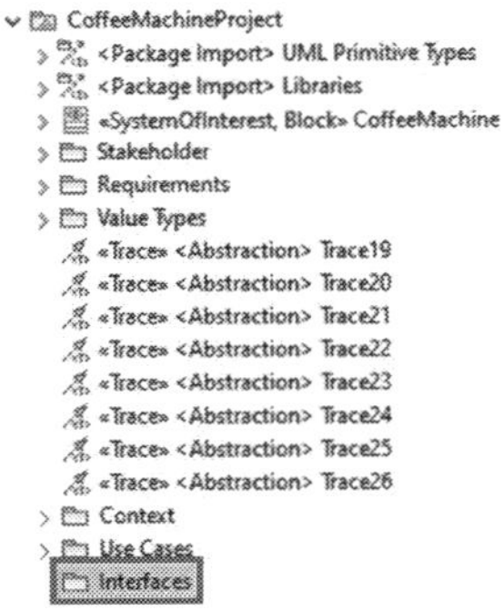

Figure 5.197. New "Interfaces" package

Select the "Inform user that cleaning and descaling is required" message and go to its Properties View. Click on the button with the plus next to the Signature property and select "Signal".

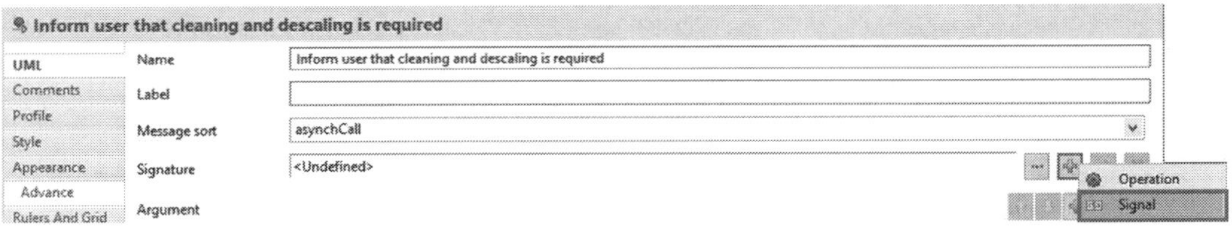

Figure 5.198. Create signal

Select the "Interfaces" package as the container for the new signal.

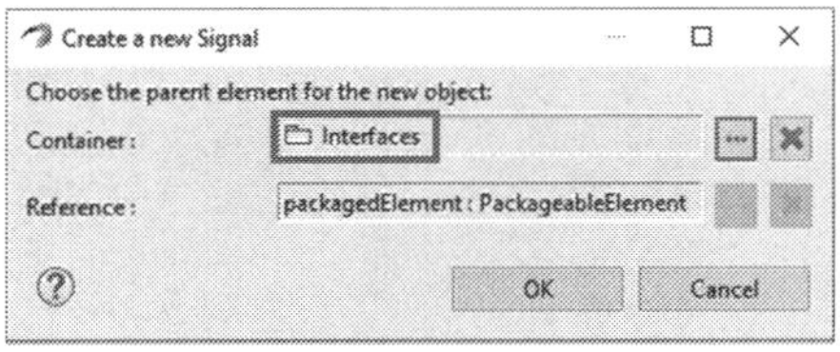

Figure 5.199. Set the container for the new signal

Name the signal "Cleaning and descaling" is required.

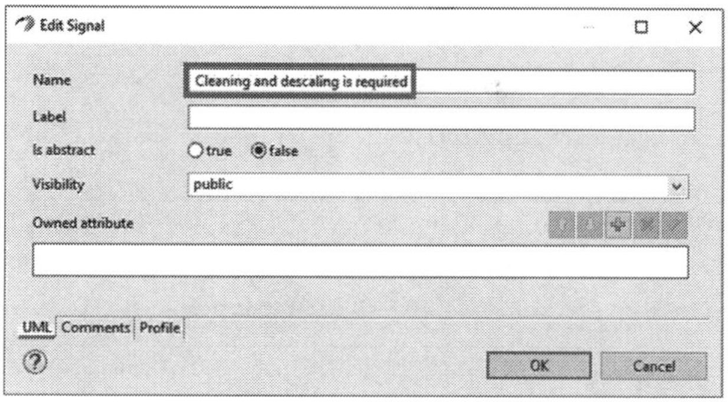

Figure 5.200. Name the new signal

Repeat the steps for all the other messages and behavior execution specifications in the sequence diagram.

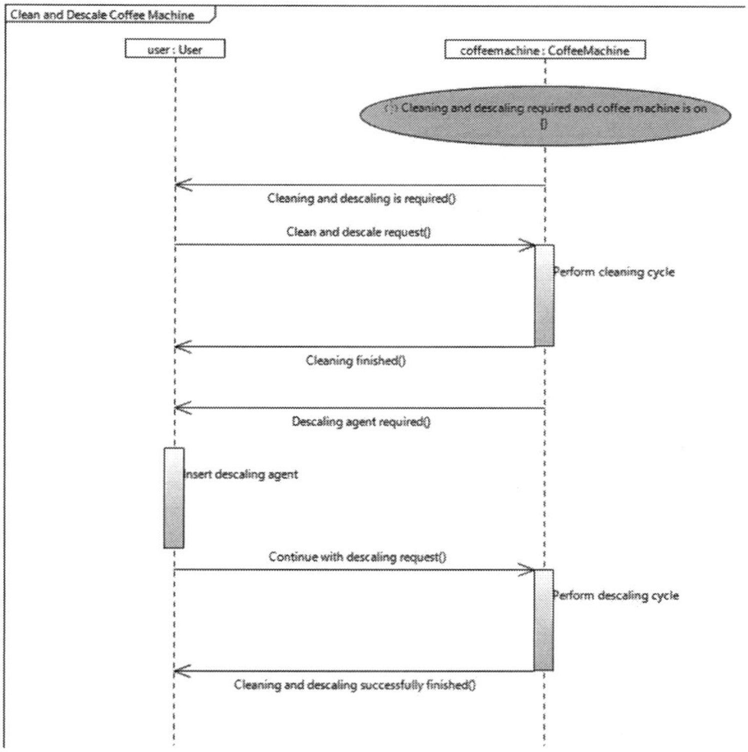

Figure 5.201. Finished sequence diagram

5.8 Analyze System States

System state analysis is a fundamental concept in systems engineering that focuses on understanding the system's current and potential future states. It provides a framework for analyzing the system's behavior as it interacts with its environment and responds to various stimuli. By understanding the system's states and state transitions, systems engineers can effectively design, operate, and maintain complex systems.

A state in systems engineering represents a condition or situation that a system can exist in at a specific point in time. It describes the current status or configuration of the system, indicating the values of relevant variables, settings, or parameters. Systems can have multiple states, and transitions between these states occur in response to events, inputs, or changes in the system's environment.

States are often defined by the values of specific variables or parameters within the system. States can be stable or unstable, depending on whether the system remains in a particular state or transitions to another. Possible changes between states are defined by transitions. System state transitions represent the changes in the system's state. They occur in response to various factors, such as external stimuli, user interactions, or internal processes. State transitions define the system's behavior and its ability to respond to changes in its environment.

System state analysis in systems engineering involves the examination and characterization of the various states that a system can exist in and the transitions between these states. This analysis is crucial for understanding the dynamic behavior of a system over time and under different conditions. The purpose of state analysis includes:

- **Understanding system behavior**: System state analysis provides a deeper understanding of how a system behaves under different conditions, scenarios, or inputs. It helps capture the dynamic nature of the system throughout its lifecycle. This understanding is crucial for designing systems that meet specific requirements and operate effectively in their intended environment.
- **Evaluating failure modes**: States analysis is essential for assessing and evaluating potential failure modes or undesirable states that the system may enter. This helps in identifying vulnerabilities and designing appropriate mitigation strategies. States analysis contributes to safety and reliability assessments by identifying critical states where failures or malfunctions may occur. Understanding these states is crucial for implementing safety measures and designing reliable systems.
- **Identifying operational states and transitions between them**: The analysis helps in identifying and defining the various operational states that the system can enter.

Operational states represent different modes of system behavior, each characterized by specific functionalities, constraints, or performance levels. State analysis involves specifying the conditions or events that trigger transitions between different states. Understanding these transitions is crucial for predicting and managing the system's behavior as it responds to external stimuli or internal changes. By modeling the system in different states and modes, analysts can describe how the system responds to various inputs, disturbances, or changes in its environment. This modeling facilitates a more comprehensive understanding of the system's capabilities and limitations.

- **Verification and validation**: System state analysis is a key component of V&V activities. By comparing the system's actual state transitions to the expected behavior specified in the requirements, engineers can verify that the system is functioning as intended.

State analysis helps in refining system requirements by considering how the system should behave in different operational scenarios. This contributes to the development of more precise and comprehensive system specifications. State analysis is a fundamental aspect of systems engineering that focuses on characterizing the dynamic behavior of a system. By identifying, defining, and analyzing different states and transitions, engineers can make informed decisions to ensure the reliability, safety, and effectiveness of complex systems.

5.8.1 System State Analysis Pitfalls

System state analysis is a valuable practice in systems engineering, but it is essential to be aware of common pitfalls to ensure the accuracy and effectiveness of the representation. Some common pitfalls in system state analysis include:

- **Overly Complex State Machines**: Creating state machines with an excessive number of states, transitions, or sub-states especially on a single state machine diagram. Overcomplexity can lead to difficulties in understanding, maintaining, and validating the state machine. It may also increase the likelihood of errors. Use hierarchical structures to organize states and transitions. Group related states into composite states to enhance readability. Regularly review and simplify the state machine to eliminate unnecessary complexity.
- **Missing or Redundant States**: Failing to include important states or introducing redundant states that do not contribute meaningfully. Missing states may lead to an incomplete representation of the system, while redundant states can add unnecessary complexity. Conduct thorough analysis to identify essential states and eliminate redundant ones. Ensure that each state contributes meaningfully to the understanding of the system's behavior.

- **Modeling non-states as states**: A state represents a meaningful and stable condition in the lifecycle of an object. Including transient conditions or irrelevant details as states bloats the state machine. This makes it harder to see the core functionality and identify valid transitions.
- **Under-specification of transitions**: Failing to specify the conditions or triggers that cause state transitions, leading to ambiguity and potential errors. Identify and document the conditions or triggers that cause state transitions.
- **Lack of Proper Hierarchical Structure**: Neglecting the use a proper hierarchical structure to organize states and transitions. A lack of hierarchy can make it challenging to manage and comprehend the state machine, particularly for large and complex systems. Organize states into hierarchies based on their relationships and dependencies. Utilize composite states to represent higher-level abstractions and improve the organization.
- **Neglecting Concurrency**: Neglecting to consider concurrent states or transitions. Ignoring concurrency can result in an incomplete representation, especially in systems where multiple actions can occur simultaneously. Indicate concurrent states and transitions clearly. Use synchronization mechanisms or orthogonal regions to model concurrent behaviors explicitly.

By being aware of these common pitfalls and incorporating some mitigation strategies into the systems engineering process, teams can enhance the reliability, clarity, and effectiveness of the system state analysis.

5.8.2 SysML State Machine Diagrams

The main elements of a SysML state machine diagram are:

- **State**: Represented by a rounded rectangle. Captures a specific condition or situation the system can be in. Special kinds of states, composite and submachine states, enable further decomposition for hierarchical modeling.
- **Transition**: Illustrated by an arrow connecting two states. Signifies a change from one state to another. Is usually initiated by a trigger.

5.8.3 Coffee Machine States and Transitions

The coffee machine can be in the following states:

State	Description
Off	The coffee machine is powered off.
On	The coffee machine is powered on.
Idle	The coffee machine is powered on but not actively brewing coffee.
Brewing	The coffee machine is actively brewing coffee.
Error	The coffee machine has encountered an error.
Maintenance	The coffee machine is in the process of cleaning and maintenance.

Transitions between states of the coffee machine include:

From	To	Trigger
Off	On	The coffee machine is powered on.
On	Off	The coffee machine is powered off.
On	Off	The coffee machine loses power.
Idle	Brewing	The user initiates the brewing process.
Brewing	Idle	The brewing process is complete.
Brewing	Idle	The brewing process is canceled due to lack of water.
Idle	Maintenance	The coffee machine requires cleaning and descaling.
Maintenance	Idle	The cleaning and descaling are finished.
Any	Error	The coffee machine encounters an error.

5.8.4 Modeling the Coffee Machine States

The states of the coffee machine are modeled using a SysML State Machine Diagram. Add a new state machine diagram to the coffee machine by right-clicking on the "CoffeeMachine" block in the Model Explorer and selecting *New Diagram->SysML 1.6 State Machine Diagram.* Name the new diagram "Coffee Machine STM".

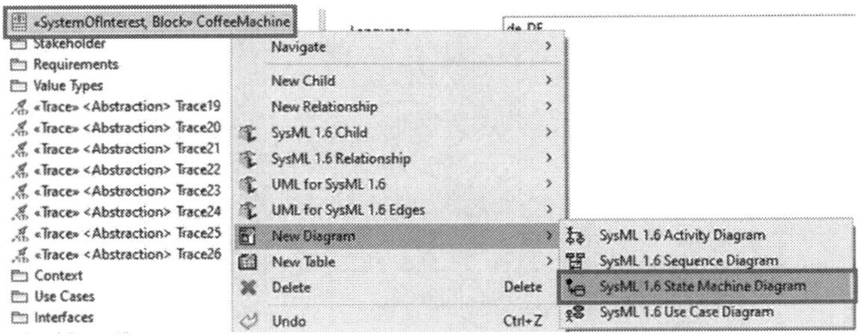

Figure 5.202. Create a new state machine diagram

Since state machine diagrams require a state machine element, Papyrus automatically creates a new state machine with a default name when the STM is created and adds it to the STM. Rename the state machine element to “Coffee Machine”.

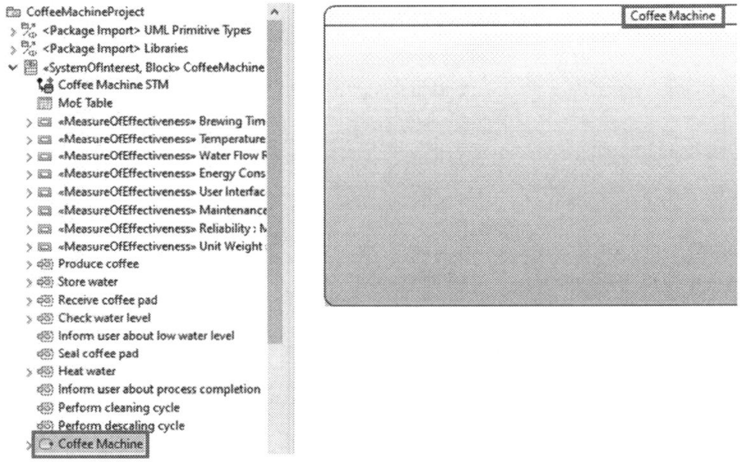

Figure 5.203. Renamed state machine “Coffee Machine”

Use the state element from the Diagram Palette to add states to the newly created STM.

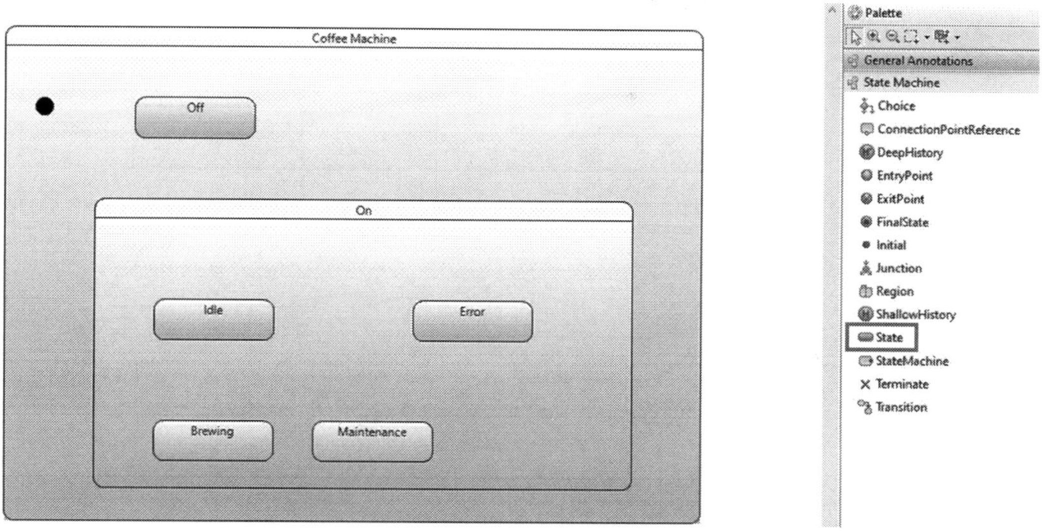

Figure 5.204. STM with new states

Add an initial node element to the STM.

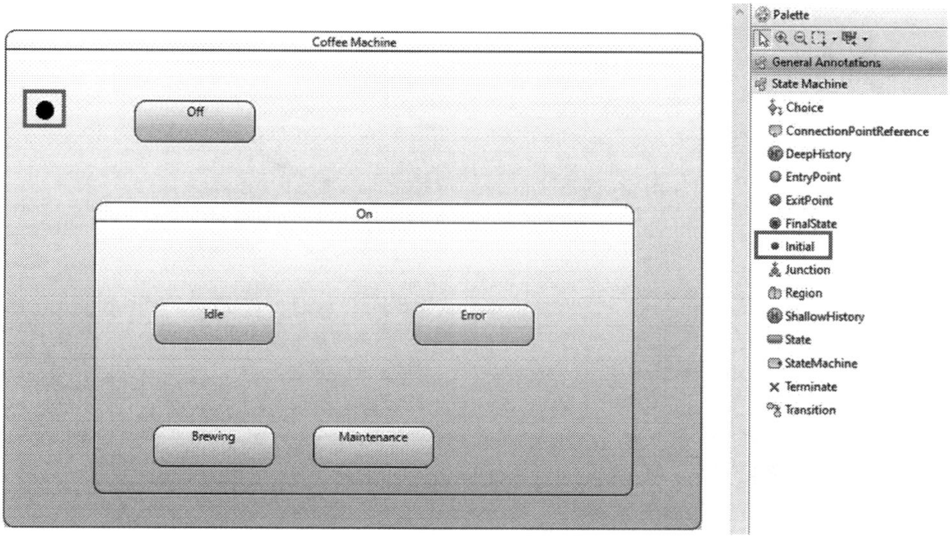

Figure 5.205. STM with Initial node

To prevent the display of text next to the start node element, go to the Properties View of the element and set the property Label to a single space or simply delete the name of the element.

Add a choice node element to the STM.

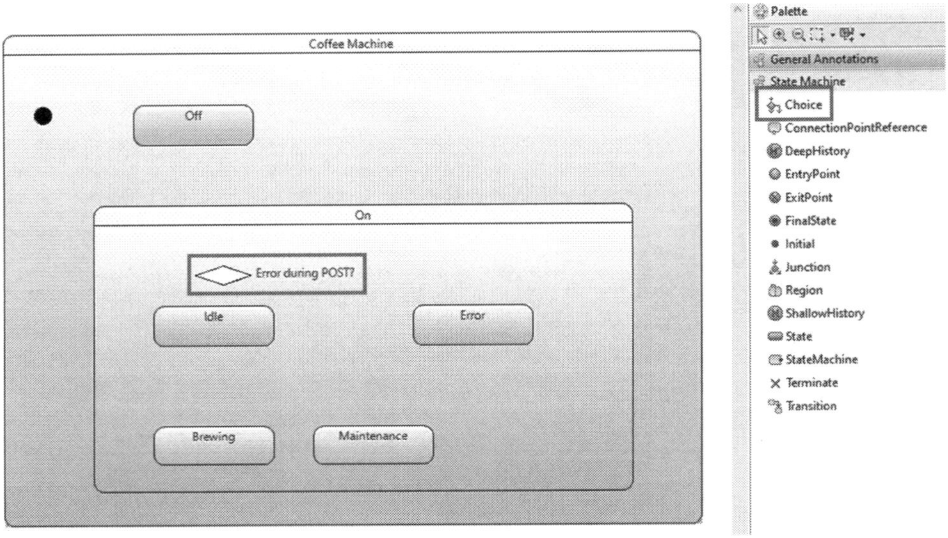

Figure 5.206. STM with choice node

Select the "Transition" element in the Diagram Palette and add the necessary transitions to the STM by clicking on the source state first and clicking on the target state second.

The transition from the "On" state to the "Error" state is a special kind of transition, a local transition. A local transition, if triggered, will not exit the source state, but it will apply to any state contained within the composite state, and these will be exited and entered, i.e., the transition can be triggered, while the coffee machine is in "Idle", "Error", "Maintenance" or "Brewing" state and when the transition is triggered the "On" state will not be exited.

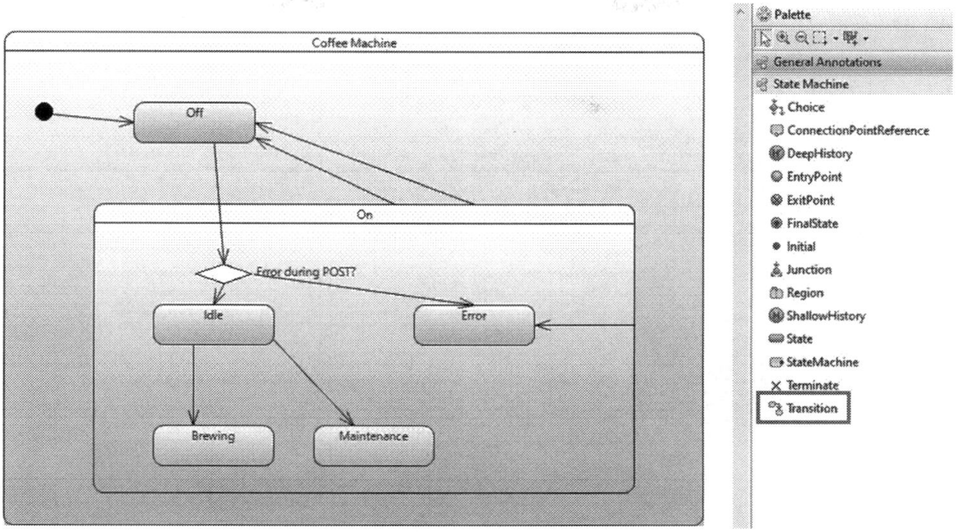

Figure 5.207. STM with transitions

The next step is to specify the triggers for the transitions. Select the transition from the "off" State to the decision node, open its Properties View, and click on the button with the plus next to the Trigger section to open the "Create a new trigger" dialog.

A transition in a state machine can have three properties: a trigger, a guard, and an action. The trigger defines when the transition can be taken. After the trigger but before the transition is actually followed, the guard of the transition is evaluated. The transition is only taken if the conditions specified by the guard are evaluated to be true. While the transition is being active, the action of the transition is executed.

Three types of elements are required to specify the trigger of a transition: a signal, an event, and a trigger. A signal is a specific type of event that represents a message sent between objects and carries data or information relevant to the message. An event is something that may occur at a specific instant in time. An event that is caused by the reception of a signal is called a message or signal event. Time events occur at a specified time or after a duration, and change events occur when a specified Boolean value becomes true. A trigger initiates a transition between states and can be caused by an event.

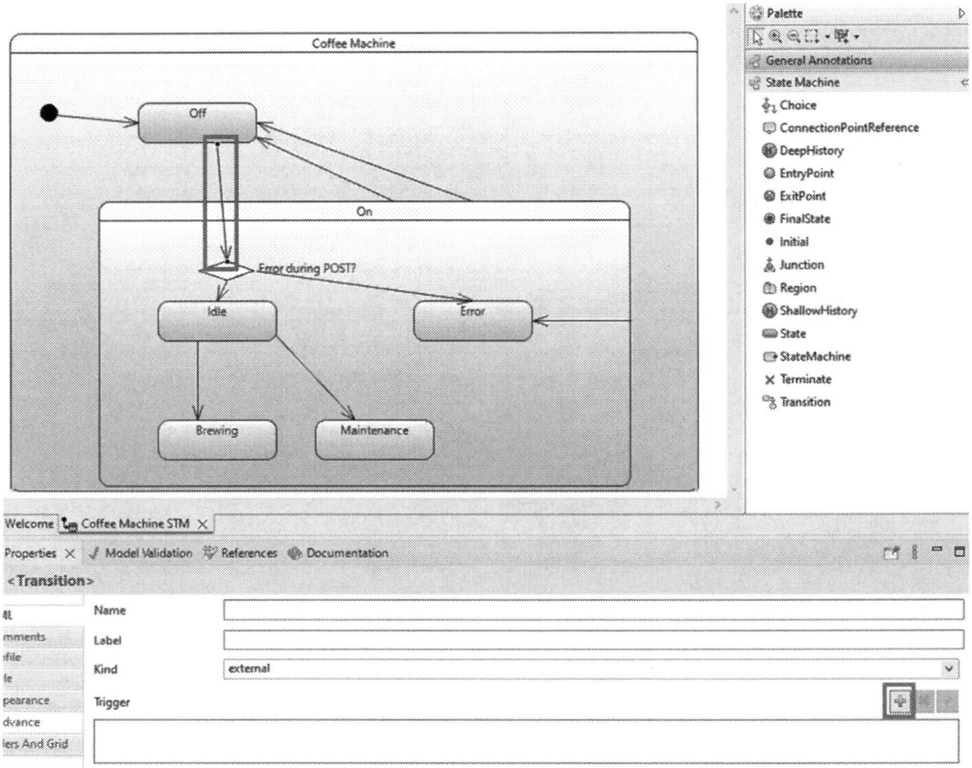

Figure 5.208. "Create a new trigger" dialog

In the dialog, click on the button with the plus next to the event property and select *SignalEvent.*

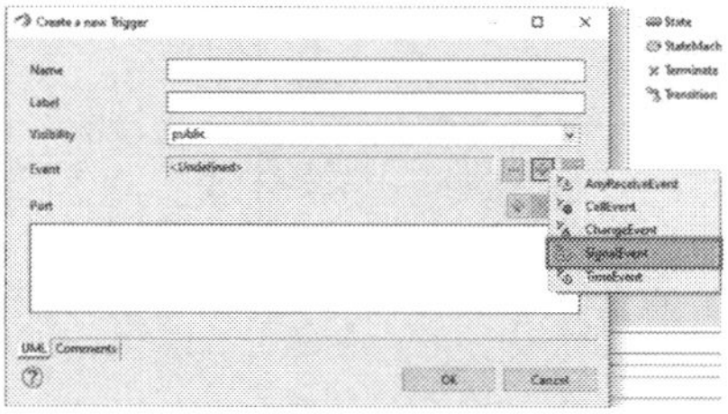

Figure 5.209. Create a new signal event for the trigger

Select the "Interfaces" package as the owner of the new signal event.

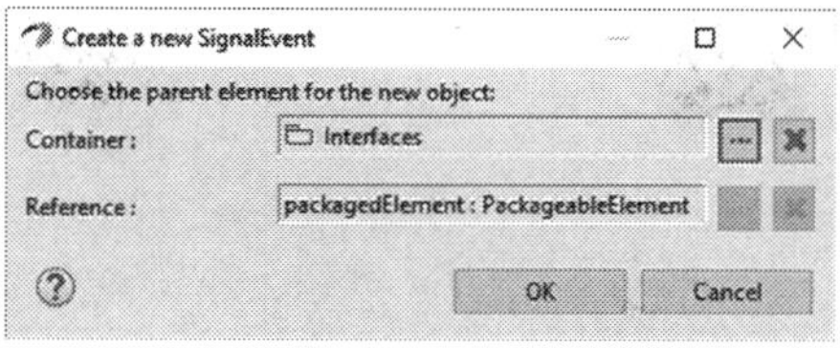

Figure 5.210. Select the owner of the signal event

Name the new signal event “Receive power on”. Then, click on the button with the plus sign next to the Signal property to create a new signal for the signal event. Set the “Interfaces” package as the owner of the new signal as well.

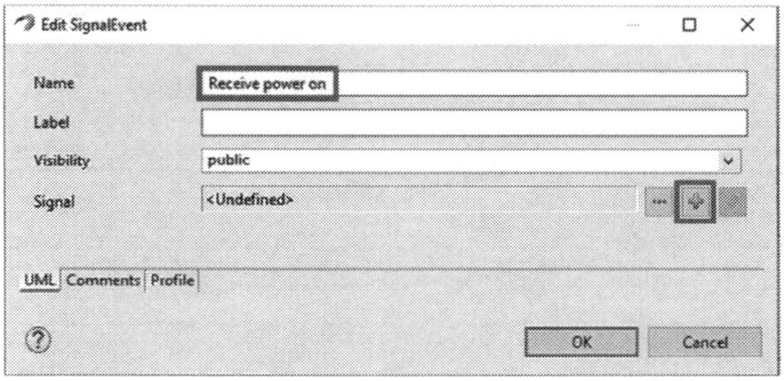

Figure 5.211. Open the “Create a new Signal” dialog

Name the signal “Power on”.

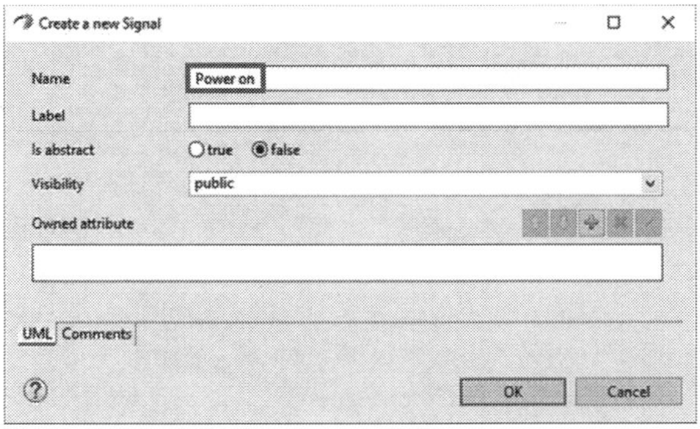

Figure 5.212. Create a new signal

Name the signal event “Receive Power on”.

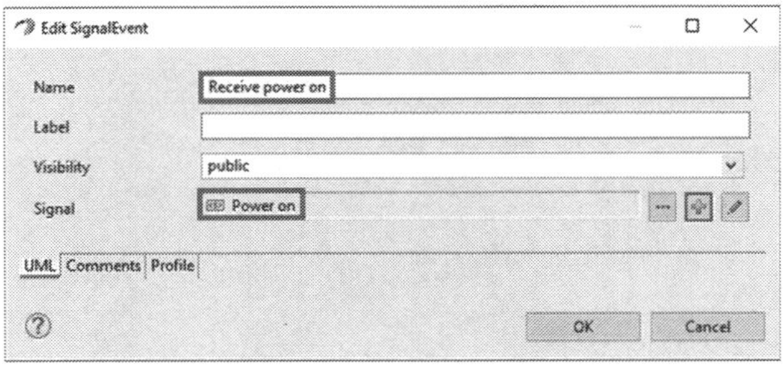

Figure 5.213. Name the signal event

The name of the signal is displayed on the transition in the STM.

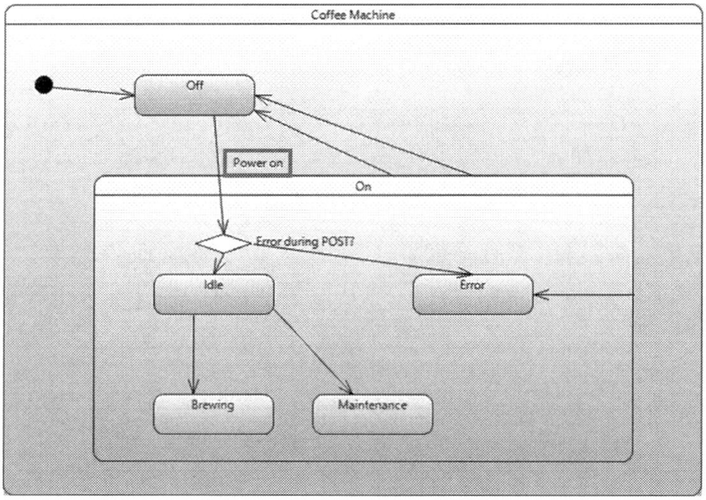

Figure 5.214. Transition with the trigger in the STM

Next, we want to represent the fact, that the coffee machine should perform a power-on self-test (POST) when it is turned on. We use a transition action to model this. The first step is to create an activity that represents the POST function. Add a new activity to the "CoffeeMachine" block by right-clicking on the block in the Model Explorer and selecting *New Child->Activity->As ownedBehavior.* Name the new activity "Perform power-on self-test".

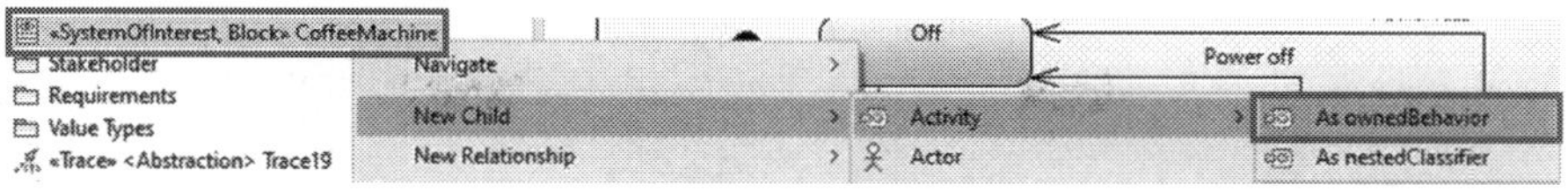

Figure 5.215. Create new activity

The new activity will be added to the "CoffeeMachine" block in the Model Explorer.

Figure 5.216. New activity "Perform power-on self-test"

To add the newly created activity as an action, select the transition from the "off" state to the decision node, open its Properties View, click on the button with the plus next to the Effect property, and select *Activity.* Call the new activity "Call perform power-on self-test".

> Note that we cannot add the activity "Perform power-on self-test" as an effect to the transition directly. Instead, we must create another new activity that is called the "Perform power-on self-test" activity. Other tools, such as the Cameo Systems Modeler, automatically perform the next steps when an activity is dropped on a transition. Papyrus does not, so we need to do it manually.

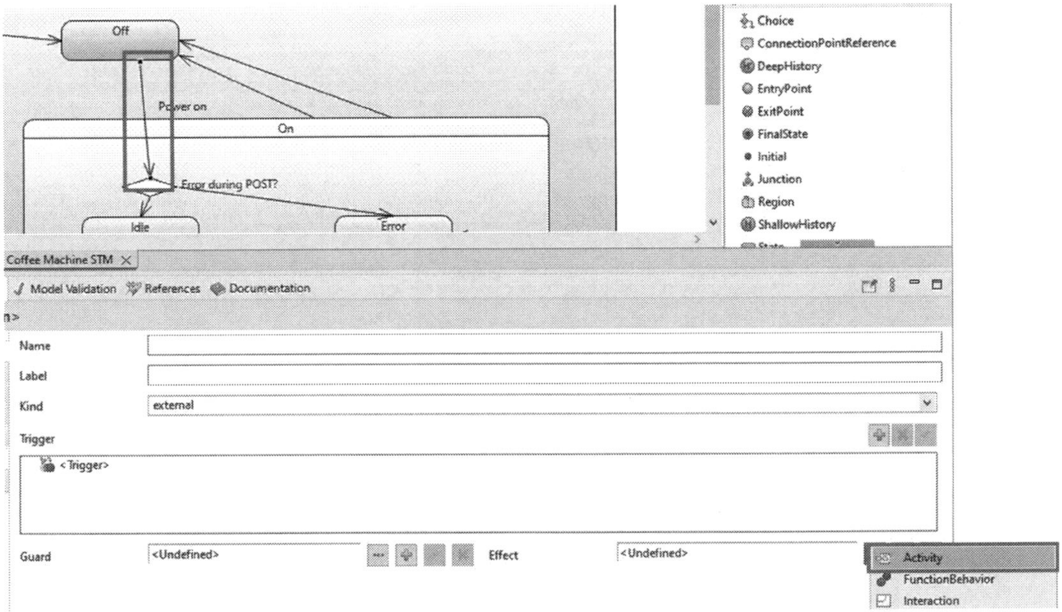

Figure 5.217. Add new activity as the effect of a transition

Select the "Call perform power-on self-test" activity in the Model Explorer and add a new activity diagram to it by right-clicking and selecting *New Diagram->SysML 1.6 Activity Diagram*. Name the new diagram "Call perform power-on self-test AD".

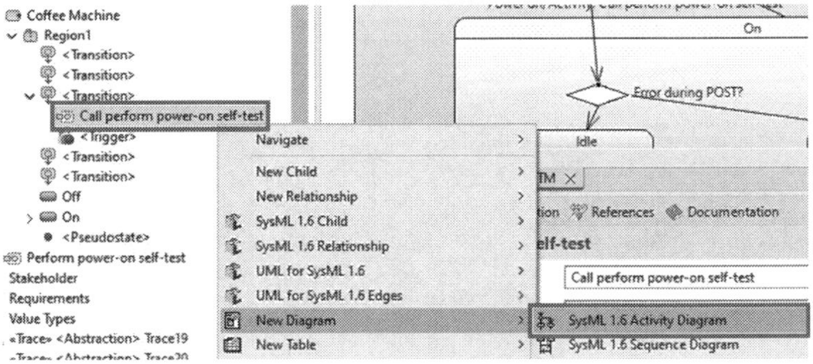

Figure 5.218. Add an activity diagram to the activity

Drag and drop the "Perform power-on self-test" activity on the newly created activity diagram and select *Drop Behavior as CallBehaviorAction.*

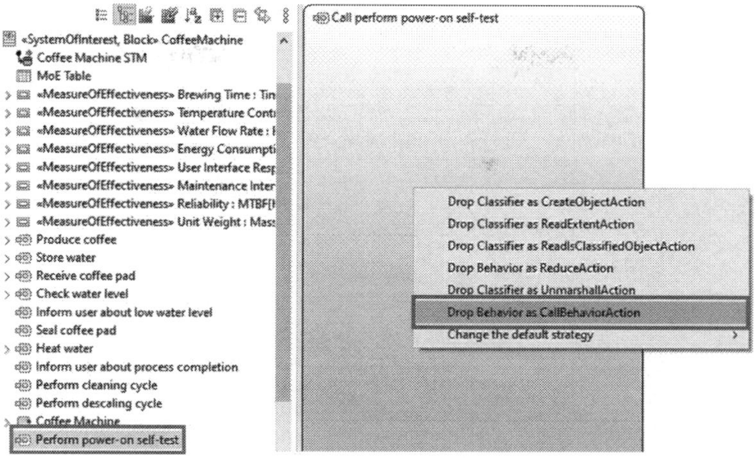

Figure 5.219. Drag and drop the activity on the activity diagram

Name the call behavior action "Perform power-on self-test". Add an initial node, an activity final node and two control flows to finish the AD.

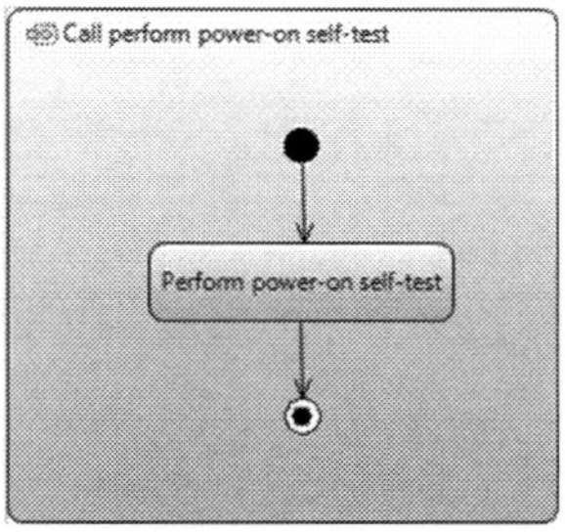

Figure 5.220. Finished activity diagram

The transition between the "Idle" state and the "Maintenance" state shall be triggered by a user request. We already created the signal element during the use case analysis of the "Clean and descale Coffee Machine" use case. Select the transition, open its Properties View, and click on the button with the plus to open the dialog for creating a new signal event.

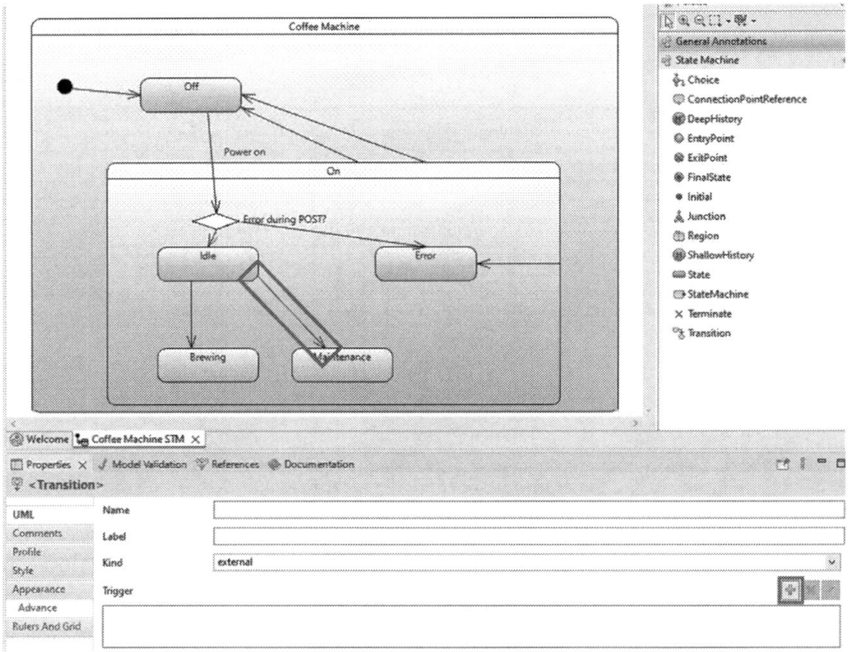

Figure 5.221. **Add a trigger to another transition**

Add a signal event to the new trigger as before.

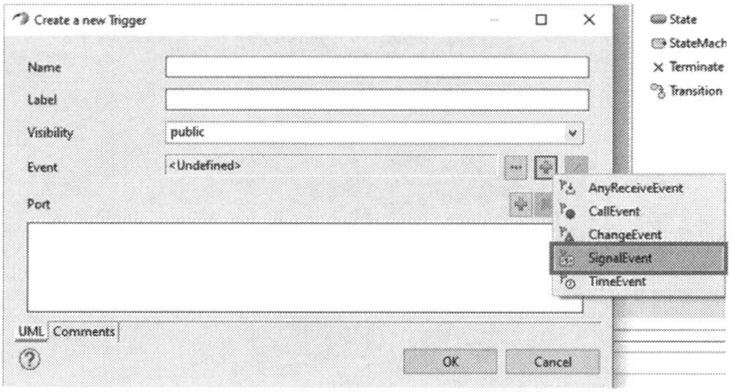

Figure 5.222. **Create a new trigger**

While creating the new signal event, instead of creating a new signal as before, we want to reuse an already existing signal. To do this, click on the button with the three dots.

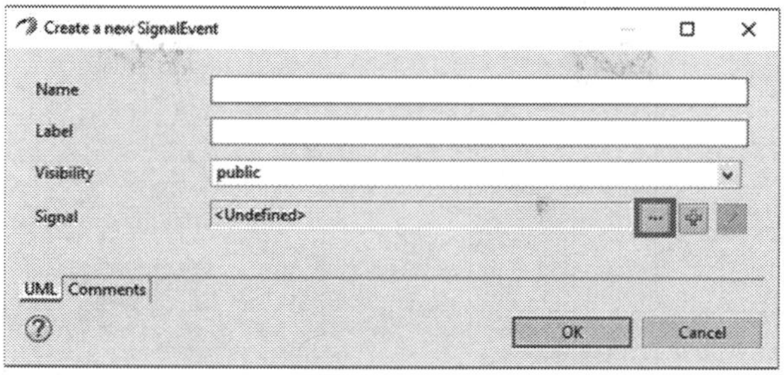

Figure 5.223. Set the signal of a signal event

Select the “Clean and descale request” signal in the “Interfaces” package.

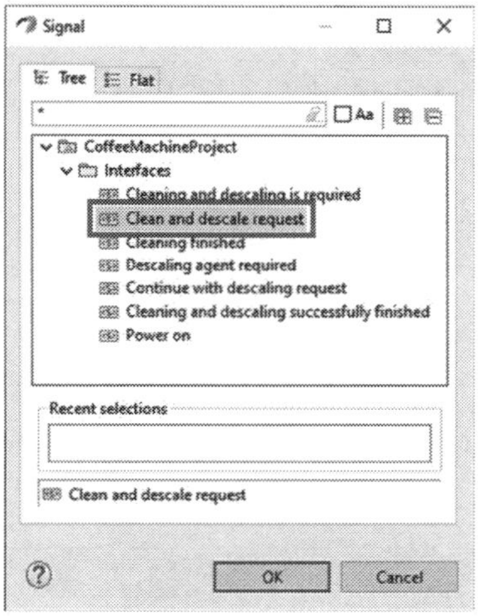

Figure 5.224. Select the correct signal

Repeat the same process to create more signals, signal events, and triggers to complete the STM.

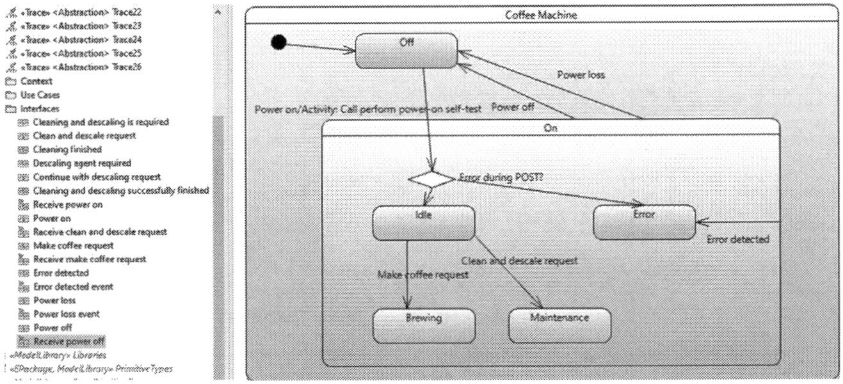

Figure 5.225. STM with all the triggers

The “Error during POST” decision node has two exits. To model, which transition from the decision node is selected, guards on the outgoing edges need to be added. Select the transition, which leads to the “Error” state, and open its Properties View. Click on the button with the plus next to the guard property and select *Constraint.*

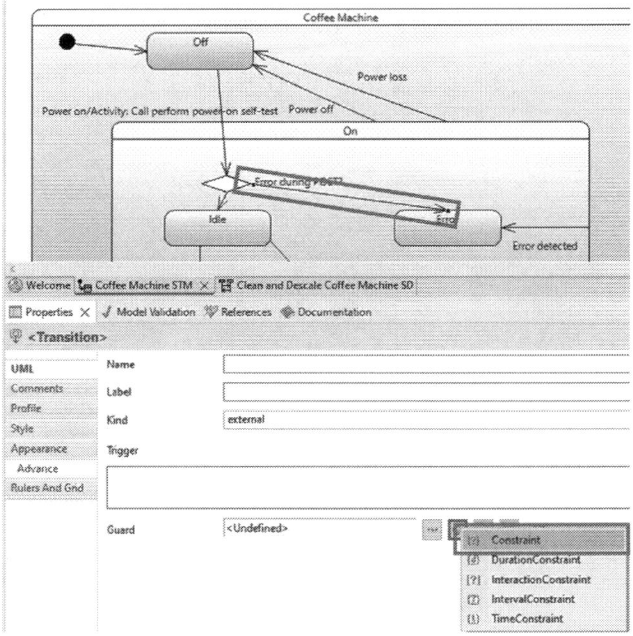

Figure 5.226. Add a constraint as a guard

To specify the constraint, click on the button with the plus next to the Specification property and select *OpaqueExpression*.

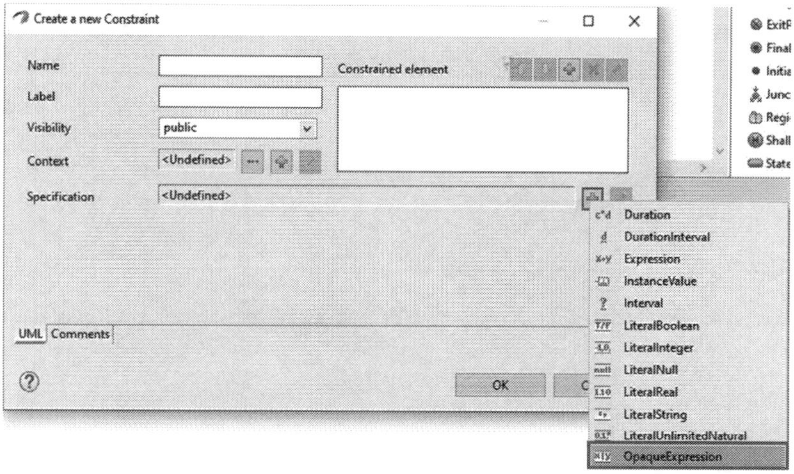

Figure 5.227. Add an opaque expression

Set the language of the opaque expression to *Natural language* and set the content of the opaque expression to "error detected".

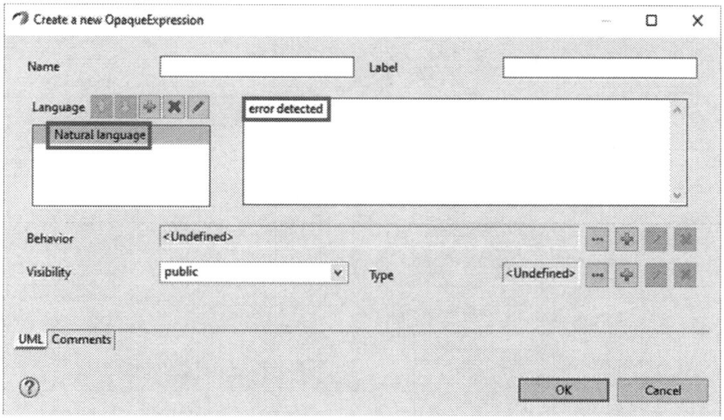

Figure 5.228. Specify the opaque expression

Repeat the same process for the transition between the "Error during POST Decision" node and the "Idle" state. Then, spend some time arranging the elements to make for a more readable diagram.

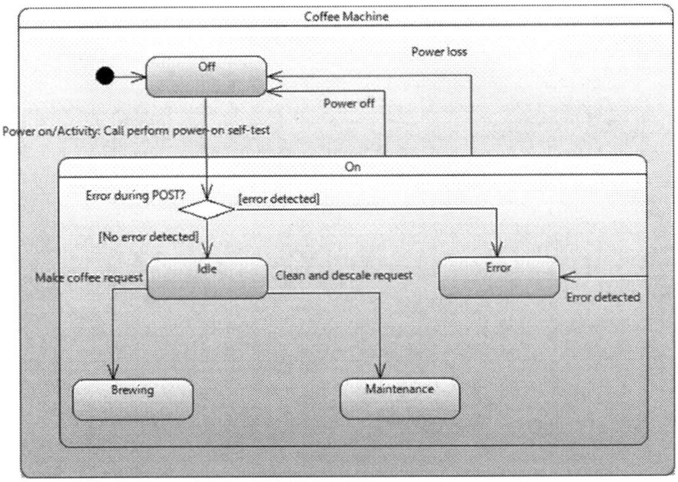

Figure 5.229. STM with two guards

Next, we want to add more details about the "Maintenance" state. To avoid cluttering the overall "Coffee Machine STM", we make use of the possibility of hierarchical state machines, i.e., a state machine is nested inside another state machine as a sub-state machine.

To find the "Maintenance" state in the Model Explorer, right-click on the "Maintenance" state on the STM and select *Navigate->Go to element (Maintenance)...->Model Explorer.*

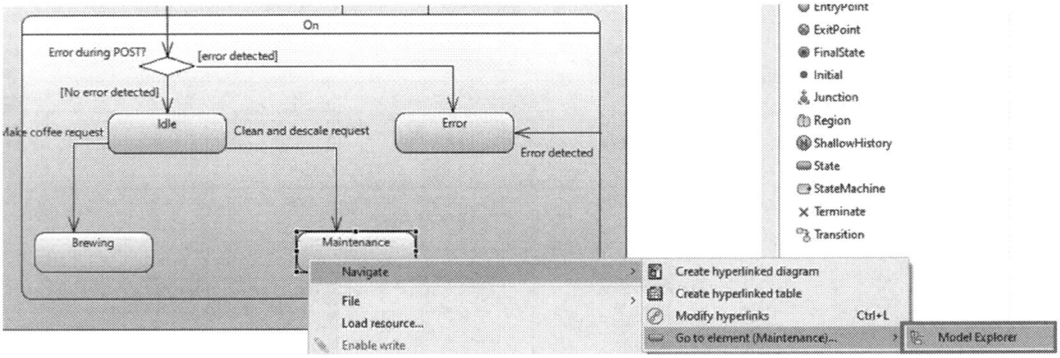

Figure 5.230. Navigate to the "Maintenance" state

To add a new state machine to the "Maintenance" State, right-click on the state in the Model Explorer and select *New Diagram->SysML 1.6 State Machine Diagram.* Name the new diagram "Maintenance STM".

Creating the new STM for an existing state to create a sub-state machine is only possible using the context menu in the Model Explorer. Other tools allow this directly on the diagram, which enables a more efficient way of working.

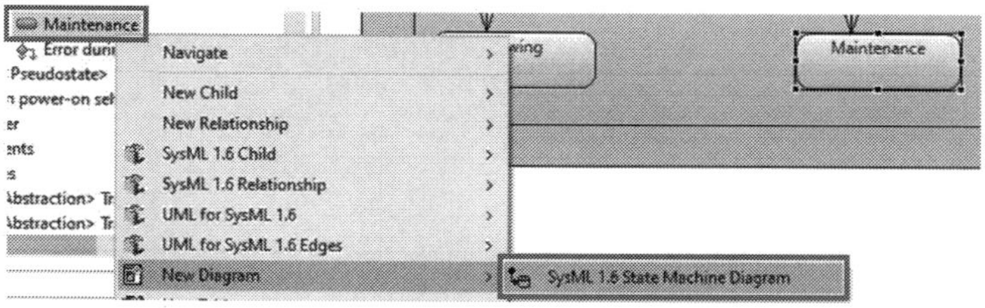

Figure 5.231. Create a new state machine diagram

Select the "ExitPoint" element from the Diagram Palette and left-click on the border of the "Maintenance" State. Name the exit point "Maintenance Finished". The exit point will allow us to connect the end of the "Maintenance" state to a state on the upper-level STM.

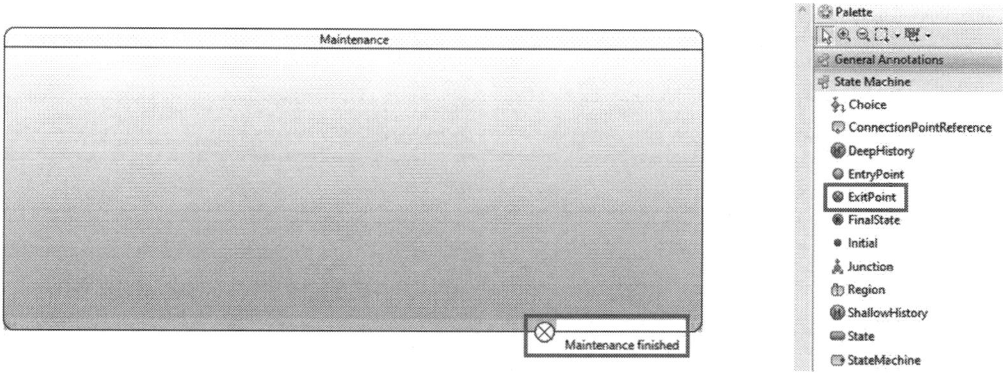

Figure 5.232. Create a new exit point

Add states, an initial node, and transitions to the STM.

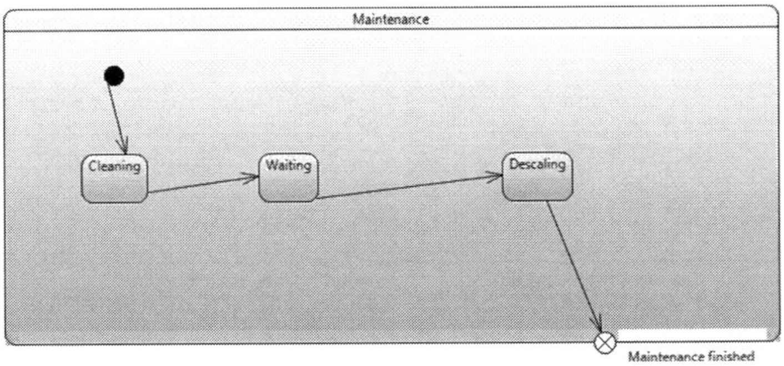

Figure 5.233. STM with basic elements

Next, we want to model that the coffee machine shall start the cleaning cycle when it enters the "Cleaning" state. Select the "Cleaning" state, click on the button with the plus next to the Do activity property, and select *Activity*.

A state can reference other behaviors like activities in three ways: as entry action, as do action, and as exit action. An entry action is executed upon entering a state. An exit action is executed upon leaving a state. A do action commences execution when the state is entered after the entry behavior has completed and executes until it completes or the state is exited. These behaviors are properties of the state and are represented within the state using appropriate notation, e.g., text within the state rectangle.

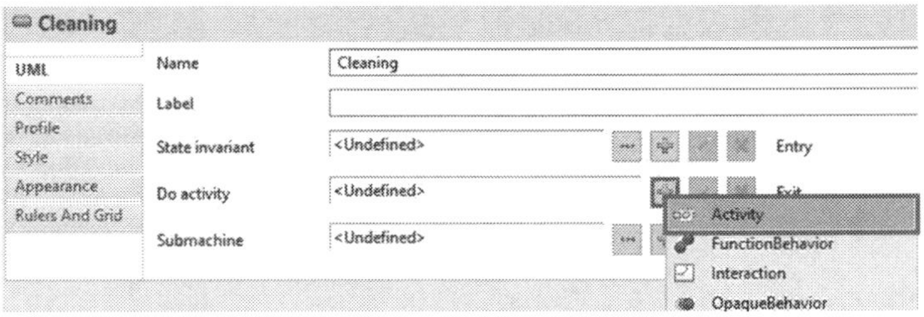

Figure 5.234. Create new activity

Name the new activity "Call perform clean cycle".

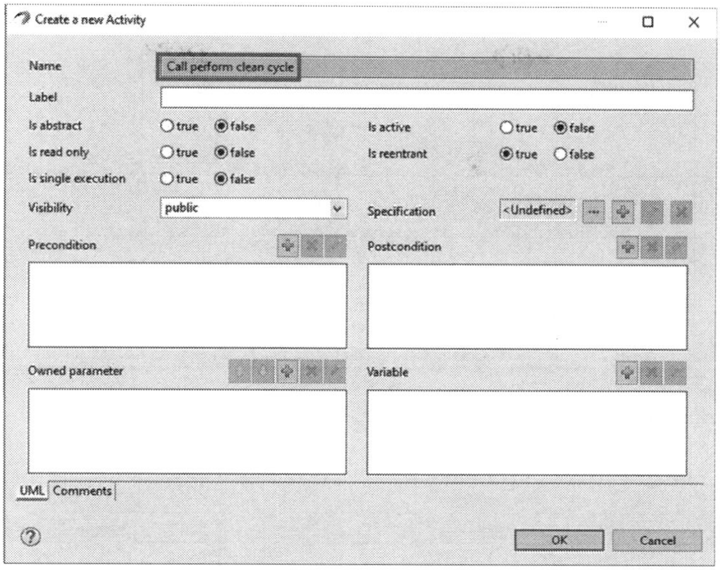

Figure 5.235. Name the new activity

As before for the “Call perform power-on self-test” activity, create a new activity diagram for the activity and add the “Perform cleaning cycle” activity from the “CoffeeMachine” block as a call behavior action.

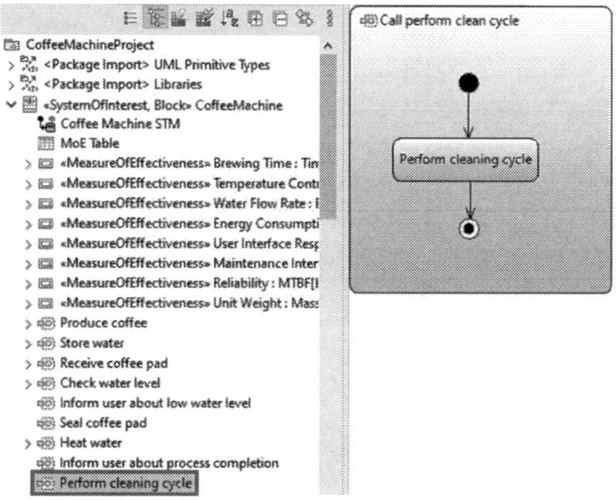

Figure 5.236. Activity diagram “Call perform clean cycle AD”

Specify the do actions and the triggers on the STM and arrange the elements to complete the STM for the “Maintenance” state.

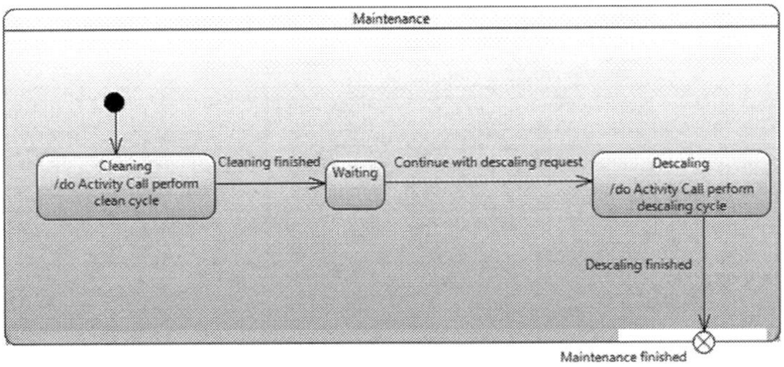

Figure 5.237. Complete STM for the "Maintenance" state

Adding the exit point to the "Maintenance STM" automatically added the exit point to the "Maintenance" state of the "Coffee Machine STM" as well.

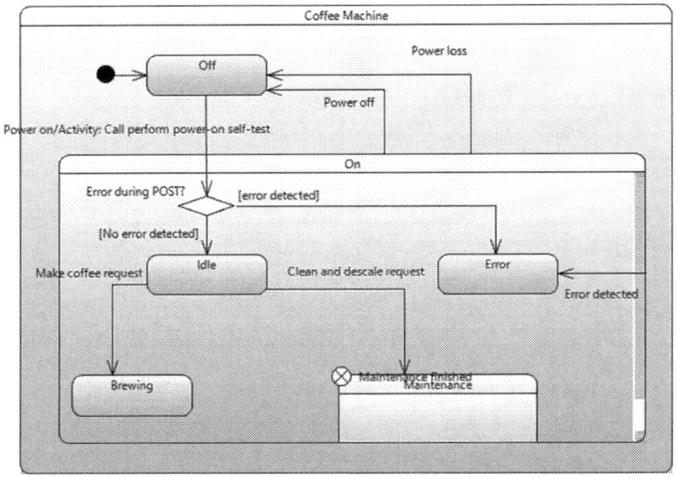

Figure 5.238. "Maintenance" state with exit point

Create a transition between the "Maintenance Finished" exit point and the "Idle" state. Note that this transition does not require a trigger. It will automatically be taken when the transition to the exit point is taken on the "Maintenance STM".

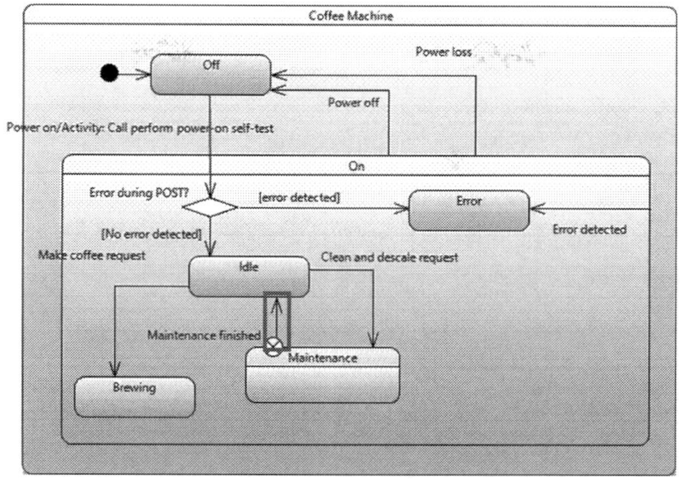

Figure 5.239. Transition from the exit point to the "Idle" state

Repeat the same process to create a sub-state machine for the "Brewing" state.

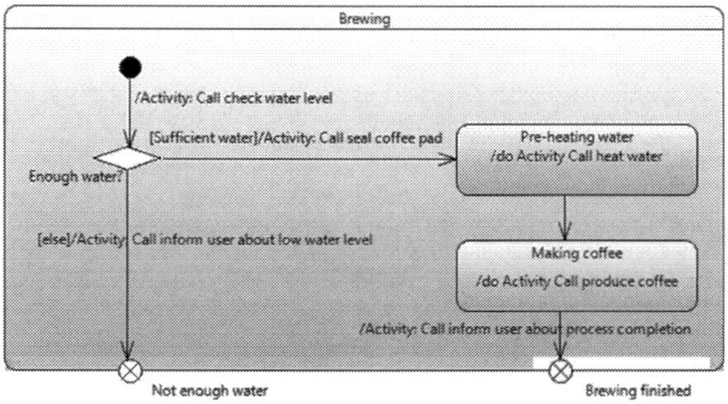

Figure 5.240. Sub-state machine for the "Brewing" state

To trigger the transition from the "Pre-Heating water" state to the "Making coffee" state, a change event is used instead of a signal event. A change event represents monitoring something, e.g., the value of a property and is triggered when a certain condition is reached. Select the transition and create a new trigger but instead of *SignalEvent* select *ChangeEvent* when adding the new event to the trigger.

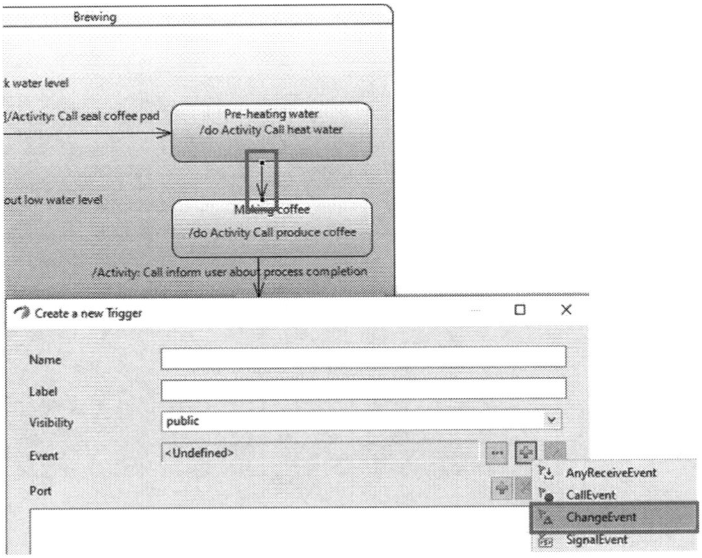

Figure 5.241. Create a new change event

Add a new change expression to this change event by clicking on the button with the plus next to the Change expression property and selecting *OpaqueExpression*.

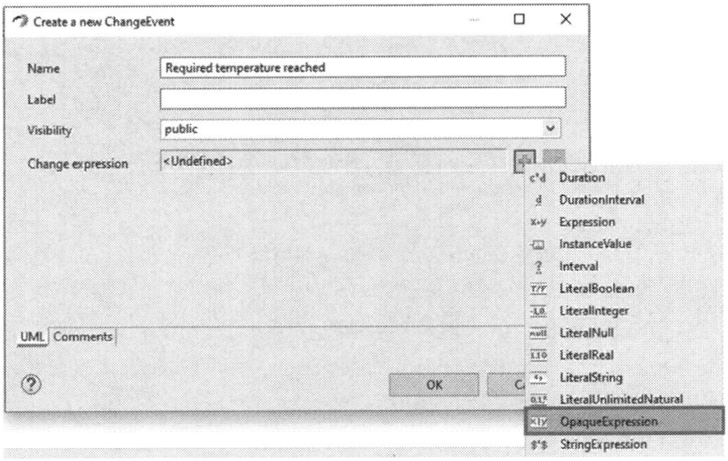

Figure 5.242. Create a new opaque expression

Set the language and specify the condition of the newly created opaque expression.

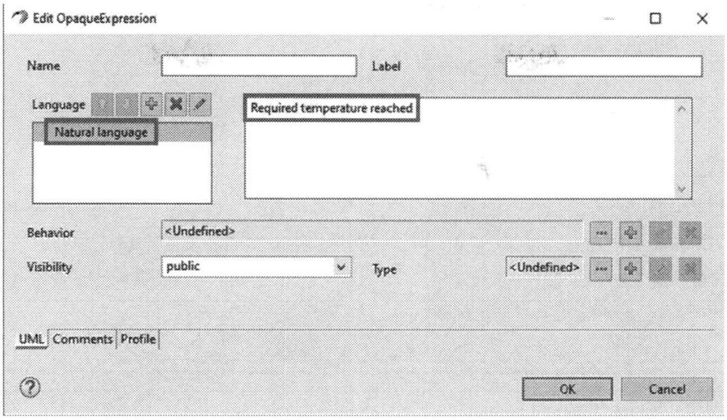

Figure 5.243. Specify the opaque expression

Note that for some reason, Papyrus does not show the opaque expression of the guard on the diagram. Instead, only three dots are shown. As a workaround, you can name the transition itself "Required temperature reached".

This finishes the "Brewing STM" diagram.

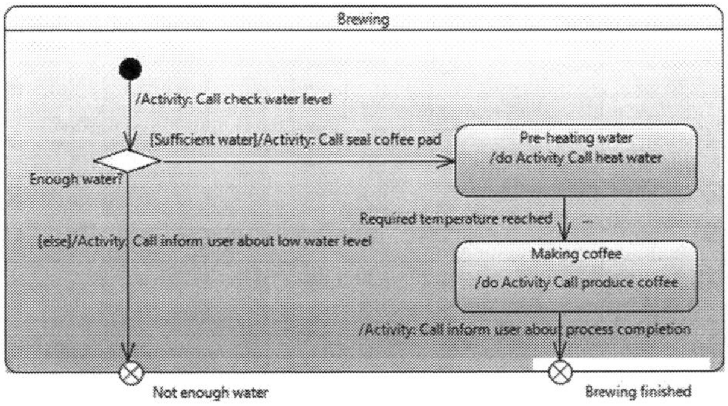

Figure 5.244. Finished "Brewing STM" diagram

Model additional transitions from the two exit points of the "Brewing" state to the "Idle" state. Route all the transitions and size the states to make a readable diagram.

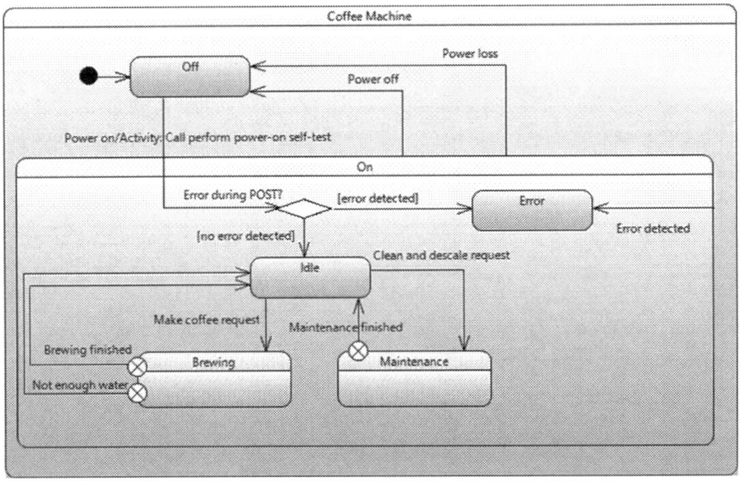

Figure 5.245. Finished "Coffee Machine STM"

5.9 Analyze and Decompose System Functions

Functional analysis is a process in systems engineering that focuses on understanding the behavior and purpose of a system without considering the specific implementation details. It involves identifying the system's functions, their relationships, and the inputs and outputs associated with each function. This level of abstraction allows for a clear separation between the analysis of functionality and the design of the implementation.

System functions refer to the capabilities, actions, or operations that a system is designed to perform to achieve its intended objectives. A system function is a fundamental unit of behavior that defines what a system does or is intended to do. Together, the system's functions enable a system to fulfill its purpose and meet the requirements specified for its development. In systems engineering, identifying and understanding system functions is a critical step in the analysis and design of a system.

System functions can vary widely depending on the type and the purpose of the system and the current level of abstraction. Examples include: For an automotive system: "Accelerate", "Brake", "Steer", or "Monitor Engine Temperature". For a communication system: "Transmit Data", "Receive Data", "Establish Connection", or "Maintain Signal Quality".

Following the recurring theme of decomposition in systems engineering, functional decomposition is a systematic process of breaking down a complex function of a system into its constituent functions or sub-functions. This hierarchical breakdown allows engineers to analyze and understand the system's organization and functionality at different levels of detail.

Function allocation assigns specific functions to specific parts. Functional decomposition provides the framework for function allocation, as it identifies the smaller functional units, component-level functions, that compose the system's function. Function allocation then assigns those component-level functions to the appropriate components within the decomposition structure. By combining decomposition and function allocation, systems engineers can create a well-defined and organized system architecture that meets the system's requirements. The benefits of functional analysis and decomposition include:

- **Increased flexibility**: A well-defined function allows the design team to explore various technologies and approaches that can be used to implement the function, leading to a potentially more innovative and adaptable system.
- **Identification of interfaces**: Functional analysis helps identify the interfaces between different components within a system. By systematically breaking down a system into its functional elements, these techniques provide a structured approach to understanding the interactions in the form of functional interfaces between components, which is, in turn, the basis for the definition of the technical interfaces between the system components. Understanding these interfaces is vital for ensuring seamless integration and interaction between system elements.
- **Risk Analysis**: As a basis for failure mode analysis, functional analysis contributes to risk analysis by identifying critical functions and assessing the potential impact of failures on the overall system performance. This is essential for designing reliable and fault-tolerant systems.
- **Support reuse**: Functional analysis and decomposition facilitate the identification of reusable functions by providing a structured approach to understanding the system's functionality. By systematically decomposing the system into its constituent functions, engineers can identify functions that are common across different systems or that can be adapted to serve similar purposes in new systems.

5.9.1 Functional Analysis Pitfalls

Functional analysis and decomposition are crucial steps in systems engineering, but like any complex process, they come with potential pitfalls. Being aware of these common pitfalls can help engineers navigate through the challenges and improve the effectiveness of functional analysis. Here are some common pitfalls in functional analysis and decomposition:

- **Insufficient understanding of system goals**: Starting functional analysis without a clear and comprehensive understanding of the system's goals and requirements can lead to incomplete or misaligned functional decompositions. Prioritize a thorough understanding of system objectives and requirements before initiating functional analysis. Engage with stakeholders and subject matter experts to gather complete and accurate information.
- **Lack of modularity**: Failing to design functions with modularity can result in a rigid and inflexible system, making it challenging to adapt to changing requirements. Encourage a modular approach to functional decomposition. Design functions as independent and reusable components, allowing for easier integration and maintenance.
- **Ignoring system boundaries**: Neglecting to define and consider the boundaries of the system can lead to either an overly broad or narrow functional analysis, impacting the accuracy of the decomposition. Clearly define the system boundaries to establish the scope of functional analysis. Determine what is included within the system and what lies outside of it.
- **Excessive detail at early stages**: Delving into excessive detail too early in the functional decomposition process can be time-consuming and may lead to a loss of focus on high-level objectives. Start with high-level functions and progressively decompose them to more detailed levels as needed. Focus on capturing the essential aspects before going into finer details.
- **Functions that predetermine the design solutions**: When functions specify technological solutions, they limit the design engineer's ability to explore alternative solutions. If a function is tied to a specific technology, the system might become outdated faster as new and better options emerge. Functional analysis should define what the system needs to do, not how it should achieve it. Acknowledge that there might be various ways to achieve the desired function. Consider leaving room for the design team to explore different options later.
- **Overlooking interdependencies**: Failing to identify and analyze interdependencies between functions can result in a fragmented understanding of the system's behavior. Explicitly document relationships and dependencies between functions. Consider how changes to one function may affect others and ensure that these interactions are well-understood.
- **Not considering non-functional requirements**: Concentrating solely on functional requirements and overlooking non-functional requirements (e.g., performance, reliability) can lead to an incomplete analysis. Integrate non-functional requirements into the functional analysis. Ensure that the system's design addresses both what it does (functions) and how well it does it (non-functional aspects).
- **Lack of iterative review**: Failing to conduct iterative reviews and refinements of

the functional decomposition throughout the development process can result in an outdated or inaccurate representation. Schedule regular reviews with stakeholders, subject matter experts, and the development team. Allow for feedback and updates to the functional decomposition based on evolving requirements and insights.

- **Failure to document assumptions**: Not documenting assumptions made during the functional analysis can lead to misunderstandings and misalignments between the development team and stakeholders. Clearly document any assumptions made during the analysis. Share these assumptions with relevant parties to ensure a shared understanding of the analysis context.

By being mindful of these common pitfalls and actively addressing them throughout the functional analysis and decomposition process, engineers can enhance the quality and relevance of their system designs. Regular communication with stakeholders, thorough documentation, and a structured approach to analysis are key elements for successful functional decomposition.

5.9.2 Coffee Machine Functional Decomposition

The starting point for the functional analysis of the coffee machine functions is the system functions that were already identified during the use case analysis. These functions have been modeled as activities.

Note that the system functions identified during the use case analysis steps typically do not include all of a system's functions since the focus during the use case analysis is on the interaction with the user. Other functions that have a more hidden or supporting role might have been overlooked. During the functional analysis step, it is, therefore, common to identify additional system functions, which can be added to the list of system functions and included in the analysis and decomposition step.

To provide an overview of all the system-level functions that have been identified during the use case analysis, add a new table to the model and name it "System Level Functions Table". Drag and drop all of the activities of the "CoffeeMachine" block and the "User" actor that were created during the use case analysis onto the table. Open the "Select Columns" dialog and add the properties "context", "inputParameters", and "outputParameters" to the right side.

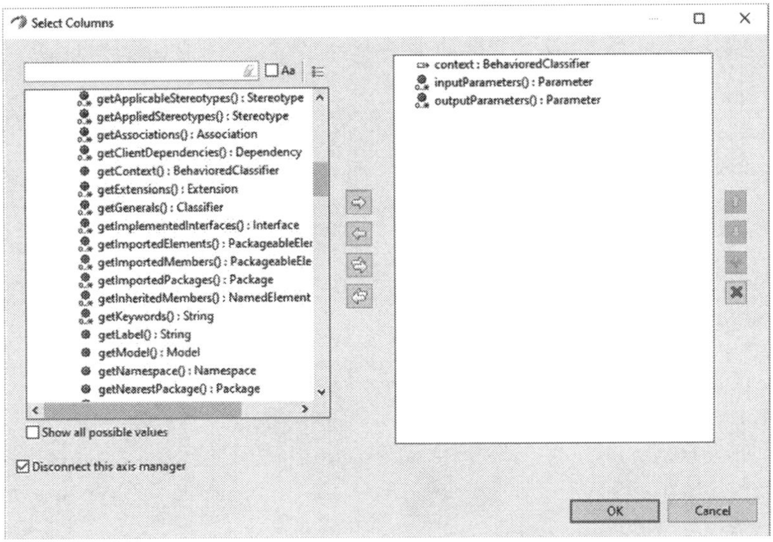

Figure 5.246. Select columns for "System Level Functions Table"

Rename the columns to "Performed by", "Inputs," and "Outputs".

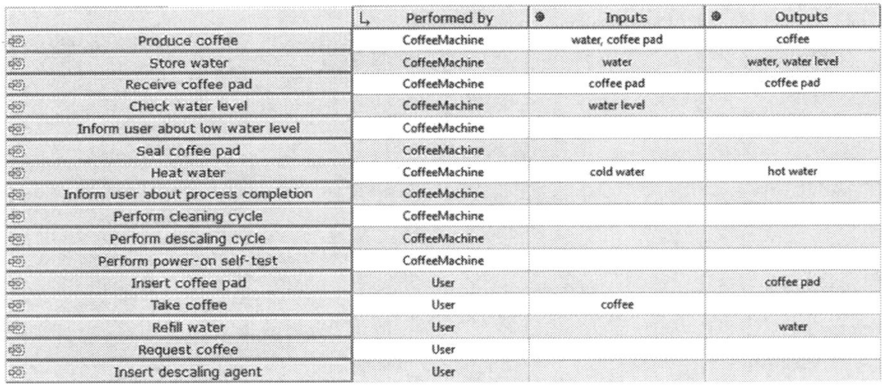

	Performed by	Inputs	Outputs
Produce coffee	CoffeeMachine	water, coffee pad	coffee
Store water	CoffeeMachine	water	water, water level
Receive coffee pad	CoffeeMachine	coffee pad	coffee pad
Check water level	CoffeeMachine	water level	
Inform user about low water level	CoffeeMachine		
Seal coffee pad	CoffeeMachine		
Heat water	CoffeeMachine	cold water	hot water
Inform user about process completion	CoffeeMachine		
Perform cleaning cycle	CoffeeMachine		
Perform descaling cycle	CoffeeMachine		
Perform power-on self-test	CoffeeMachine		
Insert coffee pad	User		coffee pad
Take coffee	User	coffee	
Refill water	User		water
Request coffee	User		
Insert descaling agent	User		

Figure 5.247. "System Level Functions Table"

Note that the table is incomplete as of yet. Especially, some inputs and outputs of the identified functions are missing. One goal of the functional analysis is to refine the current model and add missing information.

Now, we want to explore and refine the "Heat water" function. Add a new activity diagram to the "Heat water" activity by right-clicking on the activity and selecting *New Diagram->SysML 1.6 Activity Diagram*. Name the new diagram "Heat water AD".

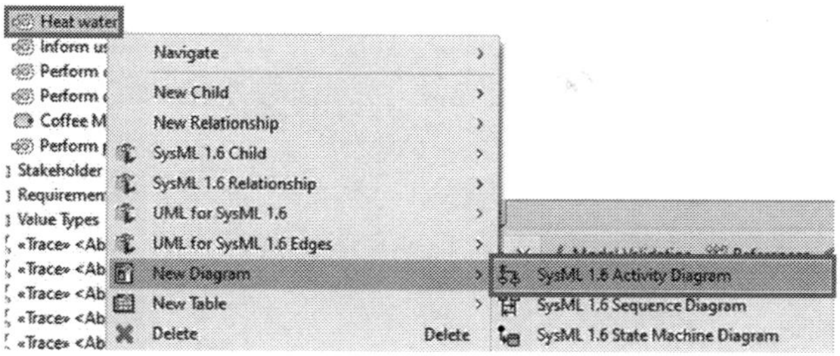

Figure 5.248. Create a new activity diagram

Select the "Call Behavior Action" element from the Diagram Palette and click on the AD. This automatically opens a dialog, which enables the creation of a new activity that is called by the call behavior action. Set the name of the new activity to "Distribute power" and its owner to the "Heat water" activity.

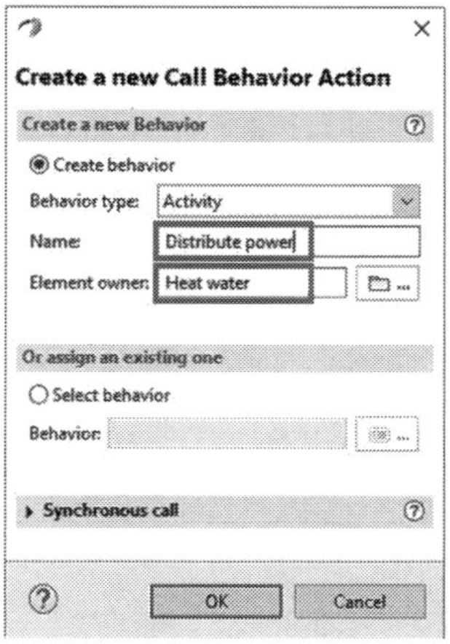

Figure 5.249. "Create a new call behavior action" dialog

Note how the activity "Distribute power" is added to the "Heat water" activity in the Model Explorer. The new call behavior action is added to the AD and the Model Explorer. Name it "Distribute power".

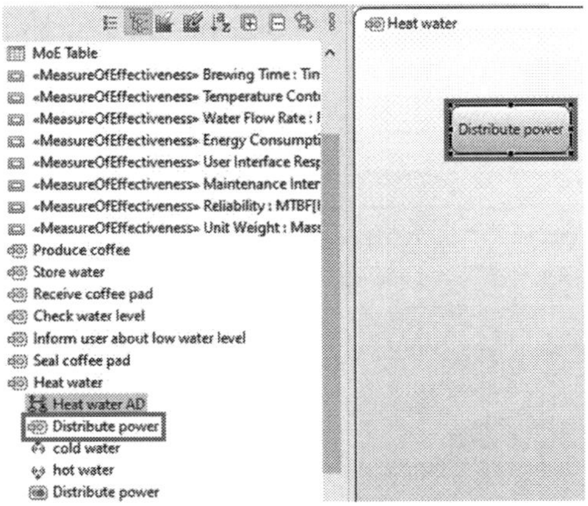

Figure 5.250. Call behavior action on the AD

Select the "Output Pin" element from the Diagram Palette and add a new output pin called "power out" to the "Distribute power" function.

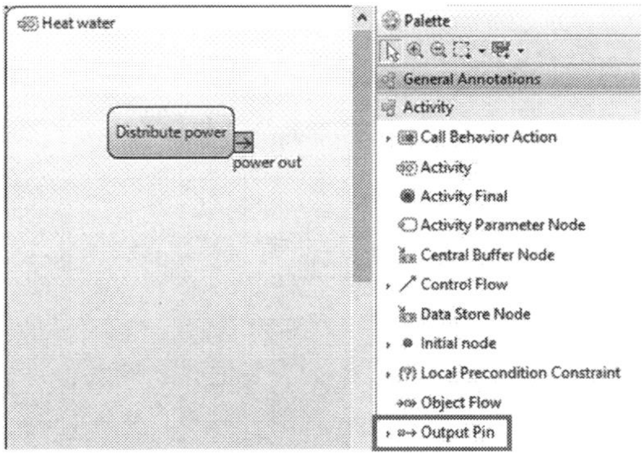

Figure 5.251. Create a new output pin

Select the "Input Pin" element from the Diagram Palette and add a new input pin called "power in" to the "Distribute power" function.

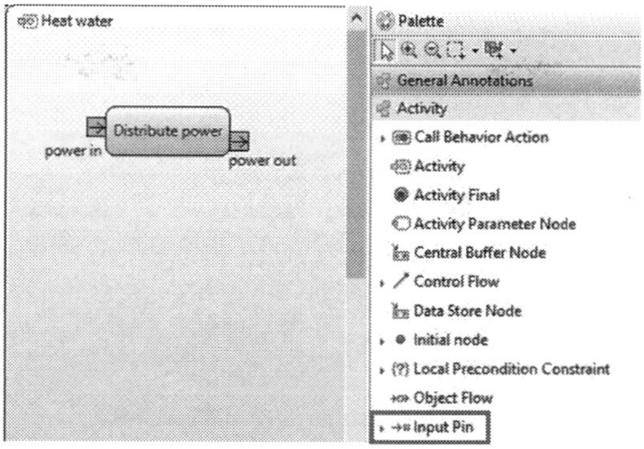

Figure 5.252. Create a new input pin

Repeat the process for the other functions and add input and output pins to all the activities.

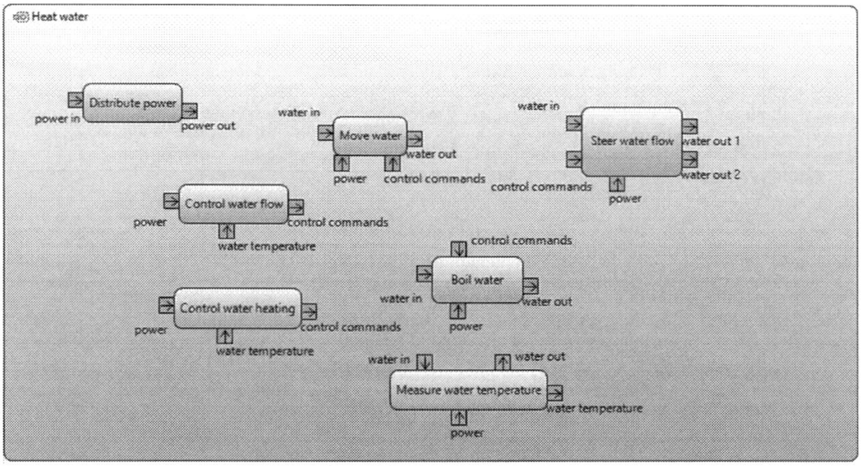

Figure 5.253. Activity diagram with functions, inputs, and outputs

Next, select the “cold water” input parameter of the “Heat water” activity in the Model Explorer and drag and drop it onto the border of the “Heat water” activity in the AD. Papyrus will create a new activity parameter node that is linked to the parameter with a default name.

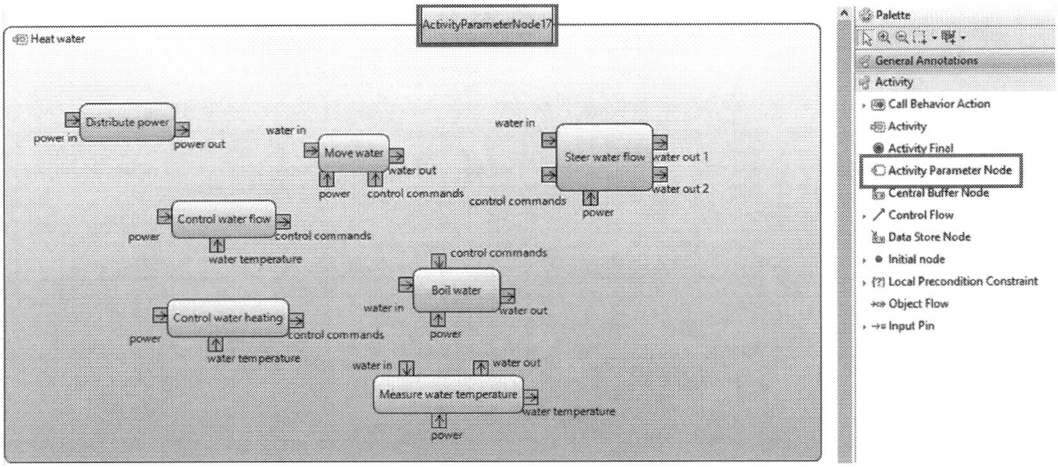

Figure 5.254. Create a new activity parameter node from the input parameter

Rename the activity parameter node “cold water”. Then, repeat the same process for the “hot water” output parameter. Add a new input parameter called “power” to the “Heat water” activity by clicking on the button with the plus sign next to the Owned parameter property in the Property View of the “Heat water” activity.

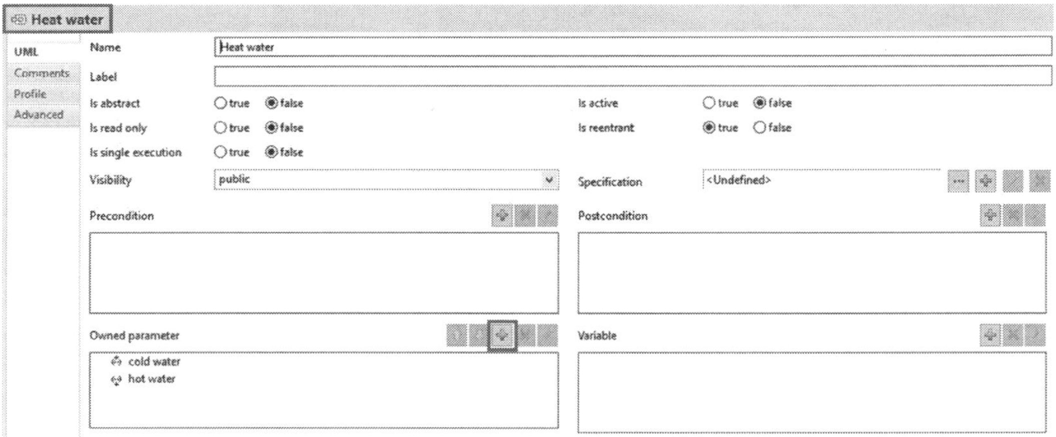

Figure 5.255. Create a new activity parameter

Add the newly created “power” input parameter as an activity parameter node to the “Heat Water AD” as well.

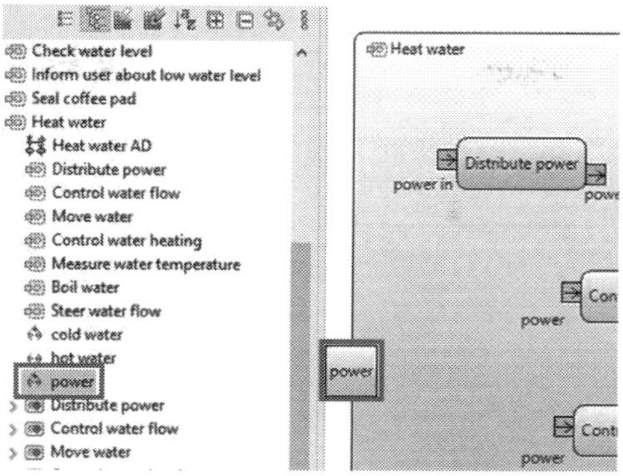

Figure 5.256. "Heat water AD" with activity parameter nodes

To finalize the "Heat Water AD", use the Object Flow element from the Diagram Palette to connect the inputs and outputs. Then, tidy up the diagram and organize the layout to make it more readable.

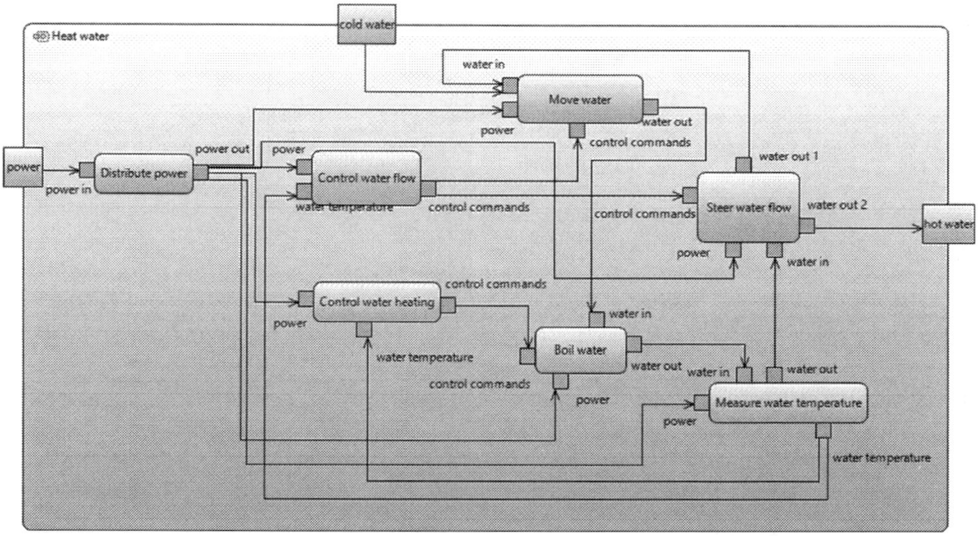

Figure 5.257. Finished "Heat water AD" activity diagram

Note that the input parameter "power" that was added to the "Heat water" function during this analysis now also appears in the "System Level Functions Table".

	Performed by	Inputs	Outputs
Produce coffee	CoffeeMachine	water, coffee pad	coffee
Store water	CoffeeMachine	water	water, water level
Receive coffee pad	CoffeeMachine	coffee pad	coffee pad
Check water level	CoffeeMachine	water level	
Inform user about low water level	CoffeeMachine		
Seal coffee pad	CoffeeMachine		
Heat water	CoffeeMachine	cold water, power	hot water
Inform user about process completion	CoffeeMachine		
Perform cleaning cycle	CoffeeMachine		
Perform descaling cycle	CoffeeMachine		
Perform power-on self-test	CoffeeMachine		
Insert coffee pad	User		coffee pad
Take coffee	User	coffee	
Refill water	User		water
Request coffee	User		
Insert descaling agent	User		

Figure 5.258. Updated "System Level Functions Table"

The last step is to create a new package called "Component Functions" in the model and move all the newly created activities from under the "Heat water" activity into the new package.

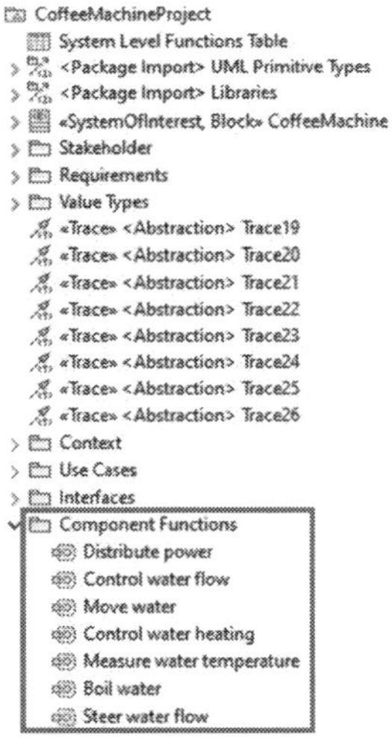

Figure 5.259. New package containing the component level functions

5.10 Define Initial System Concept

Defining the initial system concept is the first step on the way to the definition of the system architecture and follows the recurring principle of decomposition in systems engineering.

The system architecture serves as the blueprint that defines the system's fundamental structure and organization. It encompasses the arrangement, composition, and interactions of the system's components, interfaces, and connectors, providing a conceptual framework for its overall design and development. A sound system architecture holds immense significance in achieving the overall goals of systems engineering. A well-defined internal architecture is not merely a technical artifact; it is the bedrock upon which the success of the entire system engineering effort rests. Its importance transcends individual phases of development, permeating through the entire systems engineering lifecycle. Reasons for the significance of the system's architecture include:

- **Clarity and understanding**: A well-defined system architecture provides a clear and structured representation of the system, enabling stakeholders to comprehend its design, components, and interactions. This facilitates effective communication, collaboration, and decision-making throughout the development process.
- **Requirement analysis and optimization**: System architecture guides the analysis of functional and non-functional requirements, ensuring that the system's design aligns with stakeholder expectations and addresses critical performance, reliability, and security considerations.
- **Design feasibility and trade-offs**: Architecture helps identify potential feasibility issues and steers trade-offs early on, enabling informed decisions regarding technology choices, design approaches, and resource allocation. This minimizes the risk of costly rework and ensures that the system can be implemented effectively.
- **Development and implementation**: Architecture serves as a roadmap for the development and implementation of the system, guiding the selection of technologies, the allocation of tasks, and the coordination of development efforts. It provides a framework for ensuring consistency and adherence to design principles.
- **Testing and validation**: A well-defined architecture provides a basis for comprehensive testing and validation, ensuring that the system meets its functional and non-functional requirements. A good architecture emphasizes modularity, where the system is broken down into smaller, independent components. This modularity allows for isolated testing of individual components before integration. This simplifies the testing process and facilitates the identification and rectification of defects early in the development lifecycle, minimizing the risk of costly post-deployment problems.
- **Maintenance and evolution**: A sound system architecture facilitates the maintenance and evolution of the system throughout its lifecycle. It enables the system to adapt to changing requirements, technological advancements, and evolving user needs, ensuring its long-term viability and relevance.
- **Reusability and scalability**: Architecture principles promote the use of reusable components and modules, enhancing the system's adaptability and scalability. It enables

the system to be extended with new features and capabilities without compromising its overall performance and maintainability.

- **Subcontracting**: By clearly outlining the functionalities and technical requirements of each component, a well-defined architecture facilitates the identification of components that could be more efficiently or cost-effectively developed by a subcontractor with specialized skills or manufacturing capabilities. Clearly defined interfaces between components simplify the process of managing subcontracted components and ensure seamless integration with internally developed parts.

During the definition of the initial system concept, the hierarchy of the system is defined. All the system components are identified and it is defined how many of each component are used and what the role of each component is in the system architecture. Components are the fundamental building blocks of a system, representing individual modules or subsystems responsible for specific tasks or functions. They encapsulate the functionality and data associated with those tasks, enabling modularity, reusability, and maintainability. Each component operates independently and contributes to the overall system's functionality and behavior. Modularity is a cornerstone of systems engineering that promotes the creation of independent and self-contained components. Each component has a well-defined purpose and encapsulates its behavior and interactions with other components. Modularization enhances the maintainability, reusability, and testability of the system by enabling developers to work on individual components without affecting others.

5.10.1 Coffee Machine Components

Decomposing the coffee machine into system components involves identifying and organizing the various parts or modules that work together to fulfill the identified use cases and system functions. Here is an initial high-level decomposition of the coffee machine into system components based on the identified use cases and functions:

Name	Description
Power Supply	Distributes power to the other components.
Display	Displays information to the user.
Controller	Manages and directs the behavior of other components of the coffee machine.
Temperature Sensor	Measures the temperature of the water.
Water Heater	Heats water using electric energy.
Water Tank	Contains water.
Water Pump	Pumps the water through the hydraulic tubing.
Push Button	Enables the user to input requests.
Water Level Sensor	Measures the level of water in the water tank.
Valve	A three-way valve that can route the flow of water.
Power Switch	Enables the user to turn the coffee machine on and off.

5.10.2 Modeling the Initial System Concept of the Coffee Machine

Everything that needs to be done to model the initial system concept was already done and explained in previous steps such as the system context analysis. Create a new package called "Components". In the new package create a BDD called "Coffee Machine Decomposition BDD". Add the "CoffeeMachine" Block to the diagram and then create new blocks for all the components of the system. Use the directed composite association to make the newly created components parts of the coffee machine.

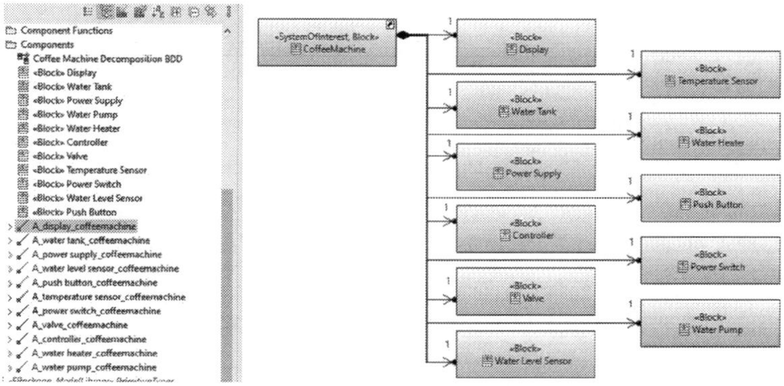

Figure 5.260. Coffee machine decomposition and components

Make sure that the multiplicity of all the parts is set to 1.

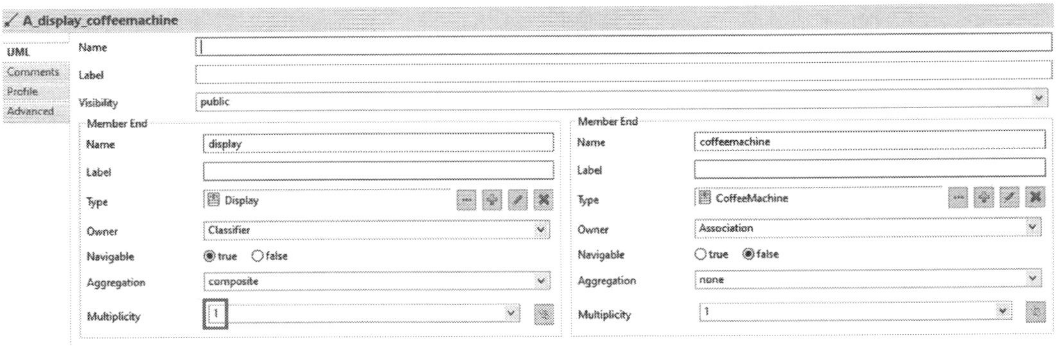

Figure 5.261. **Set the multiplicity to 1**

5.11 Allocate Functions to System Components

In systems engineering, function allocation is the process of assigning component functions to specific system components. It is a critical step in the systems engineering process, as it helps to ensure that the system's functionality is effectively implemented and that each component contributes to the overall system's performance. The benefits of function allocation in systems engineering include:

- **Clarity and understanding**: Function allocation provides a clear and structured representation of how the system's functions are distributed among its components. This clarity facilitates communication and understanding among stakeholders, ensuring that everyone has a shared understanding of the system's architecture and functionality. By clearly defining the roles and responsibilities of each component in fulfilling the overall system functions it contributes to the development of the system architecture. Allocating functions to components supports the system architects in defining the structure of the system and the relationships between its various elements. Additionally, it enables traceability between the elements of the different levels of abstraction.
- **System integration**: Function allocation plays a pivotal role in system integration, ensuring that the system's components work together seamlessly to achieve the system's overall objectives. By assigning functions to specific components, function allocation defines the interfaces and interactions between these components, laying the foundation for successful integration. This structured approach helps to identify

and address potential integration issues early in the development process, preventing costly delays and rework. This is particularly important in large and complex systems where multiple components need to interact cohesively.

- **Resource allocation**: Function allocation facilitates resource allocation during the development process. By knowing which functions are associated with specific components, project managers can allocate resources effectively, ensuring that each component is developed and tested as per its designated functionality.
- **Support system V&V**: Function allocation provides a clear framework for V&V activities. By mapping functions to specific components, engineers can develop test cases and scenarios that target each component's functionality. This structured approach ensures that the system's functionality is comprehensively tested and validated, reducing the risk of undetected defects and ensuring that the system meets its requirements.
- **Communication**: The allocation of functions to components serves as a communication tool among stakeholders involved during the development phase of a system, especially in large corporations where multiple teams work together to collaborate in the system development effort. It helps convey how the system is structured and how each part contributes to the overall functionality. This is essential for collaboration among different teams working on the project.

In summary, allocation functions to components is a critical aspect of systems engineering that contributes to the overall success of a project by providing clarity, ensuring functionality alignment, supporting integration efforts, and facilitating effective communication among team members and stakeholders.

Two key principles that guide the functional allocation process are minimizing interaction and coupling and maximizing cohesion.

Interaction refers to the communication and item exchange between system components. Coupling is a measure of the strength of these interactions and the degree to which components are interdependent. Low coupling indicates that components are loosely coupled and have minimal dependencies on each other. Minimizing interaction and coupling offers several benefits:

- **Increased independence**: Components with low coupling are more independent and can be developed and modified without affecting other components.
- **Improved modularity**: Low coupling promotes modularity, enabling the system to be decomposed into smaller, self-contained modules.
- **Enhanced fault isolation**: Low coupling simplifies fault isolation, as issues can be more easily traced to specific components.

- **Increased reusability**: Components with low coupling are more reusable, as they can be integrated into different systems without introducing unnecessary dependencies.

Generally, cohesion refers to the degree to which the elements of a component are related and contribute to the fulfillment of the component's purpose. High cohesion between the functions that are allocated to one component indicates that the component's functions are tightly related and share a common purpose. Maximizing functional cohesion offers several benefits:

- **Improved understanding**: Components with high functional cohesion are easier to understand and maintain as their functionality is more focused and self-contained.
- **Reduced complexity**: High cohesion reduces the overall complexity of the system by organizing components into well-defined units.
- **Enhanced testability**: Components with high functional cohesion are easier to test as their functionality is more distinct and isolated.

By adhering to the principles of minimizing interaction and coupling and maximizing cohesion, systems engineers can design systems that are modular, maintainable, and easy to understand and modify. These principles contribute to the overall robustness and reliability of the system. By balancing these principles, systems engineers can achieve a well-structured and adaptable system architecture.

5.11.1 Modeling the Allocation of Functions to Components of the Coffee Machine

To support the modeling of the allocation of component functions to components, we will use an allocation matrix. Select the "Component Functions" package in the Model-Explorer, right-click, and select *New Table->SysML 1.6 Allocation Matrix.* Name the newly created element "Function to Component Allocation Matrix".

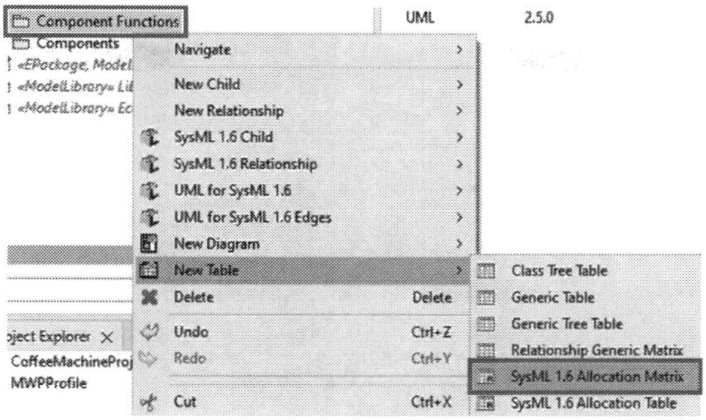

Figure 5.262. Create a new allocation matrix

Go to the Matrix tab of the Properties View. To populate the matrix automatically from the model, first, we need to define a filter for the elements that are displayed as the rows of the matrix. We want to populate the rows of the matrix with all the system components. To do this, click on the button with the plus next to the Filter property of the Rows section and select *IsStereotypedWithExpression.*

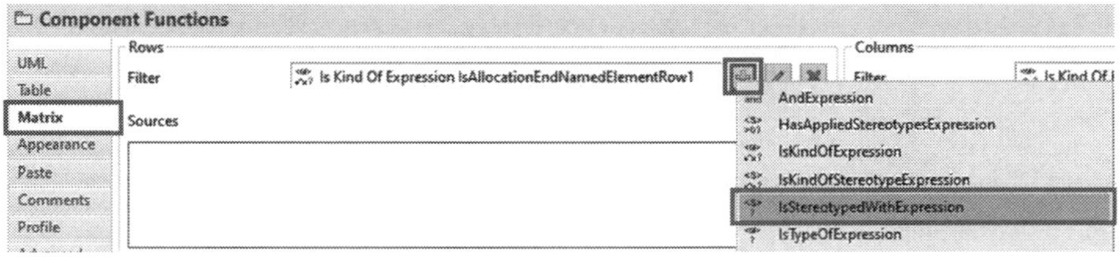

Figure 5.263. Create a new filter

In the dialog, click on the button with the three dots next to the Stereotype qualified name property.

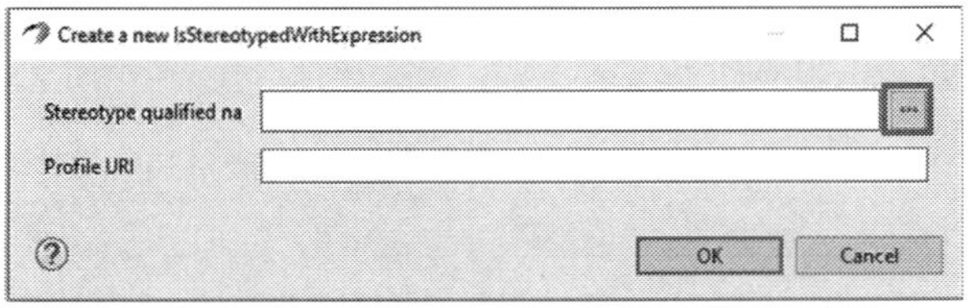

Figure 5.264. Create a new IsStereotypedWithExpression dialog

Select the "Block" element in the SysML 1.6 profile, since all the components that we want to filter for are SysML blocks.

Figure 5.265. Select "Block" in the Stereotype Explorer

Once, the block stereotype is set correctly, confirm the IsStereotypedWithExpression dialog.

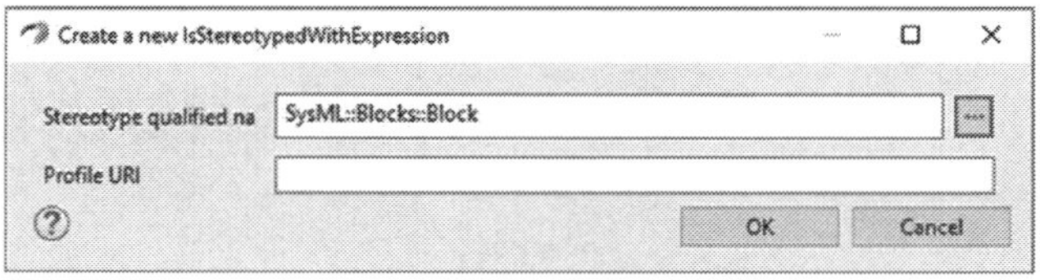

Figure 5.266. Selected stereotype

Next, we need to specify where the elements are that should become the rows of the matrix. Click on the button with the plus next to the Sources property. Papyrus automatically takes all the elements in the source packages and, once they pass the filter, adds them to the matrix.

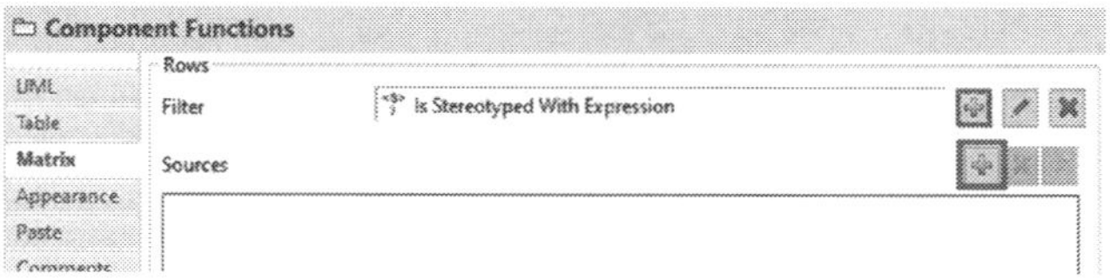

Figure 5.267. Add a source for the rows

Select the “Components” package and move it to the right side using the button with the single arrow pointing to the right.

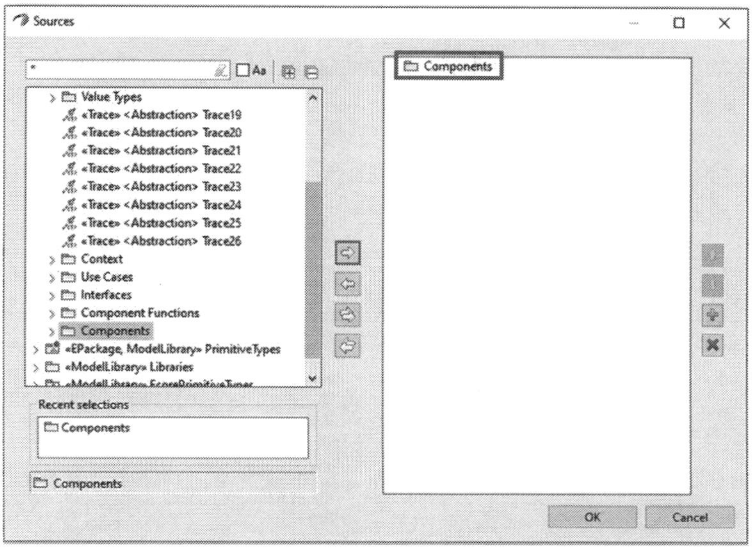

Figure 5.268. Select the “Components” package

The matrix will update and all the system components are shown as rows.

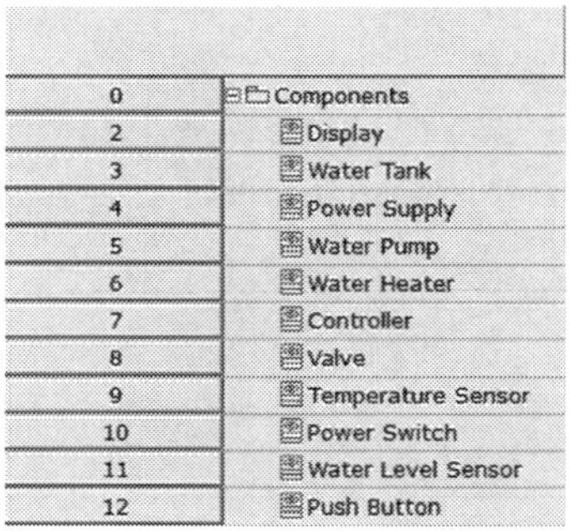

Figure 5.269. Matrix with components

Next, we want to populate the columns of the matrix with the activity elements that represent our component-level functions. To do this, click on the button with the plus next to the Filter property of the Columns section and select *IsKindOfExpression*.

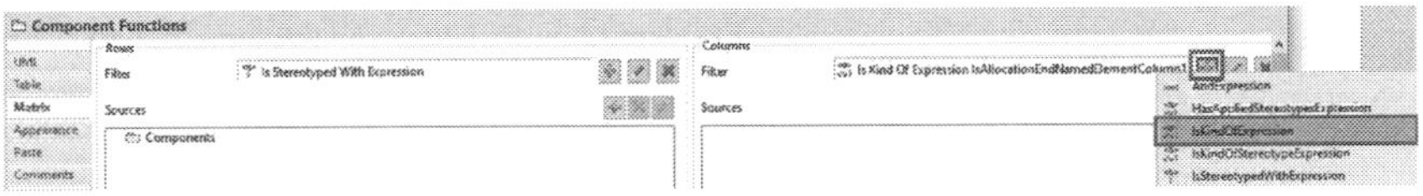

Figure 5.270. Create a new filter

Select "Activity" as the element kind that we want to filter for.

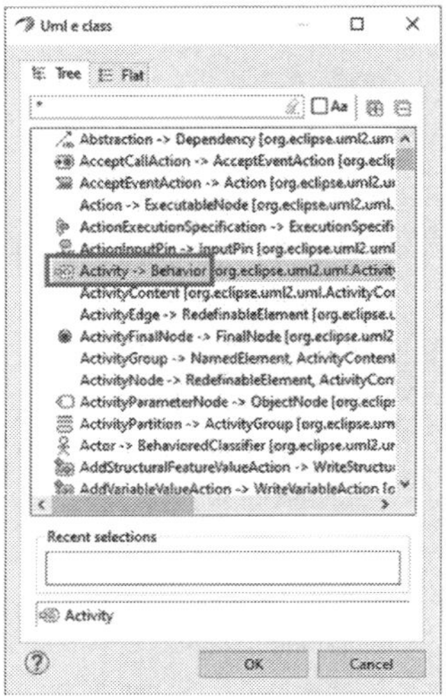

Figure 5.271. Select "Activity"

Set the "Component Function" package as the source for the columns. Then, make sure to set the Direction property to "From Column to Row" since we want to allocate functions to components.

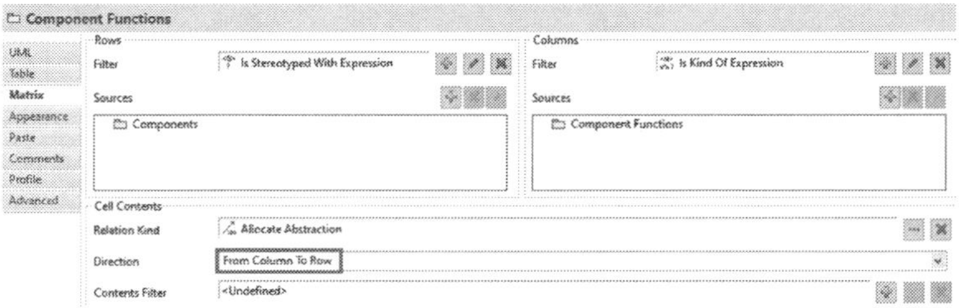

Figure 5.272. Set allocation direction

The matrix should now contain all the components as rows and all the component-level functions as columns. Since we have not created any allocations yet, all the checkboxes should be empty.

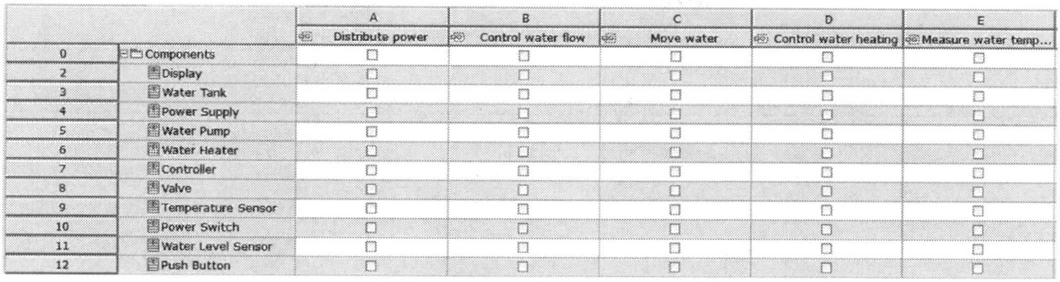

Figure 5.273. Empty matrix with columns and rows

Go to the Appearance tab and remove the indices from being displayed in the matrix. Adjust the width of the columns as necessary.

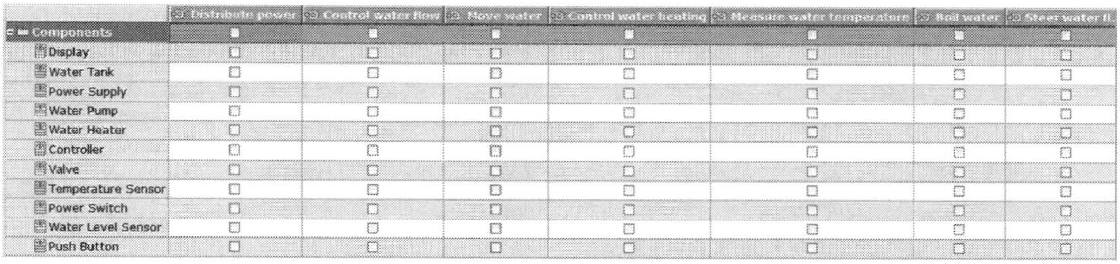

Figure 5.274. Matrix without indices

Similar to tables, the matrices are interactive components and not static views of the model. To create an allocation, first, select the column of a function that should be allocated and then, second, tick the checkbox in the row of the component to which it should be allocated. This automatically creates an allocation relationship that can be seen in the Model Explorer as well. Now, allocate all the functions to the appropriate components.

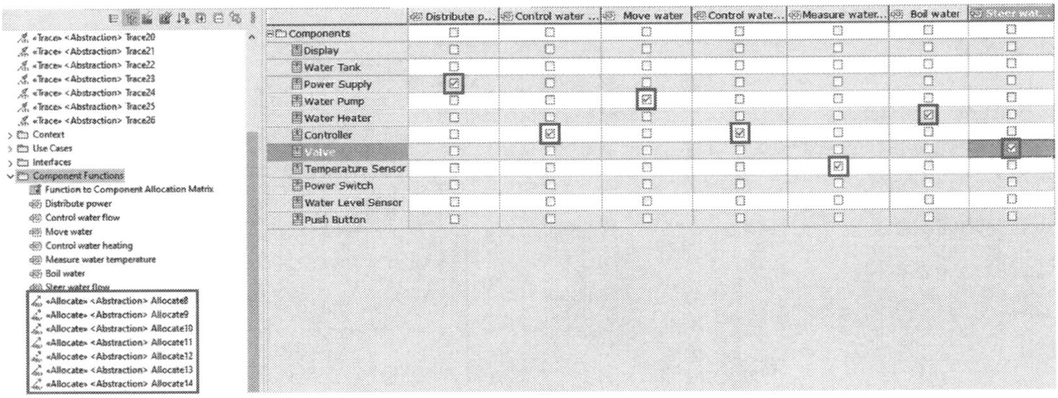

Figure 5.275. Allocate functions to components using an allocation matrix

The matrix can be exported to Excel in the same way as a table.

	A	B	C	D	E	F	G	H
1		Distribute power	Control water flow	Move water	Control water heating	Measure water temperature	Boil water	Steer water flow
2	Components	false	false	false	false	false	false	false
3	Display	false	false	false	false	false	false	false
4	Water Tank	false	false	false	false	false	false	false
5	Power Supply	true	false	false	false	false	false	false
6	Water Pump	false	false	true	false	false	false	false
7	Water Heater	false	false	false	false	false	true	false
8	Controller	false	true	false	true	false	false	false
9	Valve	false	false	false	false	false	false	true
10	Temperature Sensor	false	false	false	false	true	false	false
11	Power Switch	false	false	false	false	false	false	false
12	Water Level Sensor	false	false	false	false	false	false	false
13	Push Button	false	false	false	false	false	false	false

Figure 5.276. Allocation matrix exported to Excel

5.12 Define System Internal Architecture

The allocation of component functions to system components clarified the role that each component plays in achieving the overall system goals. The next step now is to define the internal architecture of the system. The internal architecture of a system serves as the blueprint that defines its structure, i.e., how its components are connected, to each other and the external interfaces of the system.

Interfaces define how components communicate and interact with each other. They specify the data formats, protocols, and messaging patterns that components use to exchange information, energy or matter and collaborate on tasks. Well-defined interfaces promote loose coupling, making systems more flexible, adaptable, and easier to maintain.

There are two different usages of the term interface: one way is to use the term interface to denominate the connection between two systems, including the boundary elements at both ends of the connection (Kossiakoff, 2020). The other way is to use the term interface to denominate only the boundary that is presented by a system for interaction with other systems and then use the term connector to denominate the connection between these two boundary points (Fosse and Delp, 2013). In this book, the second way is used: a system presents an interface, which can be used to form a connection using a connector to another system with a compatible interface. An interface can be specified by an interface specification, which describes the nature of the interface presented by a system in terms of properties and functionality.

Connectors are the pathways or links that connect components, enabling them to exchange information and collaborate on tasks. They represent the physical or logical pathways between components, including communication channels, shared memory regions, or network connections. Well-designed connectors ensure that components can communicate efficiently and reliably while avoiding bottlenecks or performance limitations.

Since we are defining the technical architecture of the system, we are mainly concerned with physical connectors. They allow for the transfer of power, data, and control signals, ensuring the seamless operation of the system. Common types of physical connectors include:

- **Electrical connectors**: These connectors transmit electrical power and signals between components.
- **Data connections**: These connectors use physical wires to transmit digital signals. Common types include ethernet cables, Inter-Integrated Circuit (I^2C) cables, or USB cables.
- **Mechanical linkages**: These connectors allow for the transfer of mechanical motion between components.
- **Hydraulic tubing**: These connectors facilitate the transfer of liquid, such as water or coffee, between components and to the outside.

5.12.1 Modeling the Coffee Machine Internal Architecture

To illustrate how SysML enables a separation of concerns, we will not model all of the aspects of the internal architecture in one diagram but, instead, will create multiple diagrams, each of which concentrates on one aspect of the system. We want to create separate IBDs for the coffee machine for the control aspect, the power aspect and the fluid aspect. First, add a new

IBD to the "CoffeeMachine" Block and name it "Coffee Machine Control IBD". Then, add the relevant part properties of the coffee machine to this IBD.

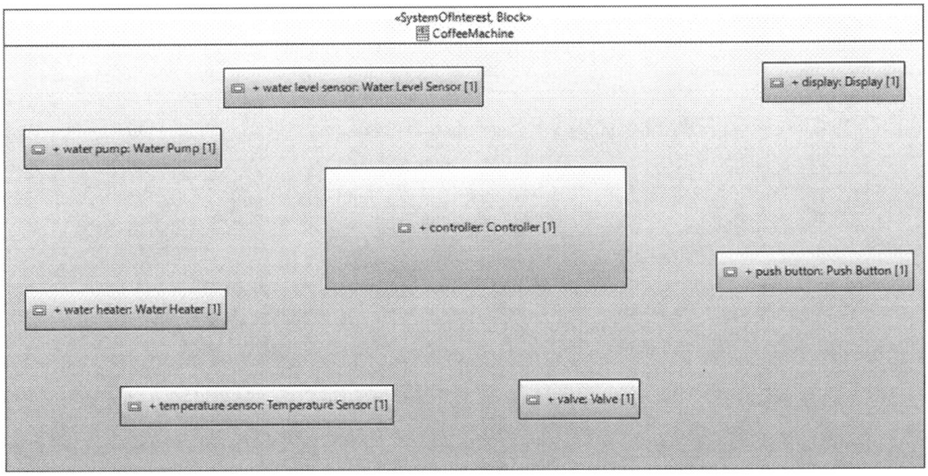

Figure 5.277. **IBD with parts**

Select the "Proxy Port" element from the Diagram Palette and left-click on the border of the controller part to add a new proxy port to the "Controller" Block and name it "display".

A port is an interaction point - an interface - at the boundary of a block through which other blocks can interact with it. SysML does not constrain what a port can represent. A port can represent any type of interaction point, similarly as a block can represent any type of object.

Flow ports and (standard) ports have been deprecated since SysML 1.4 when they have been replaced by full ports and proxy ports.

A full port is a special type of port that represents a separate element of the system that can have its own behavior, just like a part. Everything that can be modeled using full ports can also be modeled using blocks and proxy ports. An often repeated recommendation, therefore, is to not use full ports at all.

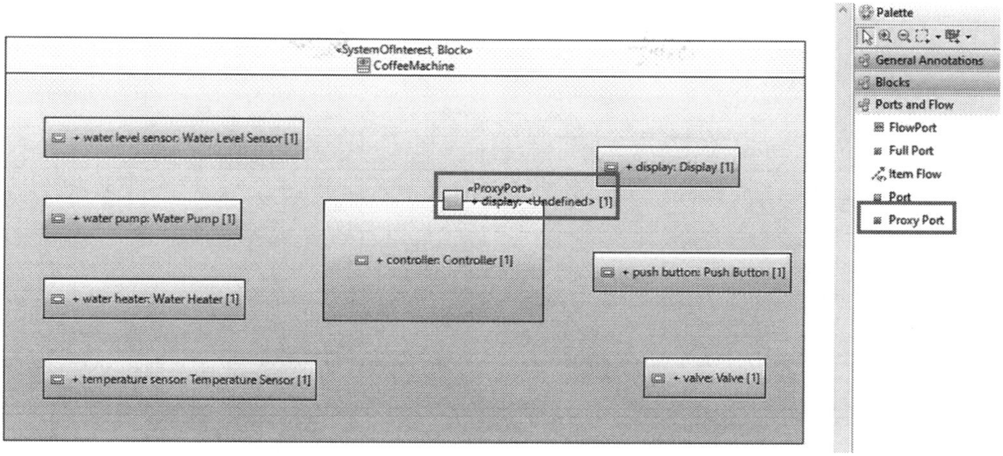

Figure 5.278. Create a new proxy port

Repeat the same process to add a proxy port to the "Display" block named "Controller". Then select the "Connector" element in the Diagram Palette, click first on the port "controller" of the "display" part and then on the "display" port of the "controller" part to connect the two ports.

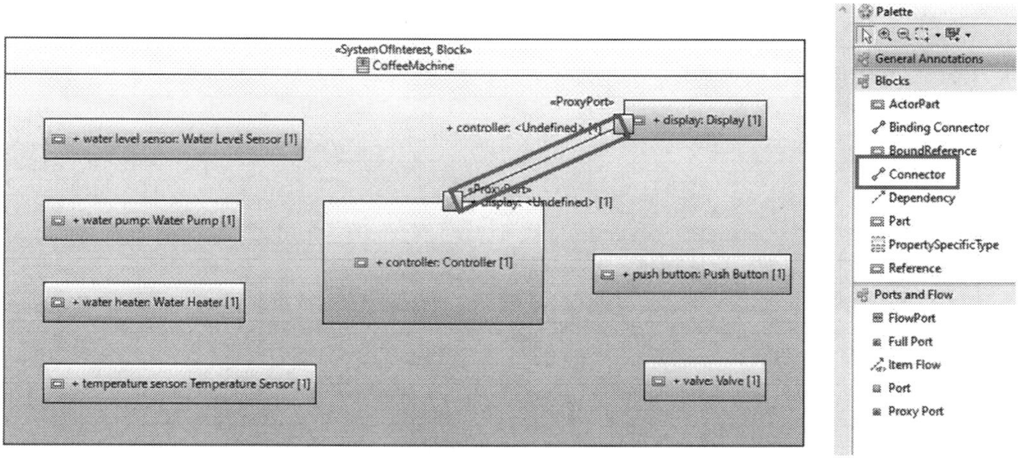

Figure 5.279. Create a new connector

It does not matter, which port is selected first when a new connector is created. connectors are undirected.

Repeat the same process for all the other parts so that all parts have a connection to the "controller" part.

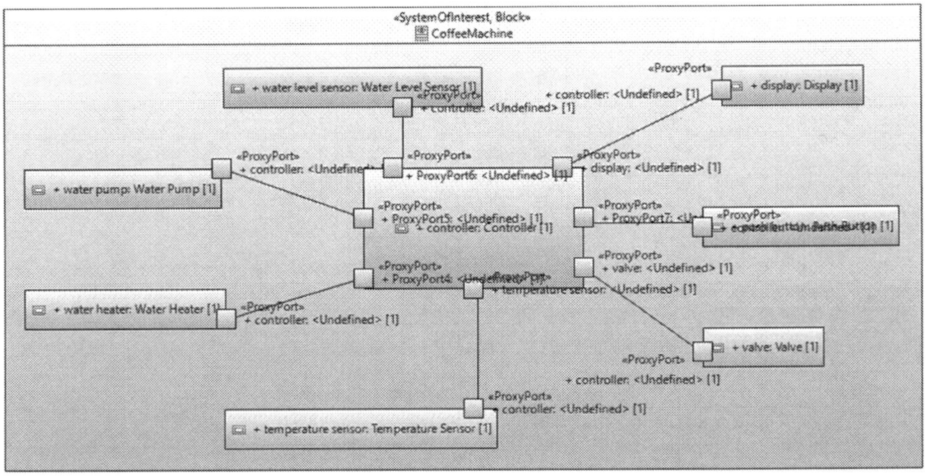

Figure 5.280. "Coffee Machine Control IBD" with all parts and connectors

As already said, ports are potential interaction points of a block. To specify the interface, i.e., to define what can pass the port, a proxy port needs to be typed by an interface block, which represents an interface specification. Create a new interface block by right-clicking on the "Interfaces" package and selecting *SysML 1.6 Child->InterfaceBlock.* Name the interface block "Controller Display".

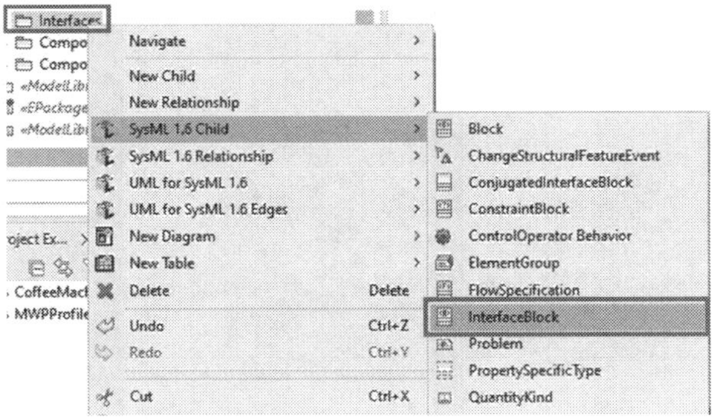

Figure 5.281. Create a new interface block

An interface block is a special kind of block for typing proxy ports. It has no behaviors or internal parts.

Repeat the same process for all the other components that are connected to the "controller" part.

Figure 5.282. New interface blocks

To type a proxy port with an interface block, drag and drop the interface block onto the proxy port. Note that the type of the port is also displayed in the IBD with the default display settings.

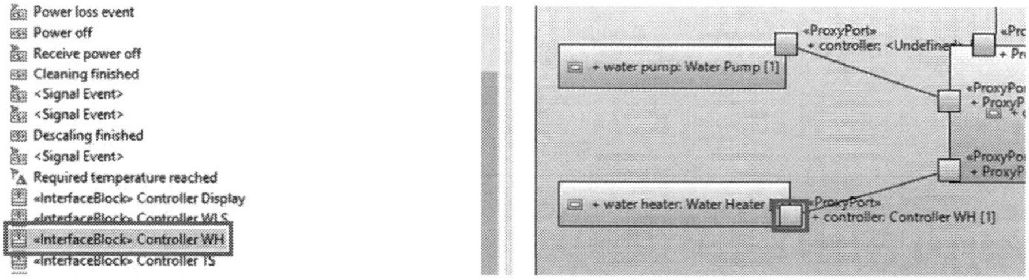

Figure 5.283. Type a proxy port with an interface block

Repeat the same process to type all the other ports in the IBD.

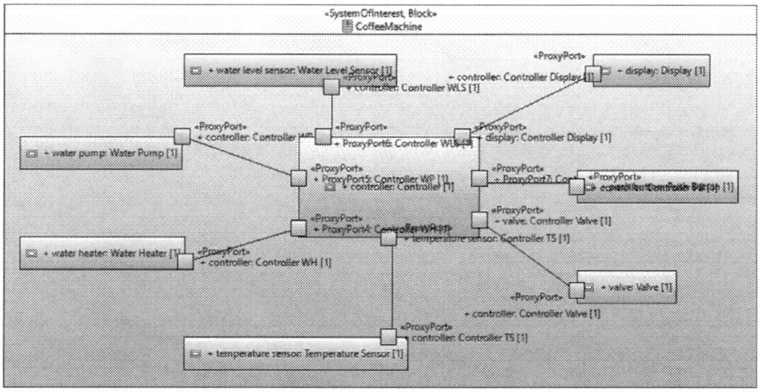

Figure 5.284. "Coffee Machine Control IBD" with typed ports

To improve the readability of the IBD prevent the "ProxyPort" stereotype from being displayed. Select all the ports on the diagram, then open the Label Manager (*Filters-*

>*Show/Hide Labels*) and de-select the "Stereotype" checkbox for all ports.

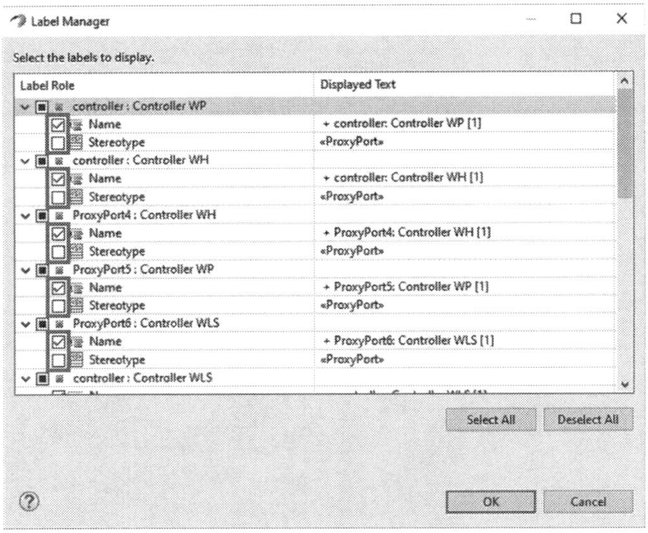

Figure 5.285. Prevent the "ProxyPort" stereotype from being displayed

Change the line style of the connectors to rectilinear and arrange the parts to make the IBD readable.

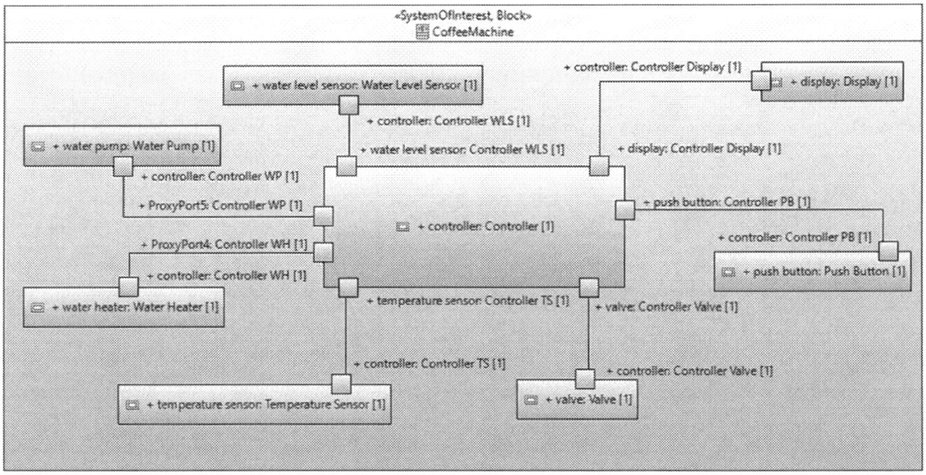

Figure 5.286. "Coffee Machine Control IBD" with improved layout

When two parts are connected through ports, it may be that they exchange the same item, just in opposite directions, i.e., one part outputs an item, and the other inputs the item. This is modeled using port conjugation. This means that the same interface block is used to type the port on both blocks, but the second port is marked as the conjugate of the first. Select the "controller" port of the "Water Level Sensor" block and go to the UML tab of its Properties View. Then, set the property Is conjugated to "true".

The Is conjugated property has been replaced in SysML 1.6 by the model element conjugated interface block, which is a specialization of interface block that has the same features as its original interface block except that its directed features and flow properties are reversed. The SysML specification states that "it is expected that tools conforming to this specification automatically create features of [conjugated interface blocks]" (OMG, 2019) to avoid the error-prone work of manually created mirrored conjugated interface blocks for interface blocks and keeping the model elements consistent after changes. Although Papyrus supports the creation of conjugated interface blocks, unfortunately, it does not have automation for conjugated interface blocks. Therefore, the nominally deprecated Is the conjugated property is used here.

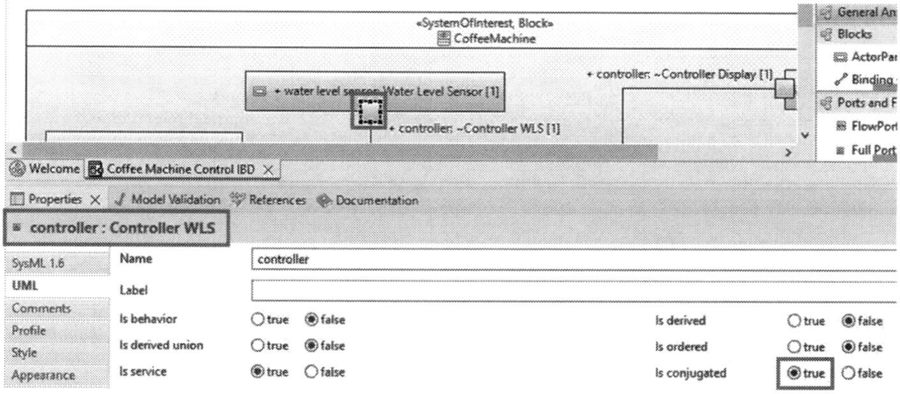

Figure 5.287. Set a port as conjugated

Repeat the same process for all the other ports on the parts that are connected to the "controller" part. Note that the conjugated ports are prefixed with a tilde symbol to show the fact that they are conjugated.

It does not matter, which side of the connector has a conjugated port. But if an interface block contains directed features, then the port on one side needs to be conjugated.

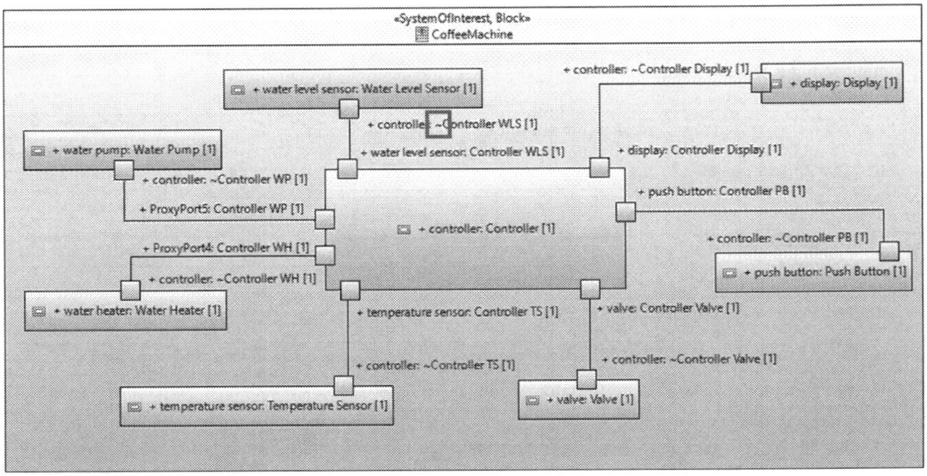

Figure 5.288. "Coffee Machine Control IBD" with conjugated ports

The next step is to define, what kind of items can pass a proxy port typed by an interface block. To this end, we add flow properties to the interface block. Right-click on the "Controller TS" interface block and select *SysML 1.6 Child->FlowProperty.*

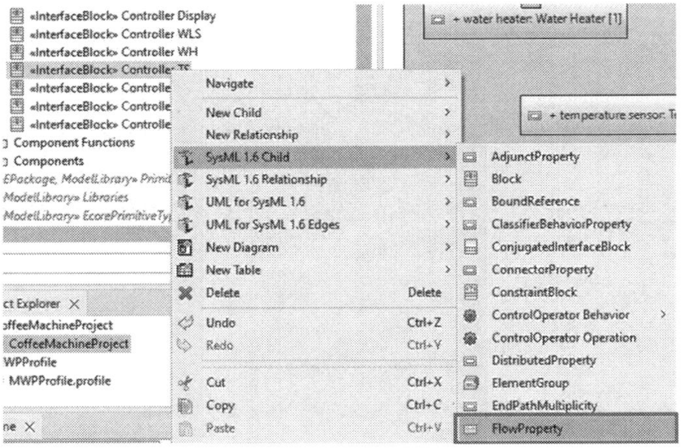

Figure 5.289. Create a new flow property

Go to the Properties View of the newly created element, name the flow property "water temperature", and set the Direction property to "In".

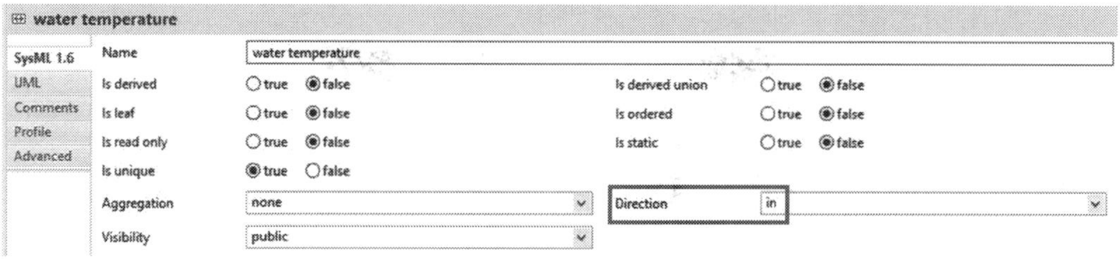

Figure 5.290. Properties of the “water temperature” flow property

To type the flow property, we can reuse some of the value types that have already been created. Go to the UML tab of the Properties View and click on the button with the three dots next to the Type property to open the “Select Type” dialog.

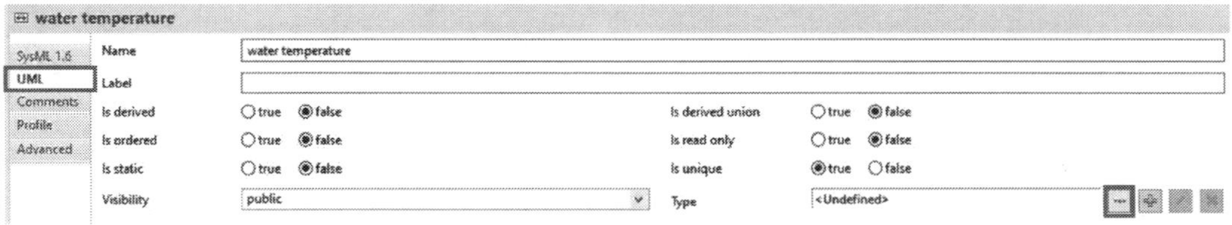

Figure 5.291. Open the “Select Type” dialog

Select the “Temperature[celsius]” value type.

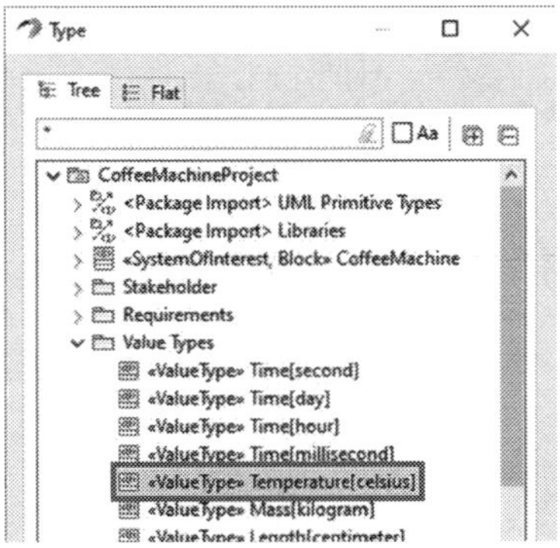

Figure 5.292. Select flow property type

Repeat the same process to create, name, and type all flow properties in all of the other interface blocks. Note that during this process it may be necessary to create new value types, such as the “Fill level[percent]” value type.

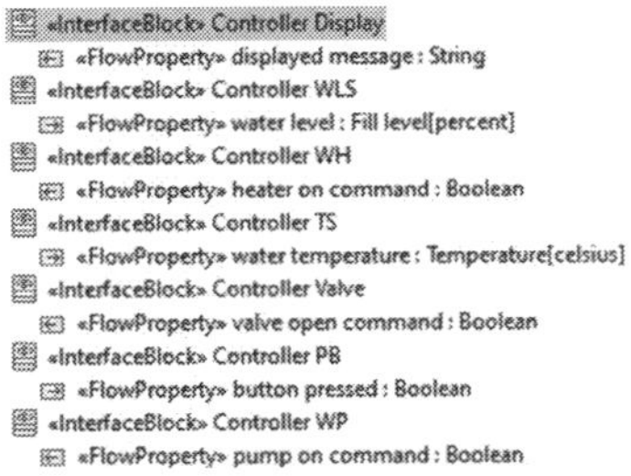

Figure 5.293. Interface Blocks with typed flow properties

Next, we want to model the coffee machine fluid interfaces. As a prerequisite, we need to create two additional components: the "Coffee Pad Cavity" and the "Coffee Nozzle".

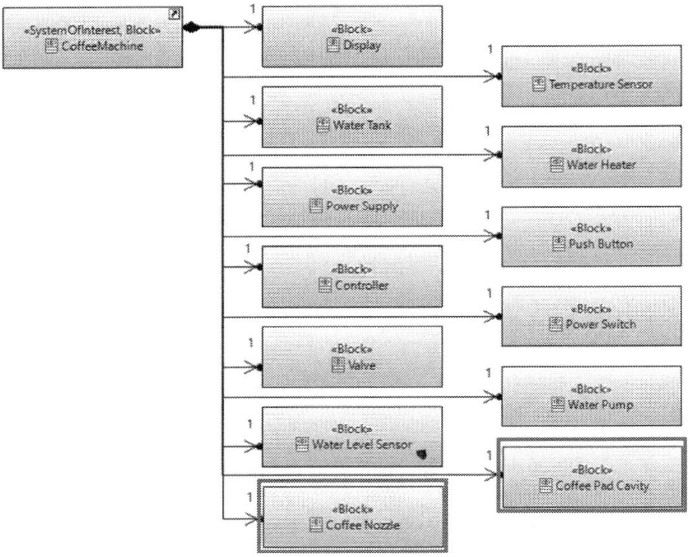

Figure 5.294. "Coffee Machine Decomposition BDD" with two new components

Create a new IBD in the "CoffeeMachine" Block and name it "Coffee Machine Fluid IBD". Repeat all the steps that have been done for modeling the control aspect of the coffee machine, to create new ports, interface blocks, and flow properties. Connect all the ports using connectors except for the port "coffee out" of the "coffee nozzle" part property and the "coffee" port of the "CoffeeMachine" block.

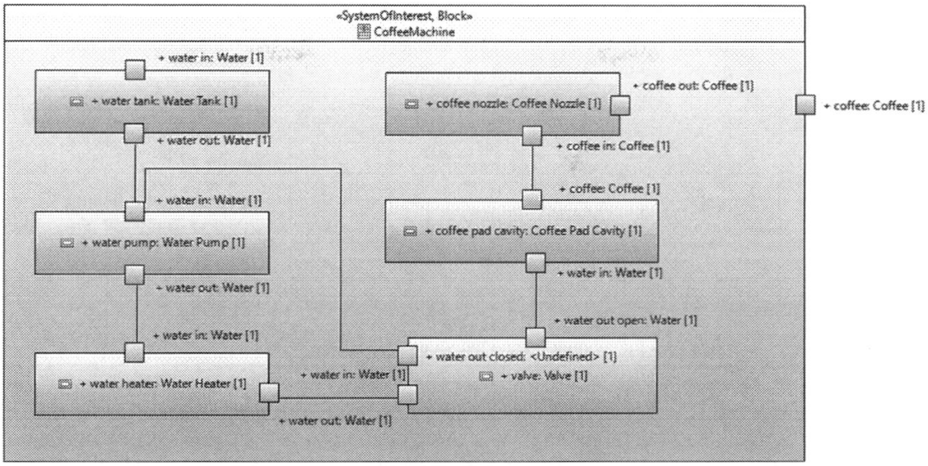

Figure 5.295. "Coffee Machine Fluid IBD"

Select the "Binding Connector" element from the Diagram Palette and create a connection between the "coffee out" port of the "coffee nozzle" part property and the "coffee" port of the "CoffeeMachine" block.

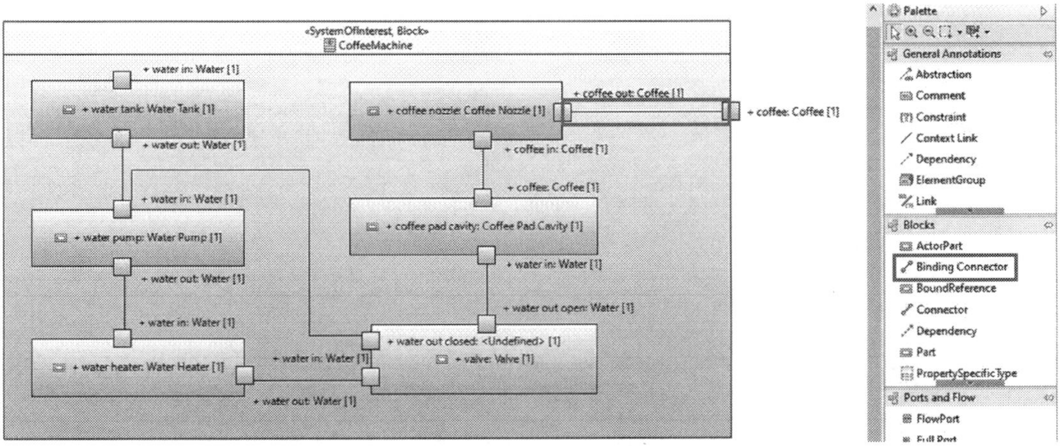

Figure 5.296. Create a new binding connector

A Binding Connector is a connector that specifies that the properties at both ends of the connector have equal values. It is used to connect ports of part properties inside a block to the ports on the boundary of the block.

Note that the direction of the flow properties for the two fluid interfaces was set to "inout" since fluids can typically flow in both directions. And since the interfaces are non-directed, no port conjugation is required on the fluid ports.

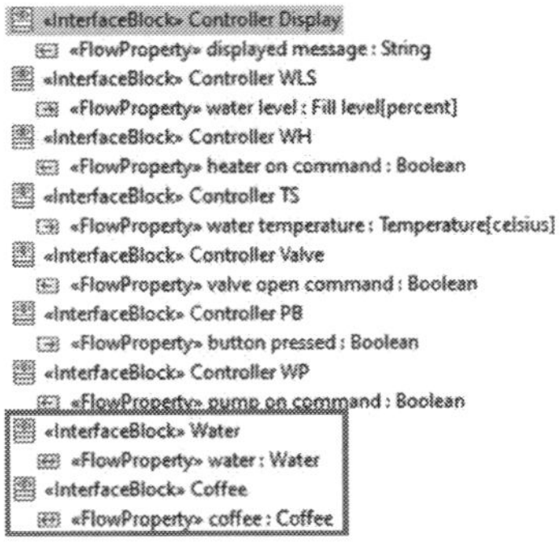

Figure 5.297. New interface blocks for the fluid interfaces

A third aspect of the coffee machine is the distribution of power inside the coffee machine. Create a new IBD called "Coffee Machine Power IBD" and repeat all of the previous steps to model the power distribution between the coffee machine components.

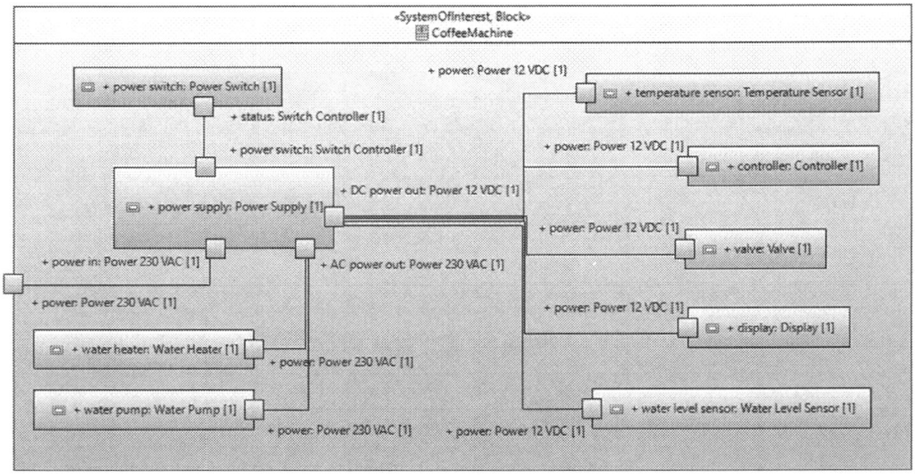

Figure 5.298. "Coffee Machine Power IBD"

5.13 Model Parametrics

Parametric modeling involves the use of parameters and relationships to describe the characteristics and constraints of a system. The SysML Parametric diagram provides a visual representation of these relationships and is a key tool in capturing and communicating complex, interrelated information. The Parametric diagram is specifically designed to represent the mathematical relationships between different parameters in a system. The benefits of parametric modeling in systems engineering are:

- **Advanced Analytics**: By expressing design variables as parameters and defining relationships between them, parametric modeling allows for the quantification of system characteristics and lays the foundation for advanced analytics. It facilitates sensitivity analysis, enabling engineers to assess how changes in one parameter affect other aspects of the system. It supports trade-offs where several designs can be compared to each other by reusing the same parametric models to analyze each of the alternatives. And, lastly, it supports optimization efforts. Engineers can explore different design alternatives and configurations to find the most efficient or effective solution.
- **Assist in V&V**: Parameters in parametric diagrams can be associated with acceptance criteria and performance specifications. This facilitates the V&V process, improves the communication between the design team and the quality assurance team, and helps ensure that the system meets its specified requirements and objectives.
- **Improve Documentation**: Parametric modeling in the form of parametric diagrams serves as a valuable form of documentation for the system. They provide a clear and concise representation of the relationships between parameters, aiding in communication, understanding, and documentation of design decisions. They provide a common language for collaboration among different engineering disciplines. It allows experts in various domains to contribute to the modeling effort and ensures that system requirements are represented in a way that is understandable across disciplines. the parametric diagrams visually represent dependencies and relationships between parameters. This visualization helps stakeholders understand the intricate connections within the system, fostering a more comprehensive and transparent communication of the system's characteristics and constraints.

Some of the commercially available tools include solvers or have interfaces to external tools that can perform calculations and solve complex equations and expressions that may arise in parametric models. This enhances their capabilities and enables the application of advanced analysis and optimization techniques. To the current knowledge of the author, Papyrus does not include a mathematical solver and has no built-in interfaces to external ones. So, it is possible to model parametric information using Papyrus but it is not possible to have the tool perform actual calculations. Several publications describe efforts to integrate Papyrus with external tools, so there might be new opportunities in the future.

5.13.1 Modeling Parametrics for the Coffee Machine

Parametric models constrain the properties of a system. Constraints are expressed as equations. The parameters of the equations are then bound to the properties of the system being analyzed in an analysis context. A constraint block is a special kind of block that packages the statement of one or several constraints so they may be applied in a reusable way to constrain properties of other Blocks.

> A constraint property is another kind of property belonging to a SysML block. It is a property that is not typed by a value type or another block but by a constraint block and represents the usage of one constraint block inside a block.

We want to model how many cups of coffee can be produced by the coffee machine. The inputs for this constraint are the volume of the water tank of the coffee machine and the amount of water the coffee machine uses per cup of coffee it produces. From these two inputs we can derive the output, the number of cups per tank refill.

As a prerequisite for the parametric analysis, we need to create the properties that characterize the system and its components as well as the value types that are required to type these properties.

Create a new value type called "Fluid Volume[ml]" along with the unit "milliliter" and quantity kind "Volume" in the "Value Types" package.

Next, add a new property element to the "Water Tank" block in the "Components" package by right-clicking on the "Water Tank" block and selecting *UML for SysML 1.6->Property.*

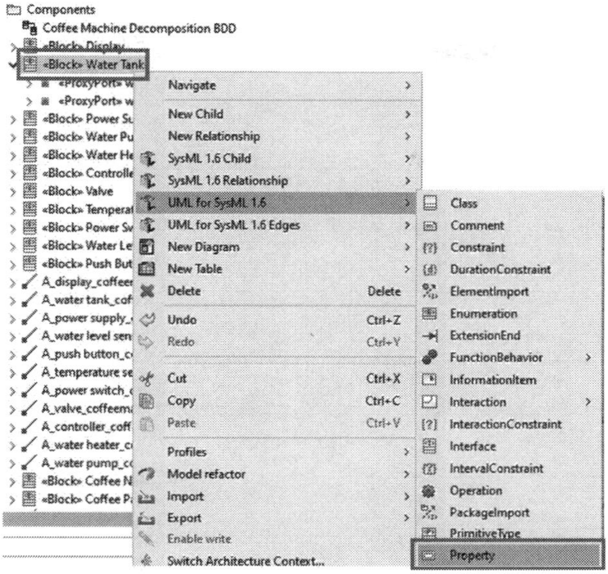

Figure 5.299. Create a new property

Name the new property "volume" then set its type to the value type "Fluid Volume[ml].

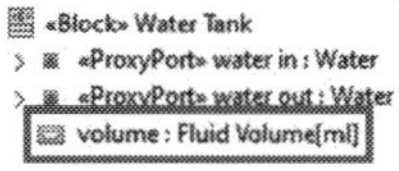

Figure 5.300. Typed property "volume"

Add another property called "Required Water per Coffee" of the same value type to the "CoffeeMachine" block.

We start the building of the parametric model for the analysis of the coffee machine by creating a new constraint block. Create a new package called "Parametrics". Then, right-click on the package and select *SysML 1.6 Child->ConstraintBlock*. Name the constraint block "Cups per Tank Refill".

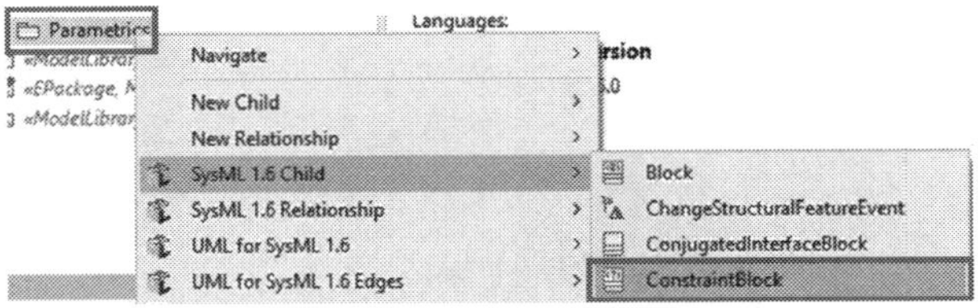

Figure 5.301. Create a new constraint block

Add a new parameter to the newly created constraint block by right-clicking on it in the Model Explorer and selecting *SysML 1.6 Child->Parameter.* Name the parameter "cups per tank refill" and type them with the value type "Integer".

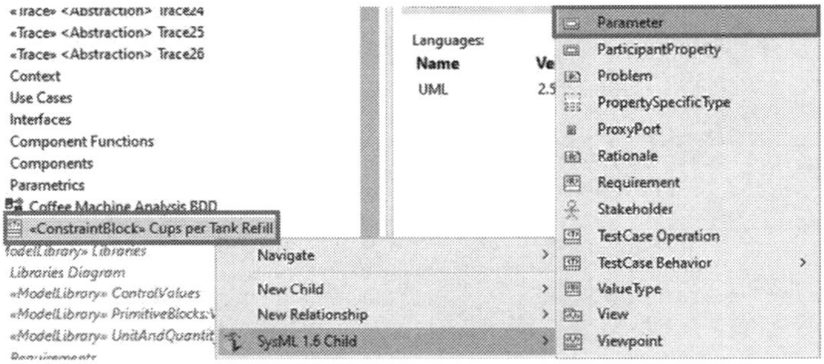

Figure 5.302. Create a new parameter

Add two more parameters - "full tank volume" and "water per coffee" - to the constraint block and type them with the value type "Fluid Volume[ml].

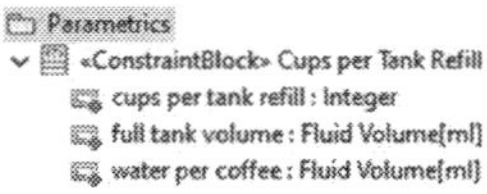

Figure 5.303. Constraint block with three parameters

Add a new constraint to the constraint block by right-clicking on the constraint block and selecting *UML for SysML 1.6->Constraint.*

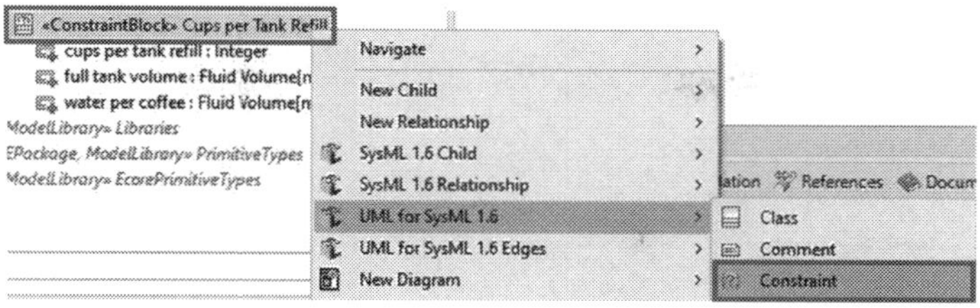

Figure 5.304. Create a new constraint

Name the constraint "Cups per Tank Refill Constraint". The constraint already has a default specification. To change it go to the Properties View of the constraint and click on the button with the pen symbol next to the Specification property to open the "Edit OpaqueExpression" dialog.

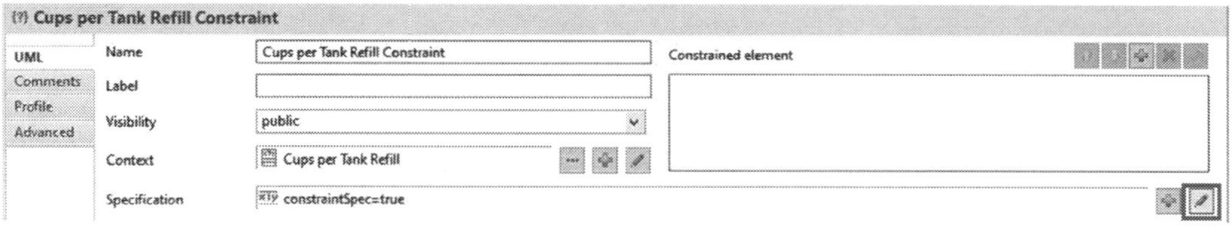

Figure 5.305. Open the "Edit OpaqueExpression" dialog

Set the language of the opaque expression to "Natural Language" and set its content to "Cups per Tank Refill = Full Tank Volume / Water per Coffee".

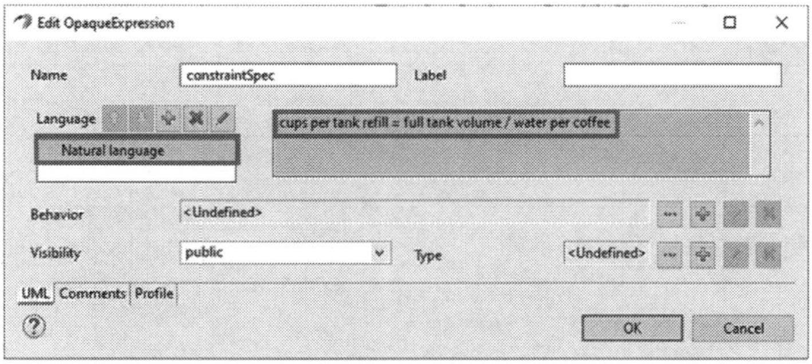

Figure 5.306. Specify the constraint

Create a new block definition diagram in the "Parametrics" package and name it "Coffee Machine Analysis BDD". Add the constraint block "Cups per Tank Refill" to the BDD and open the "Select the elements to show" dialog by right-clicking on the constraint block and selecting *Filters->Show/Hide Contents.*

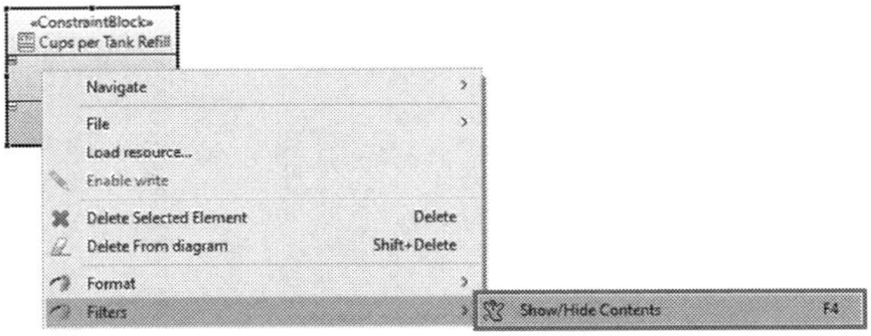

Figure 5.307. Open the "Select the elements to show" dialog

Select the constraint and the parameters.

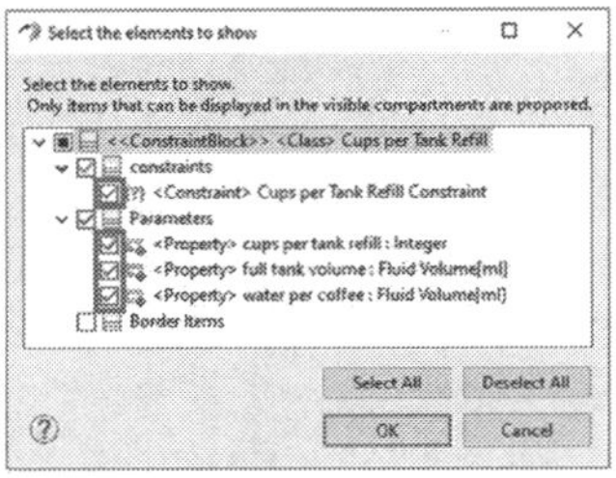

Figure 5.308. "Select the elements to show" dialog

Create a new block called "Coffee Machine Analysis Context" to the "Parametrics" package and add it and the "CoffeeMachine" block to the "Coffee Machine Analysis BDD". Then, define the constraint block and the "CoffeeMachine" block parts of the "Coffee Machine Context" block.

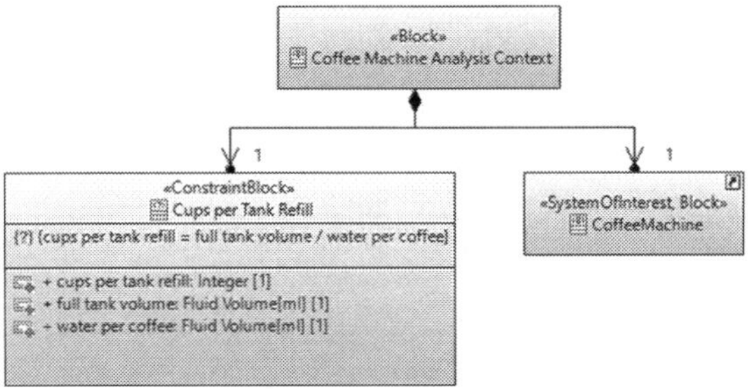

Figure 5.309. "Coffee Machine Analysis BDD"

Add a new parametric diagram to the "Coffee Machine Analysis Context" block by right-clicking on the block and selecting *New Diagram->SysML 1.6 Parametric Diagram.* Name the new diagram "Coffee Machine Analysis PD".

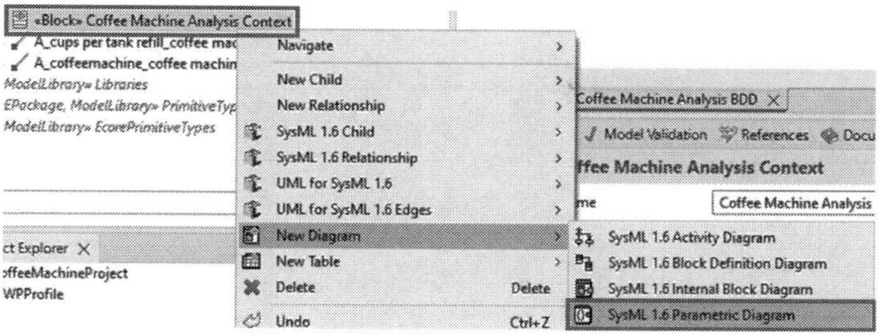

Figure 5.310. Create a new parametric diagram

Drag and drop the "cups per tank refill" constraint property of the "Coffee Machine Analysis Context" block onto the newly created parametric diagram. Select *Constraint Property drag and drop inside parametric diagram* after dropping the property.

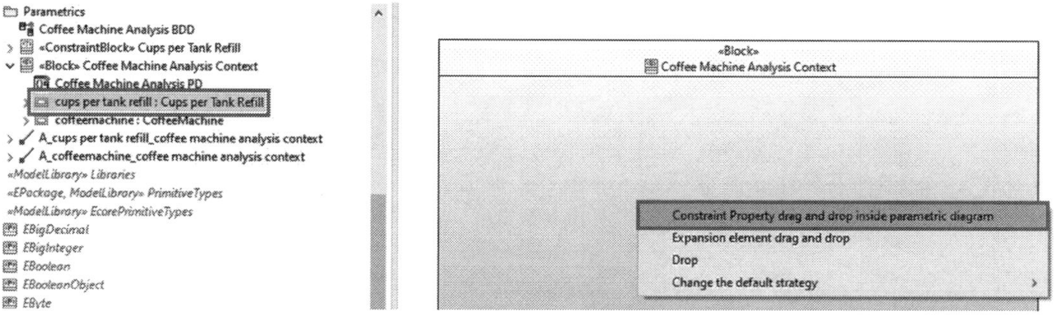

Figure 5.311. Add the constraint property to the PD

Make the constraint property bigger and open the "Show/Hide compartments" dialog by right-clicking on the constraint property and selecting *Filters->Show/Hide Compartments.*

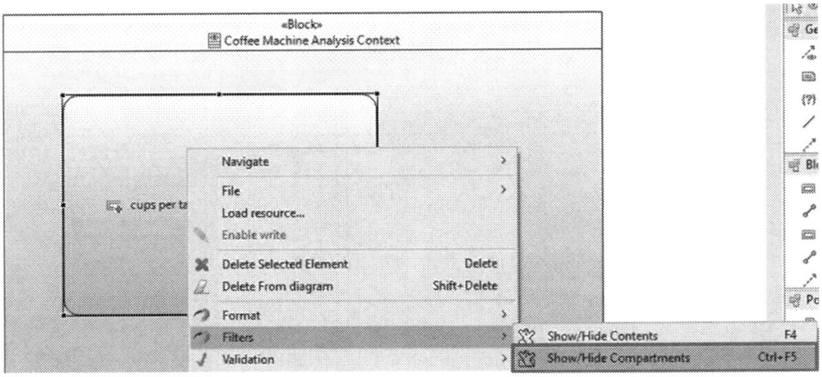

Figure 5.312. Open the "Show/Hide compartments" dialog

Check the "constraints" element.

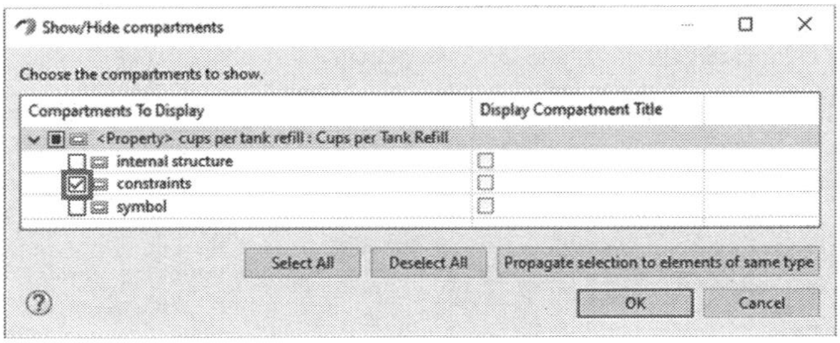

Figure 5.313. Check the constraints element

Next, we want to define which properties of the constraint block should be visible. To achieve this open the "Show/Hide Related Content" dialog by right-clicking in the constraint property and selecting *Filters->Show/Hide Contents.*

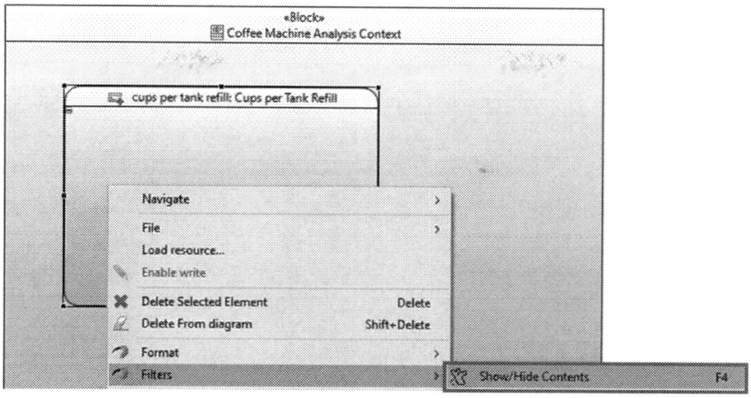

Figure 5.314. Open the "Show/Hide Related Content" dialog

Check all the elements using the "Select All" button.

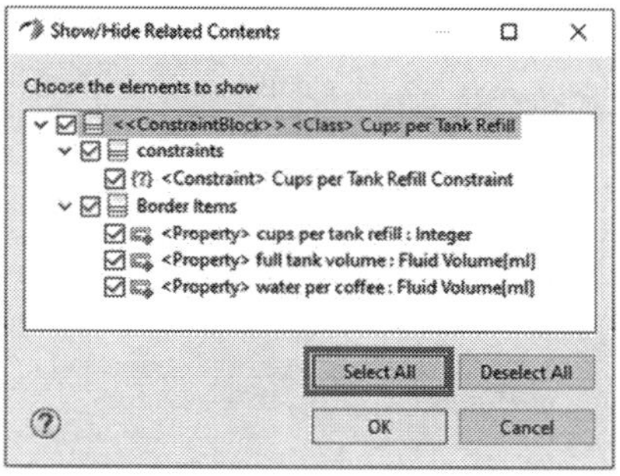

Figure 5.315. Check all the elements

The constraint and the parameters of the constraint block are now visible on the PD.

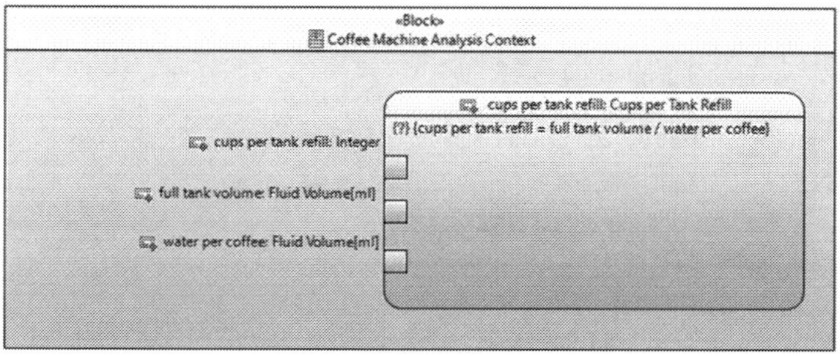

Figure 5.316. PD with constraint property

Next, add the "coffeemachine" part property to the "Coffee Machine Analysis PD" using drag and drop. Select "Expression element drag and drop" after dropping the property.

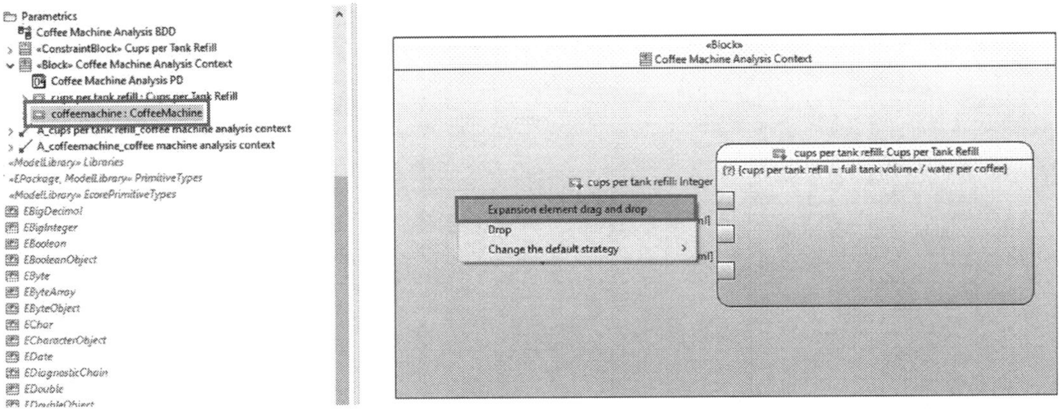

Figure 5.317. Add the part property to the PD

Open the "Show/Hide compartments" dialog by right-clicking on the part property and selecting *Filters->Show/Hide Compartments.*

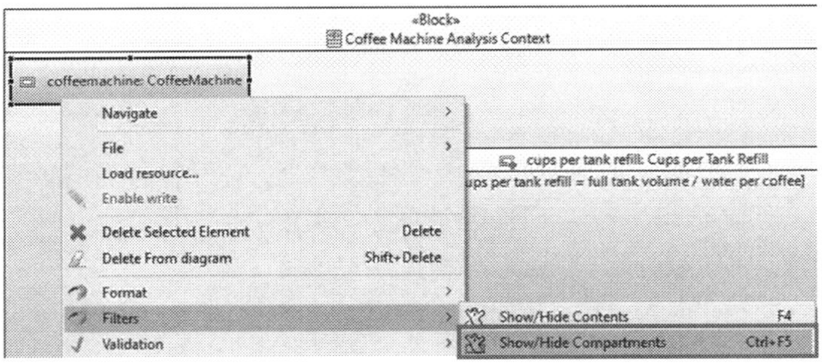

Figure 5.318. Open the "Show/Hide compartments" dialog

Since we want to look inside the "CoffeeMachine" block, check the "internal structure" element.

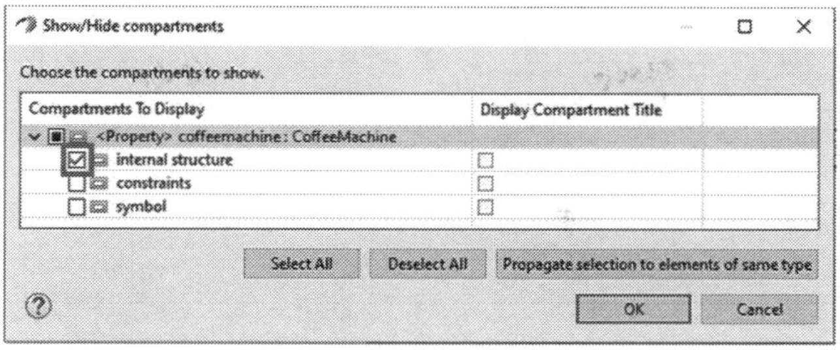

Figure 5.319. Check the "internal structure" element

Drag and drop the part property "water tank" of the "CoffeeMachine" block as well as the "Cups per Tank Refill" measure of effectiveness and the "Required Water per Coffee" property into the "coffeemachine" part.

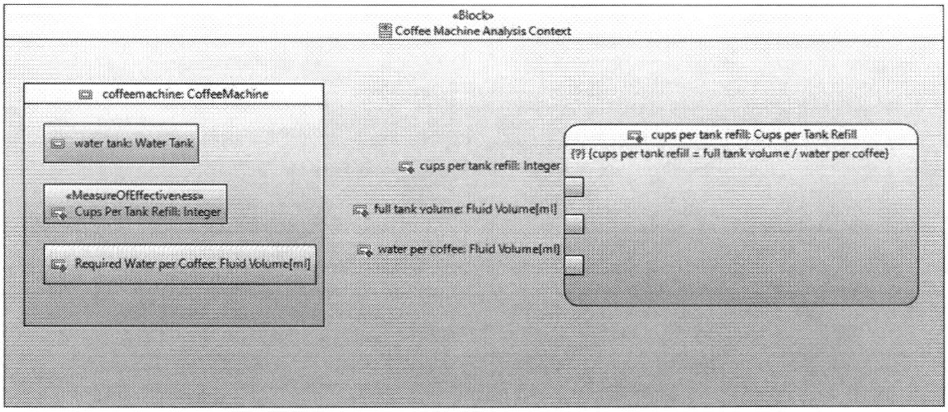

Figure 5.320. Add more elements to the PD

We want to display a property of the "Water Tank" block, its volume, on the diagram as well. So, as before we need to tell Papyrus that we want to see the internal structure of the "water tank" part. Open the "Show/Hide compartments" dialog of the "water tank" part and check the "internal structure" element.

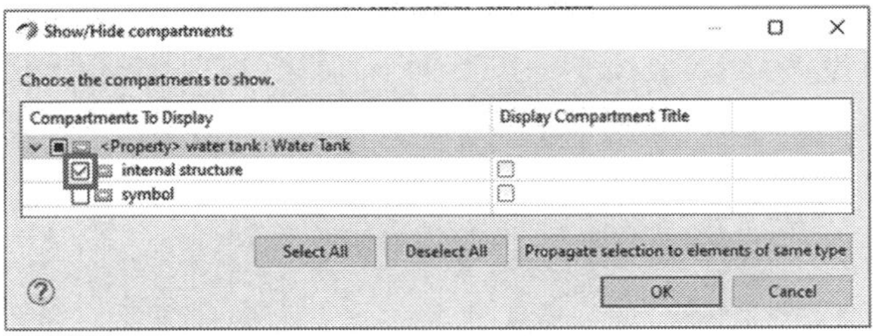

Figure 5.321. Show the internal structure of the "Water Tank" block

Then open the "Show/Hide Related Contents" dialog and select the "volume" property of the water tank.

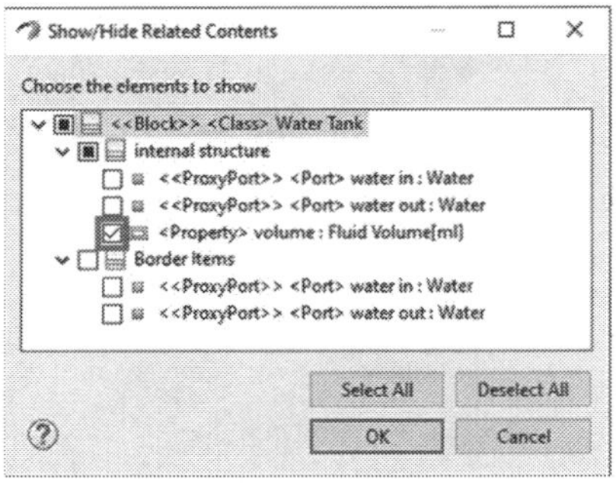

Figure 5.322. Select volume property

Now, we want to relate the parameters of the constraint block to the properties of the system model and its components. Select the "Binding Connector" element in the Diagram Palette and, first, click on the "volume" property inside the "water tank" part and then on the "Full Tank Volume" parameter of the constraint property.

Binding connectors cannot only be used for connecting ports they also allow binding constraint parameters of a constraint block to the properties of other blocks.

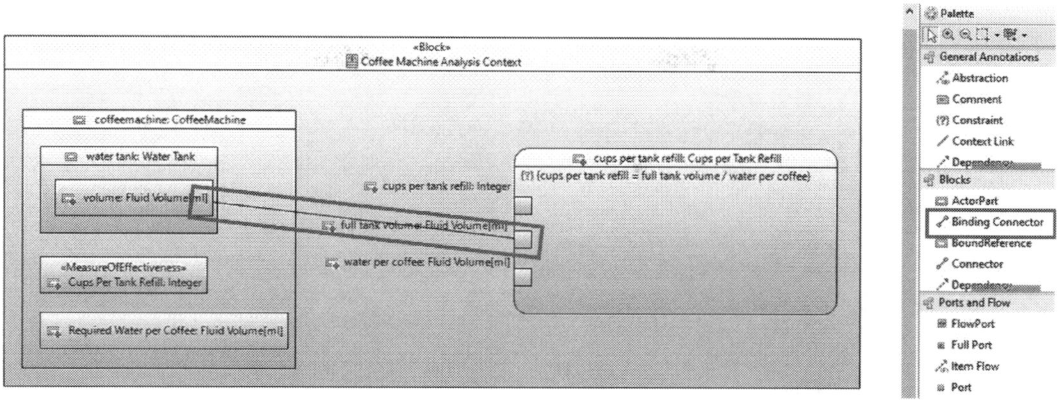

Figure 5.323. Create a new binding connector

Repeat the process for the other two parameters of the constraint property to complete the "Coffee Machine Analysis PD".

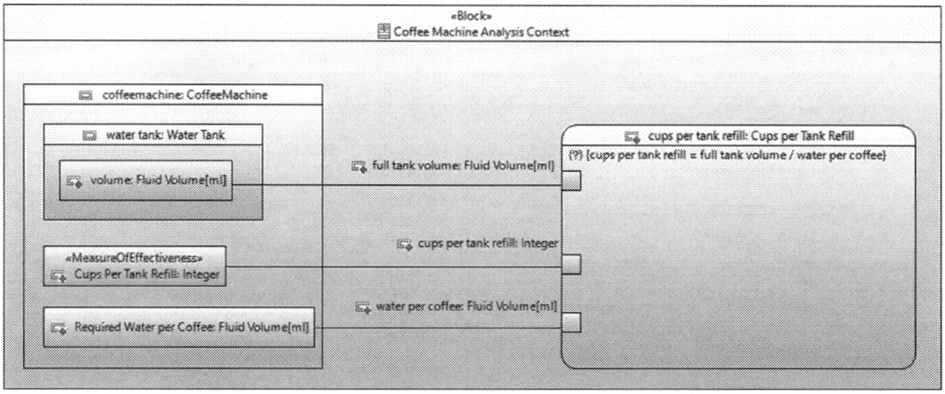

Figure 5.324. Finished "Coffee Machine Analysis PD"

Chapter 6 - Conclusion

The journey through this tutorial has provided a valuable introduction to the principles and practices of MBSE and its implementation using the Eclipse Papyrus tool and the SysML modeling language. By applying MBSE to model a seemingly simple system like a coffee machine, we have witnessed the transformative power of this approach in addressing the complexity inherent in modern systems.

Throughout the book, we have seen how MBSE has enabled us to create a holistic representation of the coffee machine. The SysML diagrams have provided a comprehensive view of the coffee machine, capturing its functional behavior, physical structure, and behavioral interactions. This kind of holistic understanding can be instrumental in identifying potential design flaws and ensuring a system's overall effectiveness.

The graphical nature of SysML can foster clear and concise communication among stakeholders, regardless of their technical expertise. The ability to share and modify models facilitates collaboration, leading to a shared understanding of the system and its requirements.

As we conclude this tutorial, it is evident that MBSE has emerged as a cornerstone of modern systems engineering. By embracing MBSE, organizations can overcome the challenges of complexity and deliver high-quality, reliable systems that meet the needs of their users. The coffee machine model serves as a testament to the power of MBSE in tackling even the most intricate systems.

As we move forward, MBSE will continue to evolve and adapt to the ever-increasing complexity of engineered systems. Its ability to provide a unified and collaborative approach to systems development will remain invaluable in ensuring the success of future endeavors.

6.1 What to Do Next?

To further expand their understanding of MBSE and SysML as well as their mastery of the Eclipse Papyrus tool, readers can embark on a journey of continuous learning and exploration:

1. Deepen expertise in SysML

   Read dedicated SysML books or online resources to gain a deeper understanding of the language's nuances and advanced concepts. Explore additional modeling elements, their possible relationships, and their applications in real-world systems engineering projects.

   Explore advanced SysML concepts: Dive deeper into the intricacies of advanced SysML concepts or SysML extensions, such as variant and test modeling, profile definition and refinement using OCL, and model reuse mechanisms.

   Analyze real-world applications: Study case studies and examples of how SysML has been successfully applied in various industries, such as aerospace, automotive, and healthcare.

   Engage with SysML experts: Connect with experienced SysML practitioners to gain insights into their modeling techniques, best practices, and challenges faced in real-world projects.
2. Enhance modeling skills with Eclipse Papyrus

   Continue practicing SysML modeling using Eclipse Papyrus. Expand the coffee machine model by adding more details to the existing diagrams or introducing new diagrams to represent additional aspects of the system's functionality or structure.

   Practice with additional modeling elements: Explore other SysML model elements and relations, such as full ports, sequence diagrams, and usage diagrams, to gain a deeper understanding of their applications.

   Utilize advanced tool features: Master the use of advanced Eclipse Papyrus features, such as model execution using the Papyrus Moka extension and documentation generation using the Papyrus Model2Doc extension, to streamline your modeling process.

   Create reusable libraries: Develop reusable SysML models and libraries to encapsulate common modeling patterns and components, boosting efficiency and consistency across projects.
3. Build a model of your system

   Embrace the hands-on approach by applying your newfound knowledge to model a system of your own choice. Choose a system that interests you and analyze its components, interactions, and requirements. Use Eclipse Papyrus to create a comprehensive model that captures the essence of your system.

   Choose a suitable system: Select a system that aligns with your interests and expertise, ensuring that you can effectively represent its complexities and challenges.

Conduct thorough analysis: Conduct a detailed analysis of the system's functionality, structure, interactions, and requirements to gather the necessary information for modeling.

Create a comprehensive model: Use Eclipse Papyrus to create a comprehensive SysML model following the MWP method that captures the essence of the system, including its physical elements, behavioral interactions, and parametrics.

4. Participate in MBSE community forums and workshops

   Engage with the MBSE community by joining online forums, attending workshops, and participating in discussions. Share your experiences, seek guidance from experienced practitioners, and stay updated on the latest advancements in MBSE practices and tools. A good starting point is the INCOSE and the local chapter in your country or region.

   Seek insights from experienced practitioners: Engage with experienced MBSE professionals in online forums and workshops to gain valuable insights, share experiences, and learn from their expertise.

   Contribute to MBSE discussions: Actively participate in MBSE discussions, share your knowledge, and contribute to the evolution of MBSE practices and tools.

   Stay Informed about MBSE Developments: Stay up-to-date on the latest MBSE advancements, standards, and emerging trends to ensure that you are at the forefront of the field.

5. Explore other MBSE tools

   While Eclipse Papyrus is a valuable tool, investigate other MBSE platforms to gain a broader understanding of the available options and their strengths and weaknesses. Experiment with different tools to determine which one suits your unique needs and workflow.

   Evaluate alternative modeling tools: Investigate other popular MBSE tools, such as Cameo Systems Modeler or Cappella, to compare their features, strengths, and potential fit for your specific needs. Most tools that require a license can be evaluated for free.

   Experiment with different approaches: Experiment with different MBSE methodologies, such as the Arcadia method, which goes hand in hand with the Cappella tool, and practices to find the ones that best suit your workflow and project requirements.

   Integrate tools effectively: Explore ways to integrate different MBSE tools to create a collaborative and unified modeling environment.

6. Stay updated with MBSE standards and trends

Keep abreast of the latest MBSE standards and trends by following industry publications, attending conferences, and participating in online communities. This will ensure that your MBSE knowledge remains current and relevant in the ever-evolving landscape of systems engineering.

Follow academic publications: Subscribe to relevant academic publications and journals to stay informed about emerging MBSE standards and trends.

Attend conferences and events: Participate in MBSE conferences, workshops, and seminars to network with professionals, learn from industry experts, and gain hands-on exposure to new technologies.

Follow the SysML v2 development: The development of the SysML v2 standard is documented openly and freely[1]. The SysML v2 Submission Team, the team working on SysML v2, has also developed a pilot implementation that allows you to "touch" SysML v2.

Follow the Papyrus development: Papyrus has a development roadmap that ranges from Papyrus adopting Sirius in the short term to implementation of a web-based version of SysML v2 through a project within the Eclipse Foundation named SysON[2]. A very early prototype of SysON has recently been published and can already be tested.

[1] https://github.com/Systems-Modeling/SysML-v2-Release
[2] https://mbse-syson.org

Appendix A - Modeling Guidelines

Implementing a SysML-based systems engineering strategy to a project can be challenging as readily available methodologies and modeling guidelines are sparse. The lack of guidelines and modeling frameworks is one of the main reasons SysML faces resistance in the industry.

To create a coherent model that covers the whole system, guidelines should be used to make sure that all aspects of the system are considered in the model. Without modeling guidelines, the model easily becomes hard to navigate and read, which makes finding potential errors harder. This issue is amplified when multiple people are working on the same model. There are several ways a system can be presented in the model, to guarantee a coherent model, conventions must be in place so that everyone working on the model follows the same modeling practices. This becomes even more important if MBSE is deployed on a wider scale in an enterprise and people are working on multiple models. If all models are built consistently along a common set of guidelines then it becomes easier for people to orient themselves in a model that they have not seen before.

A team should have a modeling standards and conventions document to guide all members of the team in modeling consistently.

The purpose of modeling guidelines is to collate this experience, then have a group of modeling experts inspect, verify, and process the resulting guidelines, and finally make them available to a wider group of developers. Without the knowledge contained in modeling guidelines, every modeler would have to rely on the extent of his own experience and start from the beginning in each new project, thus leading to regularly occurring problems in the modeling process and divergence between different modeling projects in a company or organization.

Naming conventions are helpful for several reasons:

- **Increase readability and understandability**: When names are descriptive and consistent, it is easier for stakeholders to understand what each object in the model represents.
- **Improve model quality**: By using consistent naming conventions, models can be created more quickly and easily, which can lead to fewer errors and a better overall quality of the system.

- **Facilitate collaboration between stakeholders**: When stakeholders use the same naming conventions, it is easier for them to communicate about the model and avoid confusion.
- **Aid in model maintenance**: When names are descriptive and consistent, it is easier to update and modify models without introducing errors.
- **Promote reuse**: Consistent naming conventions make it easier to reuse SysML models in other projects.

Diagramming guidelines are helpful for several reasons:

- **Effective communication**: Diagramming guidelines help to ensure that diagrams are clear, concise, and easy to understand. This facilitates communication between stakeholders, promoting shared understanding and collaboration.
- **Consistent representation**: Diagramming guidelines promote consistency in the representation of elements and relationships within diagrams. This allows for a more coherent and organized visual representation of the system.
- **Efficient representation**: Diagramming guidelines help to optimize diagram layout and content, ensuring that diagrams convey information effectively without being cluttered or overwhelming.
- **Standardization**: Diagramming guidelines establish a standard for representing system elements and relationships, making it easier to exchange and interpret diagrams across different teams and projects.
- **Repeatable use**: Well-defined diagramming guidelines ensure that diagrams can be created consistently, making it easier to reuse them for different purposes or to compare different design iterations.
- **Maintainability**: Diagramming guidelines facilitate the maintenance and evolution of diagrams as the system design changes.
- **Training**: Diagramming guidelines provide a framework for training and onboarding new team members, ensuring that they can create and interpret diagrams effectively.
- **Professionalism**: Diagramming guidelines enhance the professionalism of the modeling effort, demonstrating a commitment to clear and consistent communication.

A prescribed model structure is helpful for several reasons:

- **Faster orientation**: Anybody using the model already knows the structure from training or previous projects which is why they can extract relevant information quicker.
- **Modular and reusable models**: A prescribed model structure encourages the creation of modular models that can be reused across different projects or phases of the same project.

In short, modeling guidelines help create more readable, maintainable, and collaborative SysML models.

MWP Modeling Guidelines

Since the MWP method is rather simple, there is only a limited set of conventions and guidelines.

The following set of diagramming guidelines is used throughout this book:

- Do not use the diagram grid.
- Keep diagrams organized and uncluttered.
- Use a consistent scale for all diagrams.
- Try to keep transitions and flows from crossing each other.
- Limit the size of the diagram to one page.

Additionally, some naming conventions are used throughout this book.

General naming conventions:

- Use expressive short names.
- Do not write complete sentences, only strings of words.
- Do not use underscores but use spaces to improve readability and allow word wrapping in the tools.

Naming conventions for properties (part properties, value properties,...): Start with a lowercase letter to distinguish usage from definition.

Naming conventions for classifiers (blocks, value types, requirements,...): Start with a capital letter to distinguish definition from usage.

Naming conventions for diagrams and tables: Add the type of the diagram to the end of the name as a postfix to enable easier identification.

Appendix B - Installation

This appendix explains the necessary steps to install Papyrus on Windows. Since Papyrus is an open-source tool, it is free to download and use. The installation of Papyrus on a Windows machine does not require administrator rights.

Go to the Eclipse Papyrus homepage[1].

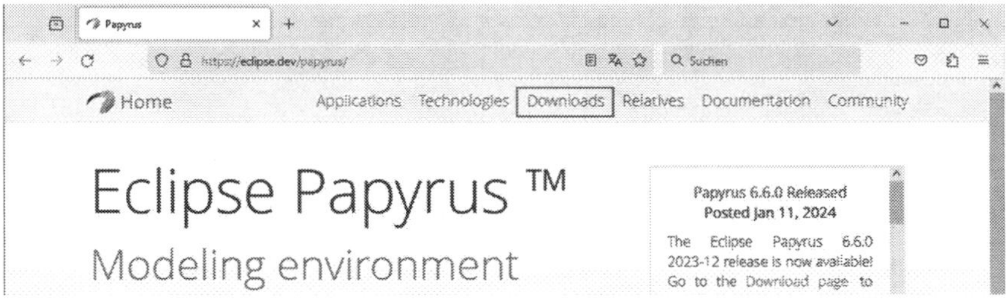

Figure 1. Papyrus homepage

Click on "Downloads", then scroll down to the section labeled "Latest released RCP" and select the version that is appropriate for your platform.

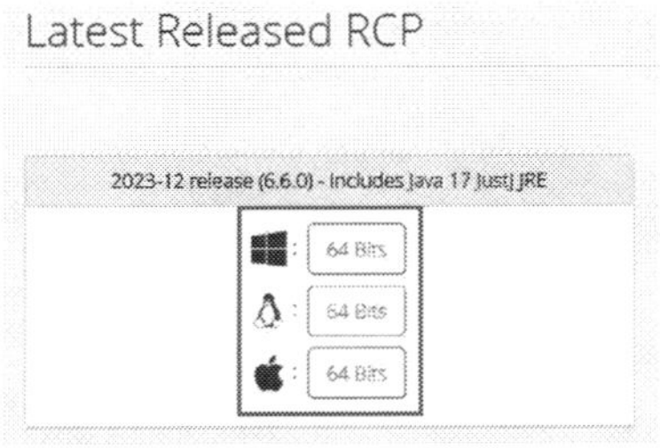

Figure 2. Papyrus downloads page

The download will be a zip file (papyrus-2023-06-6.6.0-win64.zip is version 6.6.0 2023-12 release for the Win64 platform) that needs to be extracted to a directory where the user has read and write permissions, e.g., the user space.

Once the extraction is finished, Papyrus can be launched. On the Windows platform, it

[1]https://eclipse.dev/papyrus/

is launched by double-clicking the papyrus.exe file in the root of the extracted Papyrus directory.

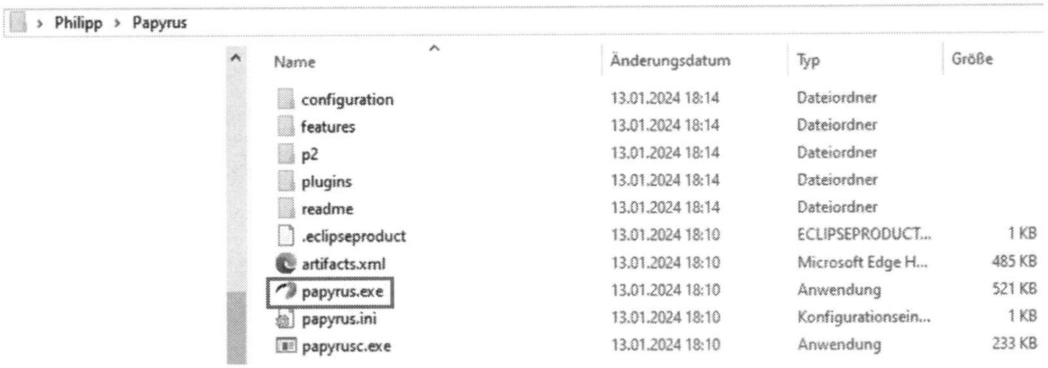

Figure 3. Extracted Papyrus folder

During the loading process, Papyrus will display a splash screen.

Figure 4. Papyrus splash screen

On startup, Papyrus requires the specification of a workspace, where all the project files are stored. The directory again needs to be a location where the user has read and write permissions, e.g., the user space. It is possible to prevent this dialog from opening on subsequent Papyrus launches by checking the "Use this as the default and do not ask again" checkbox.

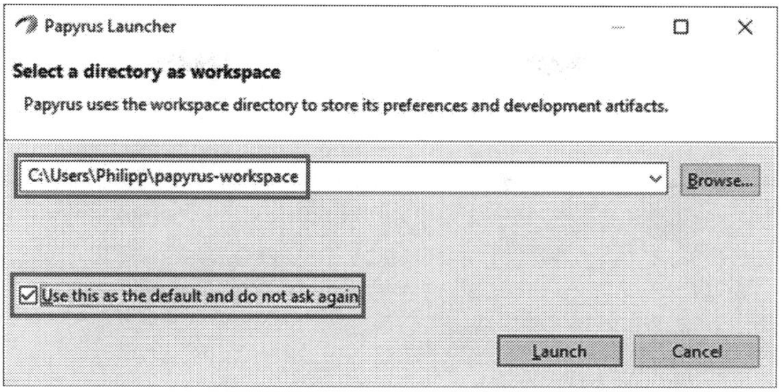

Figure 5. "Select Workspace" dialog

Once the workspace is selected, Papyrus will start with the main screen and an empty workspace.

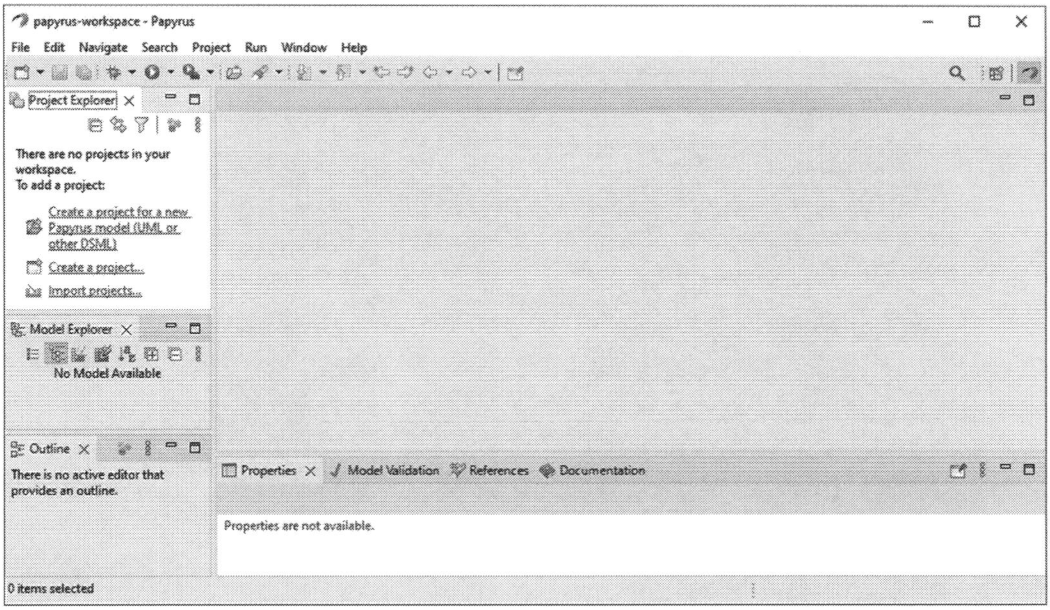

Figure 6. Papyrus main screen on the first launch

SysML 1.6 is not part of the default Papyrus installation but it can be easily downloaded from the Eclipse Marketplace. To open the Eclipse Marketplace, click on *Help->Eclipse Marketplace.*

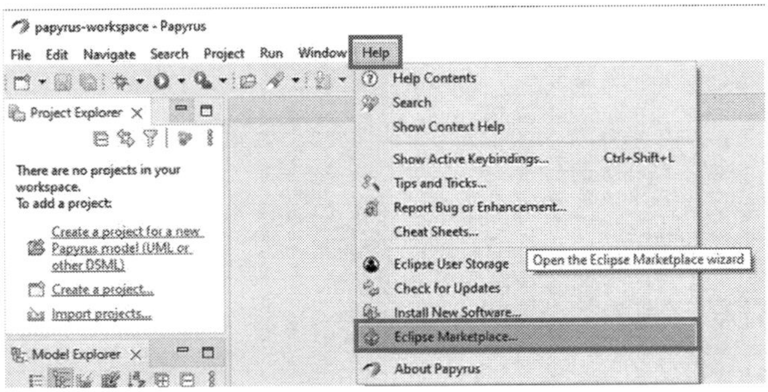

Figure 7. Starting the Eclipse Marketplace

To find the SysML 1.6 plugin search for "sysml". Then click on "Install" and follow the instructions for the installation of the plugin.

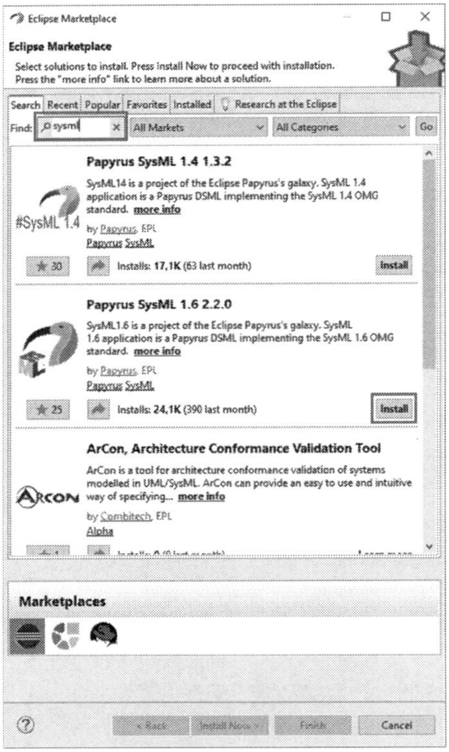

Figure 8. Install the SysML 1.6 plugin

After the installation is finished, Papyrus will prompt for a restart.

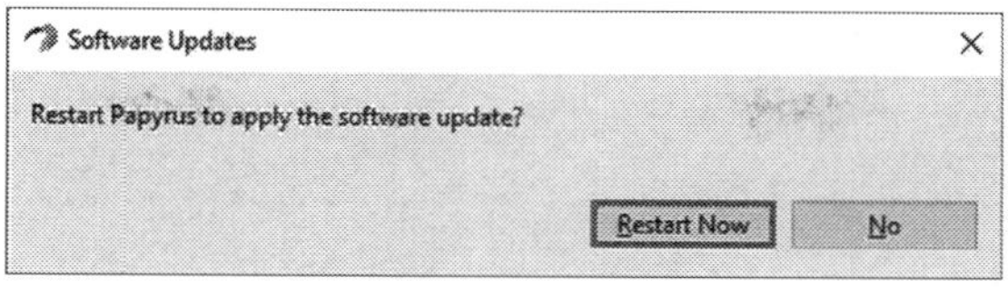

Figure 9. Restart Papyrus after finishing the installation

After the restart, Papyrus is ready to be used with SysML 1.6.

Appendix C - Collaboration

In systems engineering, configuration management (CM) is all about ensuring a system's consistency and integrity throughout its entire lifecycle. MBSE brings additional requirements and challenges to CM. MBSE models are intricate webs of information, model elements and relations between them. Effective CM for MBSE needs to consider the entire model as a unit, not just individual elements. This ensures that changes to one part do not create unintended consequences elsewhere in the model. Any changes made to one aspect of the model should be reflected consistently throughout. Since the model is the central source of truth, CM tools and processes need to handle access control, version control, conflict resolution during concurrent edits from multiple people working on the same model, and traceability of how changes impact different parts of the system.

True CM of models requires a model repository that is aware of the data that it stores, i.e., a repository that knows the structure of the model data that it stores and can differentiate between versions of each individual model element. Commercial tools often offer this as an additional product that can be used with their modeling tool. The Eclipse ecosystem has the Eclipse CDO (Connected Data Objects) project[1] but the setup of this CDO server and the integration with Papyrus is not trivial and not officially supported.

There is, however, a solution for collaboration and version management for Papyrus. Papyrus stores all models as XML (Extensible Markup Language) data stored in plain-text files. This allows using Git for version control, collaboration, and history tracking of the models. There are, however, limitations. Git is not aware, that the file that it stores contains a model. This means that concurrent editing of model files and the ensuing change analysis and merging do not work particularly well. One possibility to work around that limitation to a certain degree is to separate the model into several files so that each file only contains one package. The easiest way to avoid any problems is to prohibit concurrent editing of files and instead make sure that all repositories are synchronized before editing.

Remember that effective collaboration and version control with Git requires good communication among team members, especially when working on the same models concurrently. Always follow best practices for Git usage, such as committing regularly, writing meaningful commit messages, and reviewing changes before merging.

[1] https://projects.eclipse.org/projects/modeling.emf.cdo

To learn more about Git have a look at *Pro Git - Everything you need to know about Git* (Chacon and Straub, 2014), which is published freely under the Creative Commons license.

Publishing a Papyrus Model Using Git

This section explains how a Papyrus model can be published into an existing but empty Git repository.

Git is already integrated into Eclipse and therefore into Papyrus upon installation. Similar to Papyrus, Git provides its own perspective to Eclipse. To switch to the Git perspective, open the "Open Perspective" dialog by clicking on the "Open Perspective" button on the right side of the Menu Area.

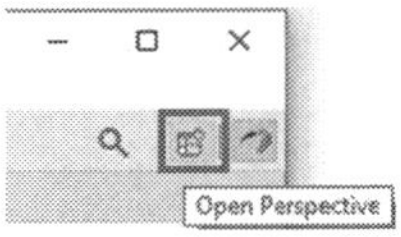

Figure 10. Open the "Open Perspective" dialog

Select "Git" and continue.

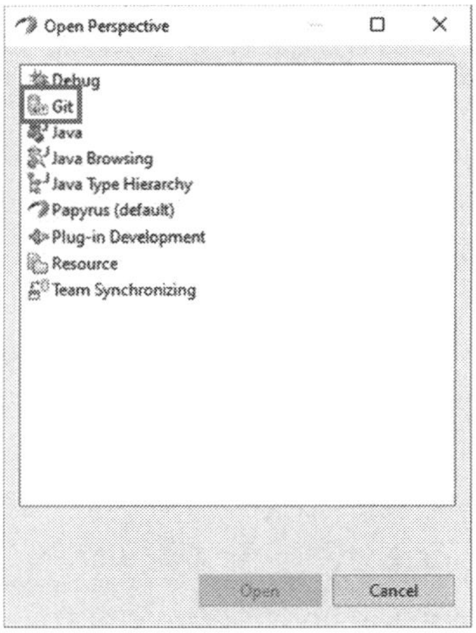

Figure 11. Select the "Git" perspective

The Git perspective opens with a Git Repositories window on the left side. This window shows a list of the currently available repositories and is empty in a new Papyrus installation. To access data in a remote repository, the repository needs to be cloned to the computer.

Click on "Clone a Git repository" in the Git Repositories window to open the "Clone Git Repository" dialog.

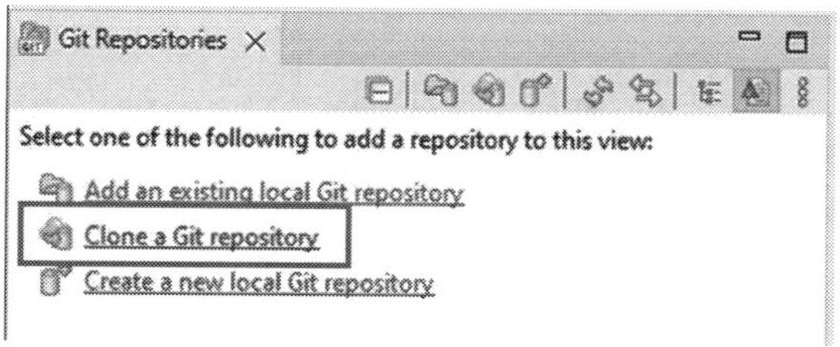

Figure 12. Open the "Clone Git Repository" dialog

Insert all the necessary information into the dialog.

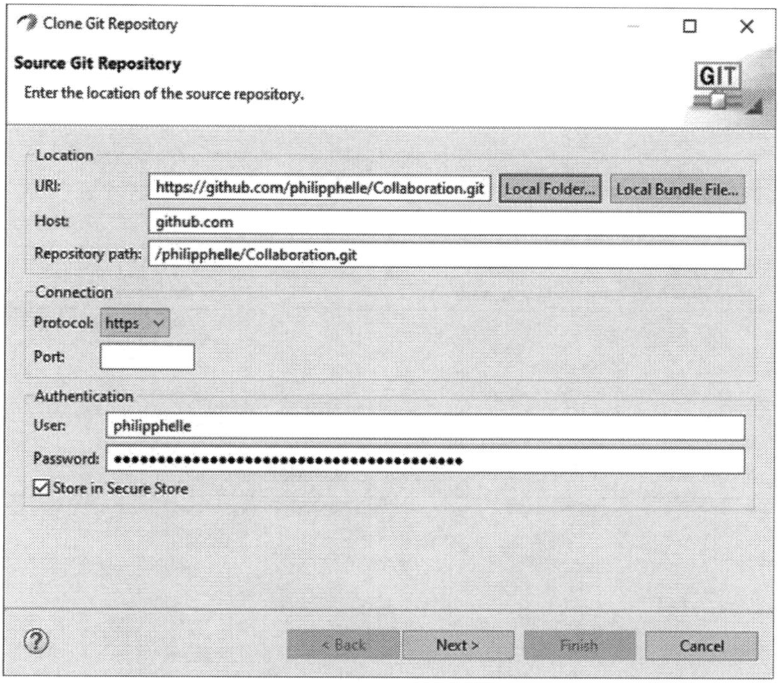

Figure 13. "Clone Git Repository" dialog

If you copy the URL to the repository in Git Hub before opening the "Clone Git Repository" dialog, then Eclipse will automatically fill some of the fields.

Since we are cloning an empty repository, the branch selection can be skipped.

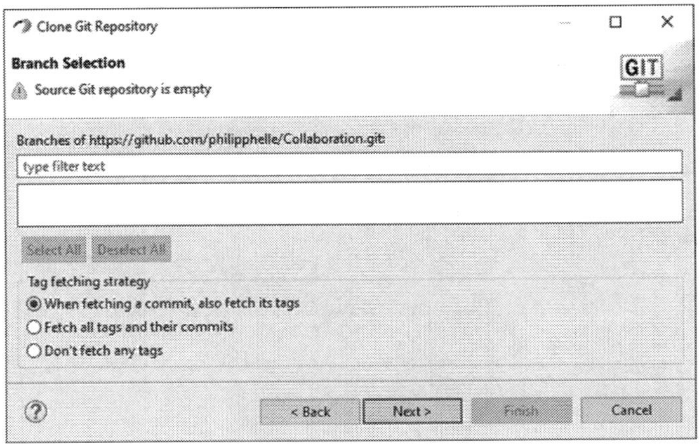

Figure 14. Skip the branch selection

Provide the name of the local directory where the newly cloned Git repository should be stored and start the cloning process by clicking on the "Finish" button.

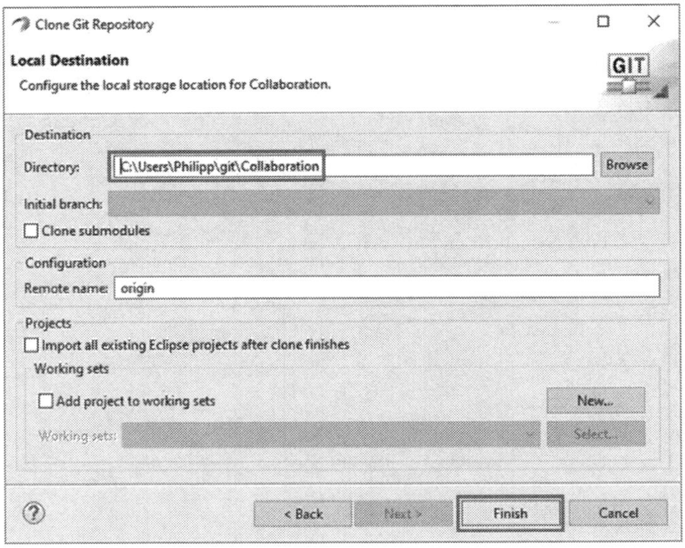

Figure 15. Set the local directory and start cloning

After the cloning process is successful, the newly cloned repository can be seen in the Git

Repositories window.

Figure 16. Cloned repository in the Git Repositories window

Now, that the repository is cloned to the local computer, the model can be moved from the local workspace directory to the local Git repository directory. Switch back to the Papyrus perspective, then right-click on the project that you want to share in the Project Explorer and select *Team->Share Project* to open the "Share Project" dialog.

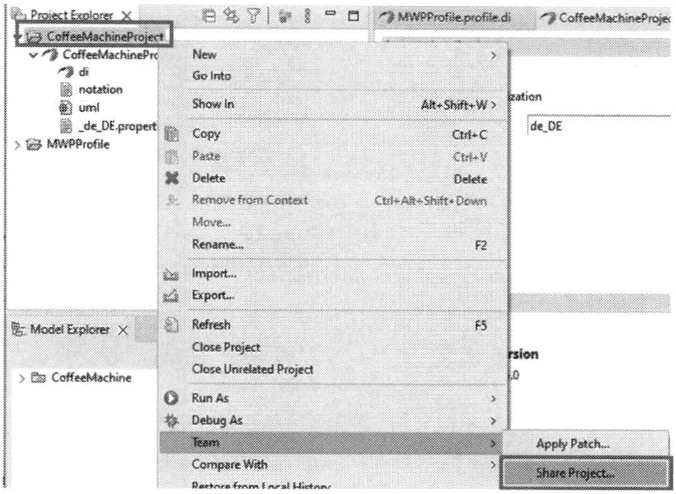

Figure 17. Open the "Share Project" dialog

Select the previously cloned repository in the drop-down list of repositories and click on the "Finish" button to start the process of moving the project to the repository directory.

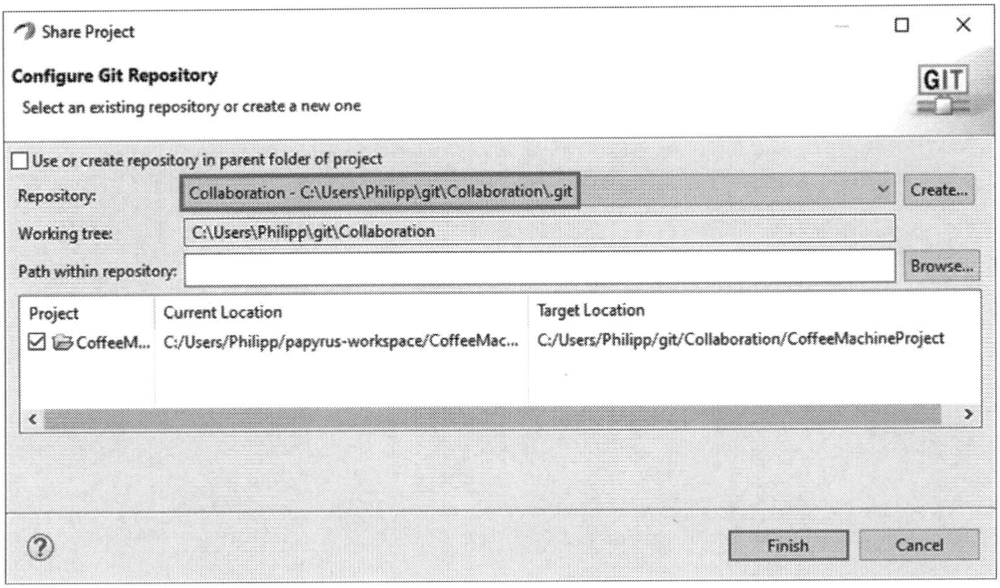

Figure 18. Select the repository and finish

The project and the files contained within the project are now marked as unversioned in the Project Explorer and the name of the repository and the current branch name is shown next to the project.

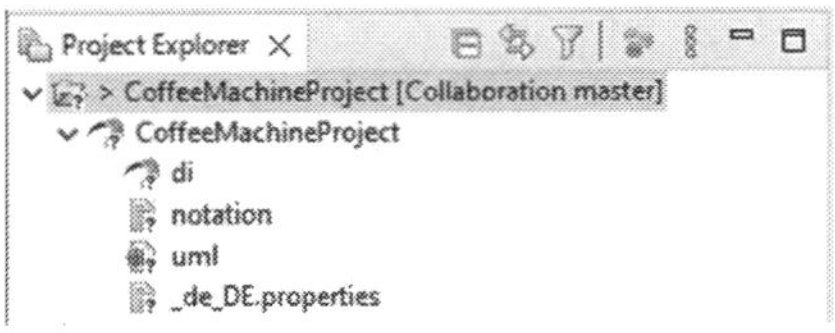

Figure 19. Unversioned elements in the Project Explorer

To add the files to the repository and push them to the remote repository, switch back to the Git perspective and go to the Git Staging view. Then, stage all the files by clicking on the button with the two plusses, write a commit message, and click on the "Commit & Push" button.

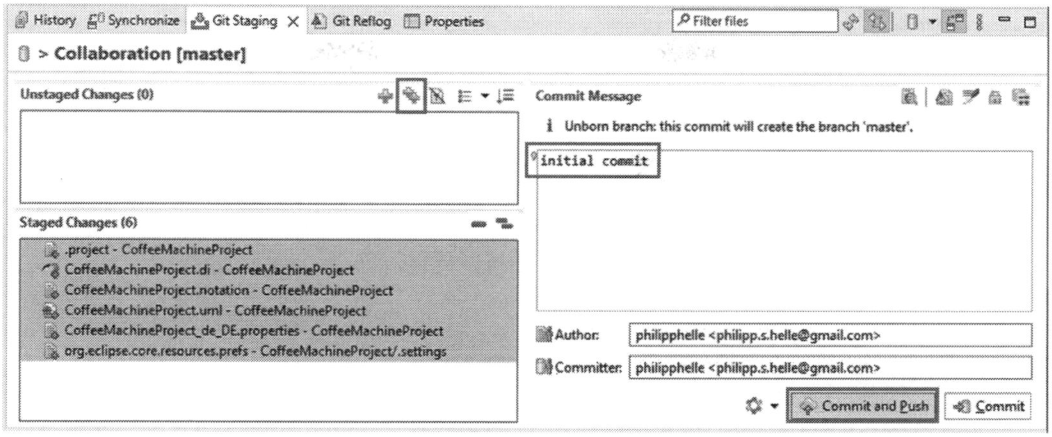

Figure 20. Commit and push

The results of the push action will be displayed in the "Push Results" dialog.

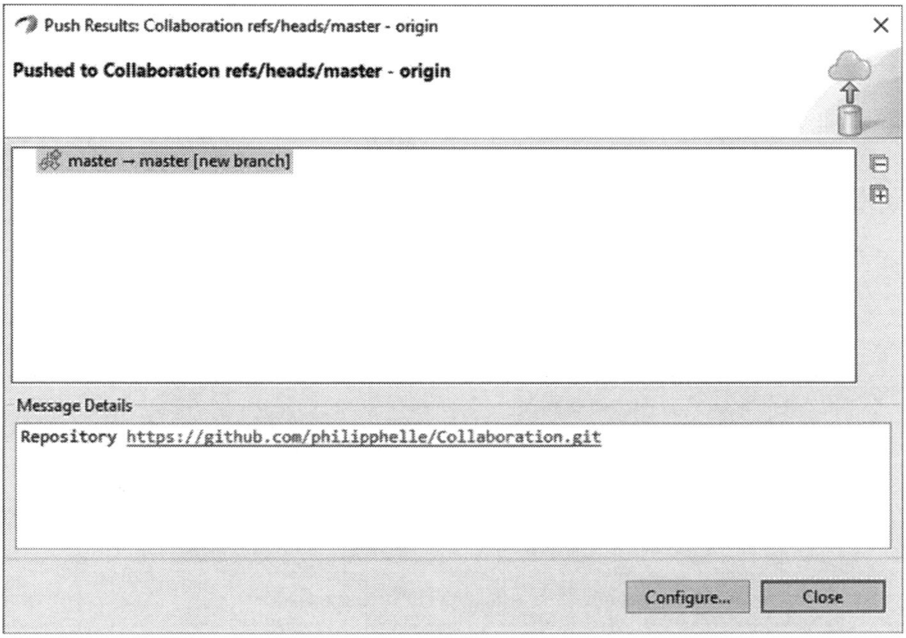

Figure 21. "Push Results" dialog

The project is now visible on the Git Hub page of the repository.

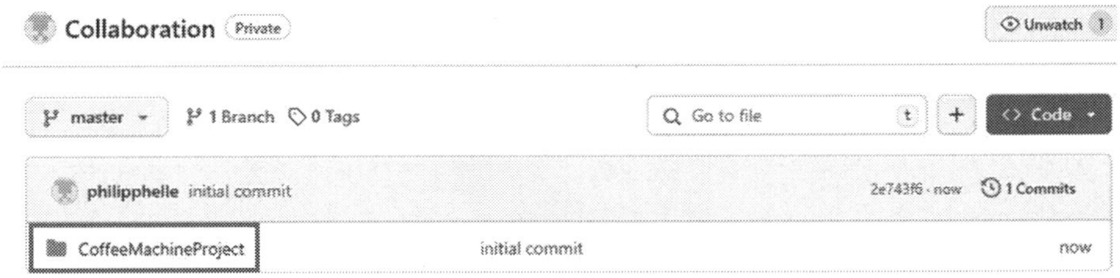

Figure 22. Git Hub page of the repository with added project

Loading a Papyrus Model From Git

This section explains how a Papyrus model that was previously published to a Git repository can be loaded into Papyrus.

Open the Git perspective and open the “Clone Git Repository” dialog as before. Now that that remote Git repository already contains data, the list of branches is not empty. Select the branch that you want to clone and continue to set the local directory as before.

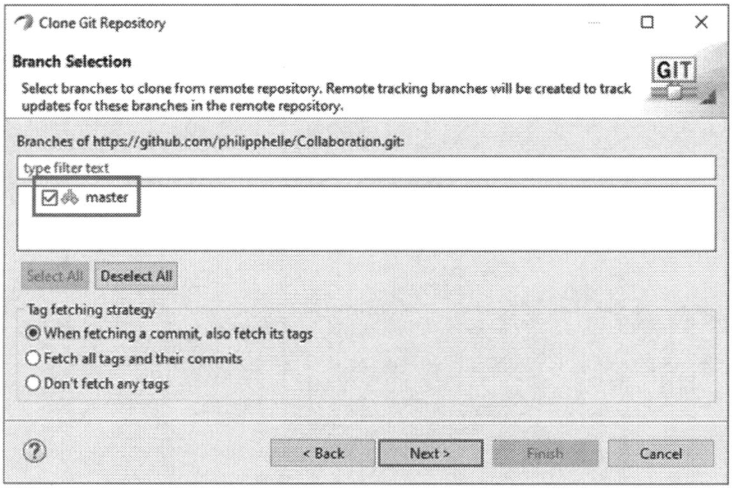

Figure 23. Select the branch to clone

After the Git repository has been cloned and is shown in the Git Repositories window, right-click on the newly cloned Git repository and select “Import Projects” to open the “Import Projects from File System or Archive” dialog.

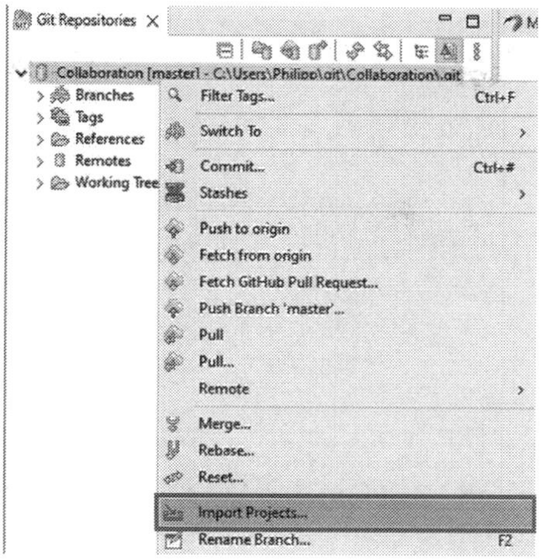

Figure 24. open the “Import Projects from File System or Archive” dialog

Select the Eclipse project that you want to import, then click the “Finish” button to start the import process.

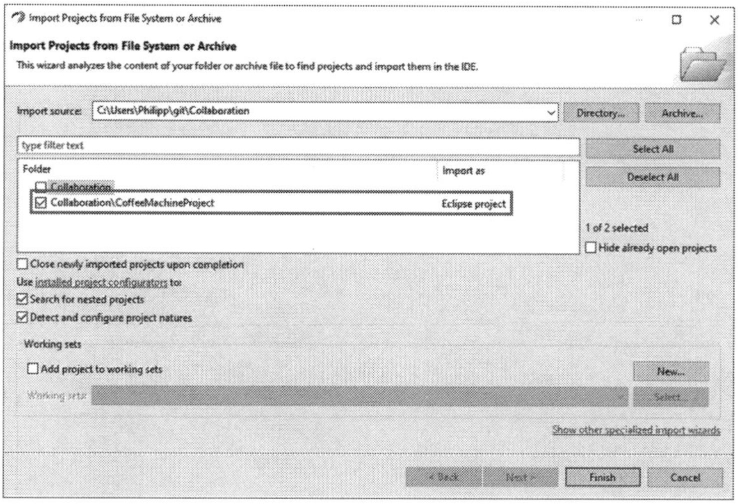

Figure 25. Select the project and start the import process

After the import process is finished, switch back to the Papyrus perspective and double-click on the Papyrus project in the Project Explorer to open it.

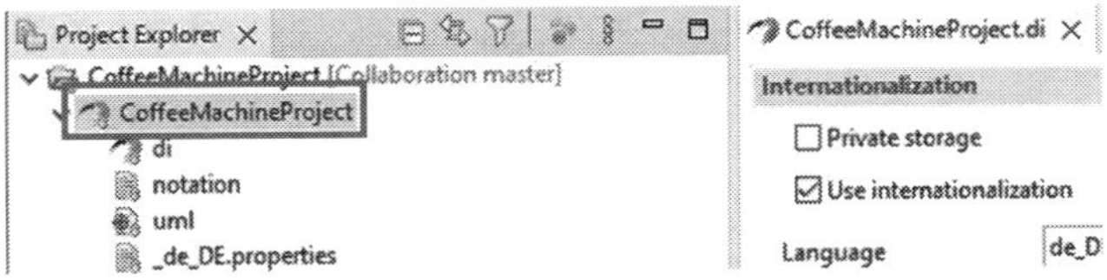

Figure 26. Open the Papyrus project

Appendix D - Profiles and Stereotypes

UML profiles provide a way to extend and specialize the basic concepts of UML to create domain-specific modeling languages (DSMLs). This is done by defining stereotypes, which are extensions of existing UML metaclasses. SysML itself is also defined as a UML profile. Stereotype definitions must adhere to the syntax and semantics of the metaclasses they extend to ensure consistency and compatibility with standard UML tools. Additionally, profiles based on UML metaclasses are relatively easy to learn for individuals with UML knowledge, as they build upon familiar concepts rather than introducing entirely new ones. This makes profiles a practical and accessible approach to creating DSMLs that cater to specific domain needs and extend the language to include additional concepts, notations, and semantics.

SysML Stereotypes

In SysML, stereotypes are a fundamental concept used to extend the language and introduce new types of elements or modify the behavior of existing ones. Stereotypes allow modelers to customize and tailor SysML to better represent specific concepts or domain-specific semantics. Stereotypes are defined using the UML extension mechanism, and they provide a way to classify and extend standard SysML elements. Stereotypes modify the behavior or appearance of the stereotyped elements and may introduce additional properties or constraints. Stereotypes often come with tagged values, which are additional properties or attributes associated with the stereotyped elements. Tagged values provide a way to extend the standard properties of SysML elements. Typically, stereotypes are also used to modify the behavior of the modeling tool. In Papyrus it is, for example, possible to modify the graphical appearance of a stereotyped model element using icons. Additionally, formalized constraints in the Object Constraint Language (OCL) can be used to further refine the definition of stereotypes.

SysML Profiles

Profiles can be used to extend the SysML language to support domain-specific modeling and to customize the language to meet the specific needs of a project or organization. A profile is a collection of stereotypes and other extensions to the SysML language that are organized and defined in a structured way. Profiles extend the language by introducing new stereotypes, tagged values, constraints, and even new diagram types. Profiles are often created to address the modeling requirements of specific domains or industries. For example, if SysML does not have specific constructs for modeling aerospace systems, a profile could be created to introduce relevant concepts. Once a profile is defined, it needs to be applied to a SysML model to use the extended concepts.

A SysML methodology and a SysML profile can be used together to create a more comprehensive and effective modeling practice. The methodology provides the overall structure and guidance for the modeling process, while the profile extends the SysML language to provide the necessary modeling elements and constraints for the specific domain or project.

Creating the MWP Profile

To support the modeling activities following the MBSE with Papyrus (MWP) method, a profile was created that is accordingly named MWP Profile. This section shows how a Papyrus SysML profile is created. The goal is to show the reader how a profile is created to empower them to adapt SysML to their own needs.

To properly maintain and distribute a Papyrus profile, it can be developed and distributed as an Eclipse plugin, similar to the SysML 1.6 plugin (see Appendix B - Installation).

Profiles are defined in Papyrus projects. To create a new Papyrus project for the MWP Profile, open the File menu and select *New->Papyrus Project.* Check the Profile element in the "New Papyrus Project" dialog.

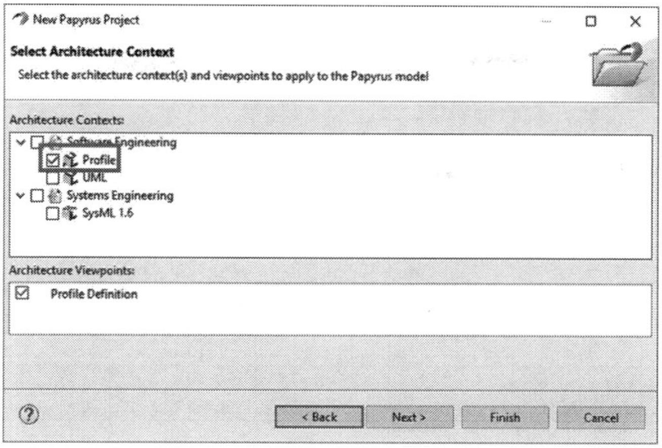

Figure 27. Create a new papyrus project for the MWP Profile

Name the new Project "MWPProfile".

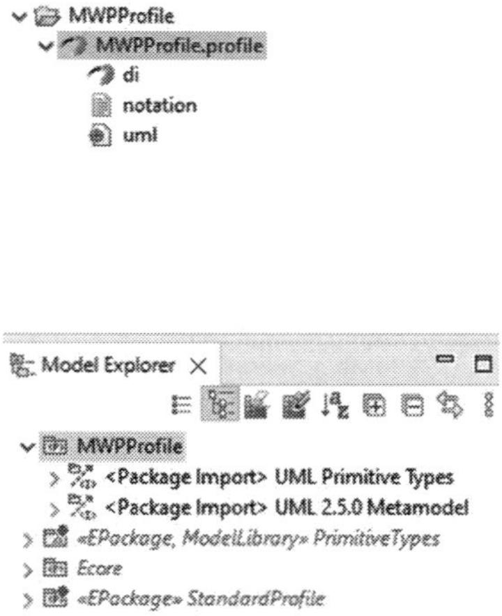

Figure 28. New project in Project Explorer and Model Explorer

Since we want to base our new stereotypes on existing SysML stereotypes, we need to import the SysML profile into the MWPProfile model. Right-click on the MWPProfile Model element in the Model Explorer and select *Import->Import Registered Profile.*

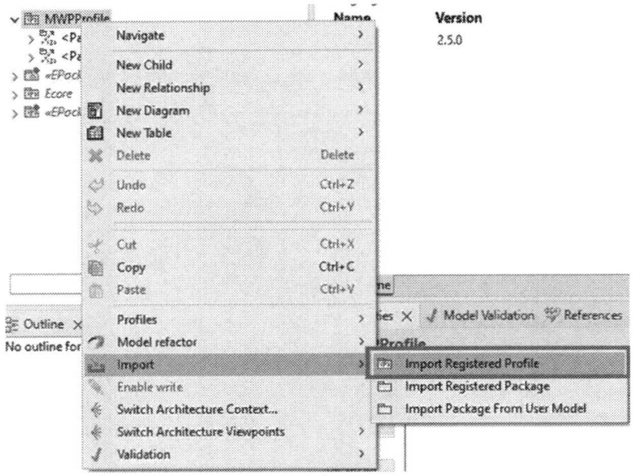

Figure 29. Open the "Import profile" dialog

Select the SysML 1.6 profile.

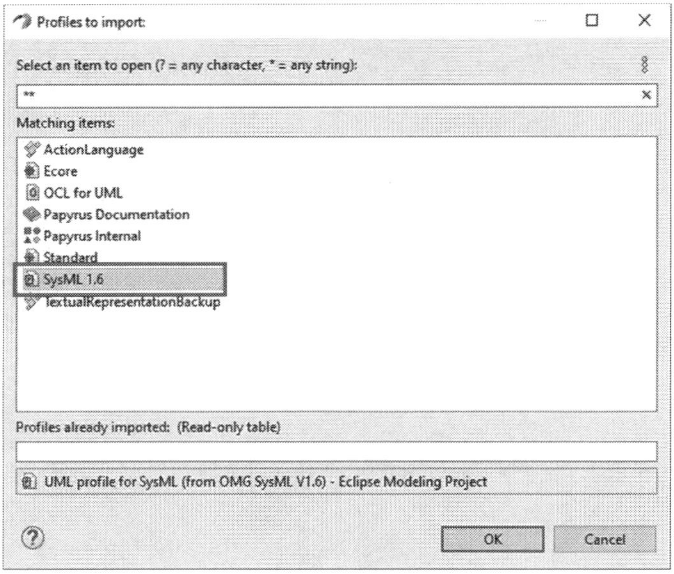

Figure 30. Select the SysML 1.6 profile

Use the "Select All" button to select all the packages of the SysML profile for import.

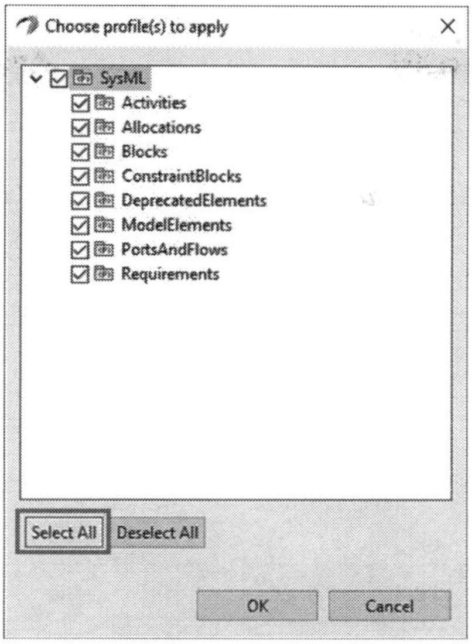

Figure 31. Select all SysML packages

Create a new profile diagram in the model by right-clicking on the model element and selecting *New Diagram->Profile Diagram.* Name the diagram "MWPProfile PrD".

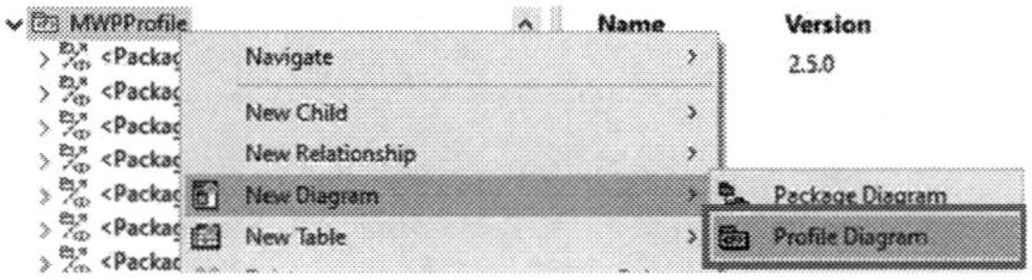

Figure 32. Create a new profile diagram

Select the "Stereotype" element in the Diagram Palette and add a new stereotype called "SystemOfInterest" to the profile diagram.

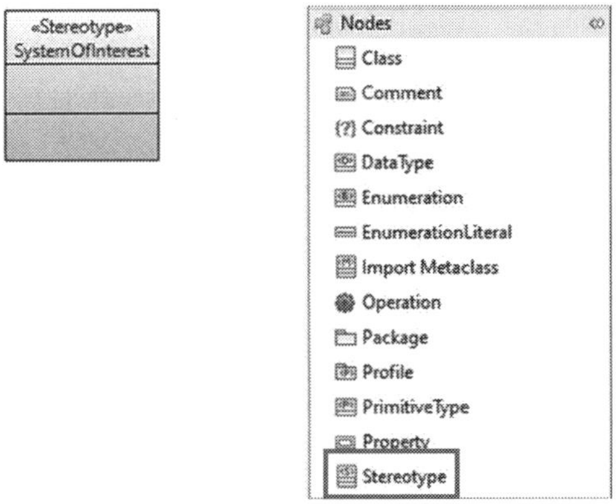

Figure 33. Create a new stereotype

The stereotype "SystemOfInterest" shall be a specialization of the "Block" stereotype. Look for the "Block" stereotype in the "SysML" package and drag and drop it onto the profile diagram.

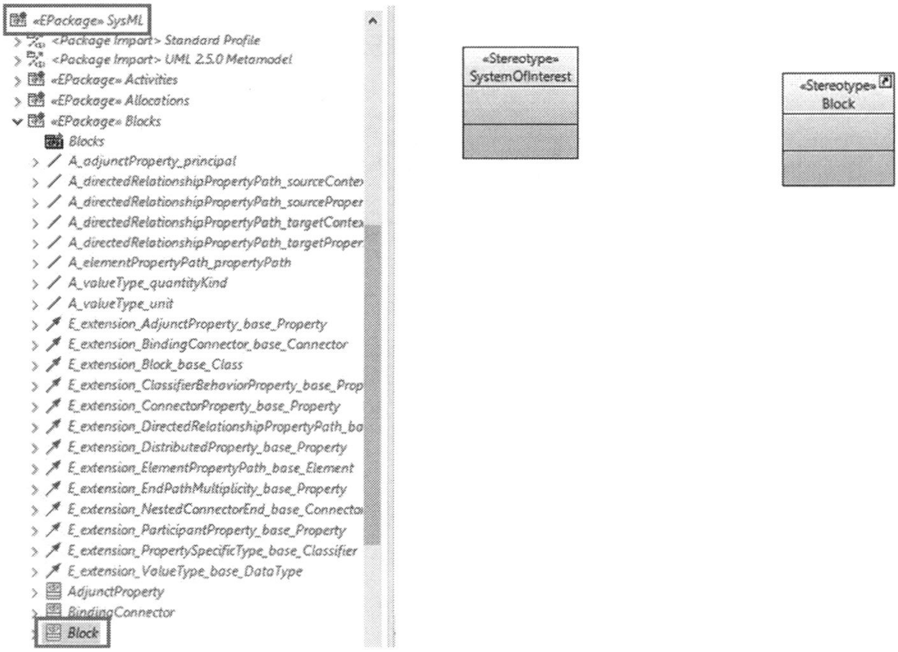

Figure 34. Add "Block" stereotype to the profile diagram

Select the "Generalization" element in the Diagram Palette and, first, click on the "SystemOfInterest" stereotype and then on the "Block" stereotype to model the relation between the two stereotypes.

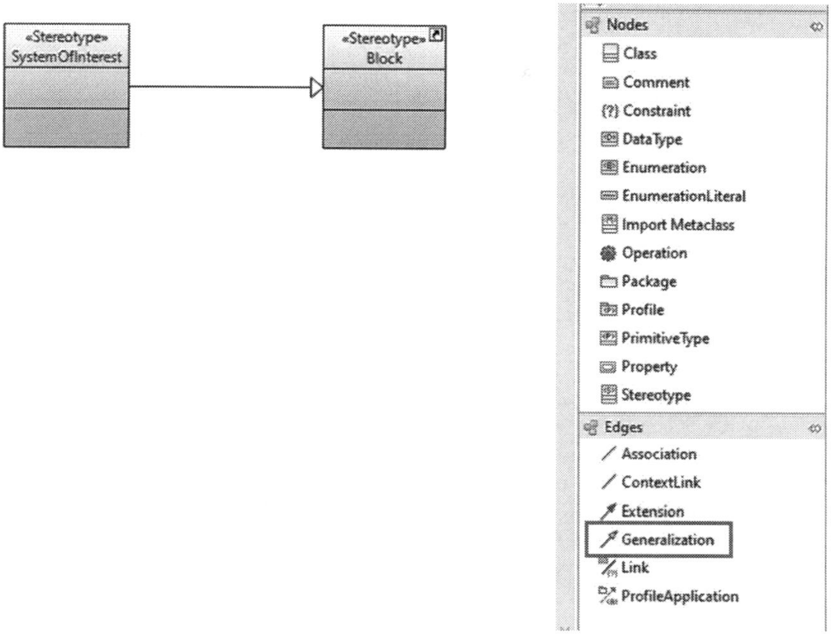

Figure 35. Create generalization

Next, we want to model the "ExtendedStakeholder" stereotype as an extension of the existing "Stakeholder" stereotype in SysML. The stakeholder shall have additional properties, such as stakeholder relevance, which can have a value of low, mid, or high. Create a new package in the "MWPProfile" model and name it "Stakeholders". Right-click on that package and select *New Child->Enumeration.*

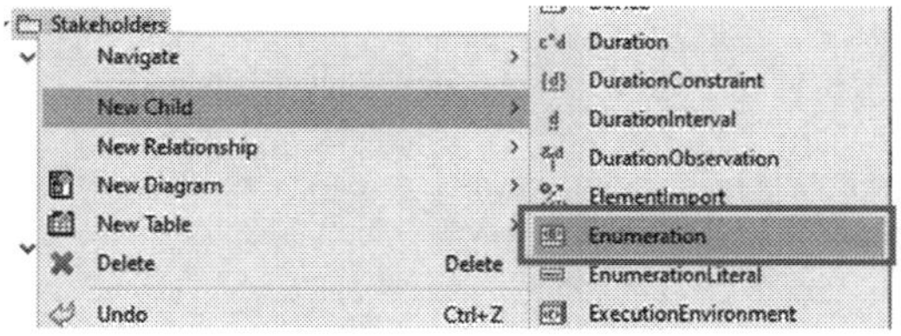

Figure 36. Create a new enumeration

Name the enumeration "StakeholderRelevance". Then add a new enumeration literal to the enumeration by clicking on the button with the plus next to the Owned literal section in the

Properties View of the enumeration.

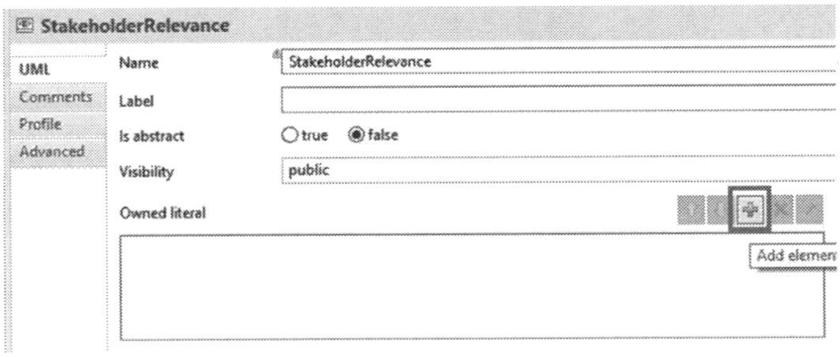

Figure 37. Create a new enumeration literal

Set the name of the enumeration literal to "LOW".

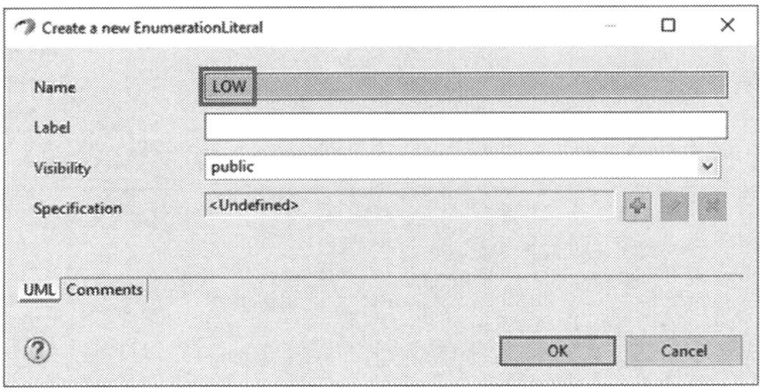

Figure 38. Name the new enumeration literal

Repeat the same process for the other two literals.

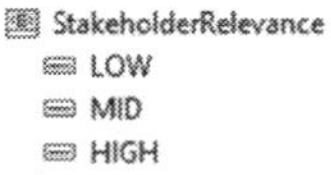

Figure 39. Finished enumeration for stakeholder relevance

Create a new stereotype called "ExtendedStakeholder" to the "Stakeholders" package. Then, add a new property to that stereotype by right-clicking on the stereotype and selecting *New Child->Property*.

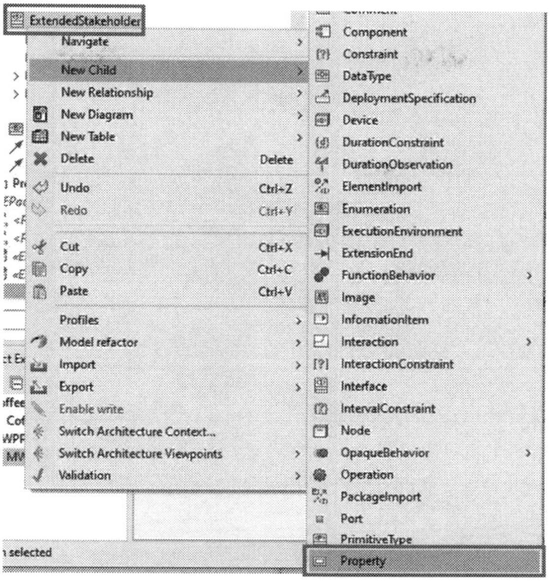

Figure 40. Create a new property

Name the new property "relevance" and type it using the "StakeholderReference" enumeration. Create an additional property called "interests" and set its type to "String". Add The "ExtendedStakeholder" to the MWPProfile profile diagram as well as the "Stakeholder" stereotype from the "SysML" package.

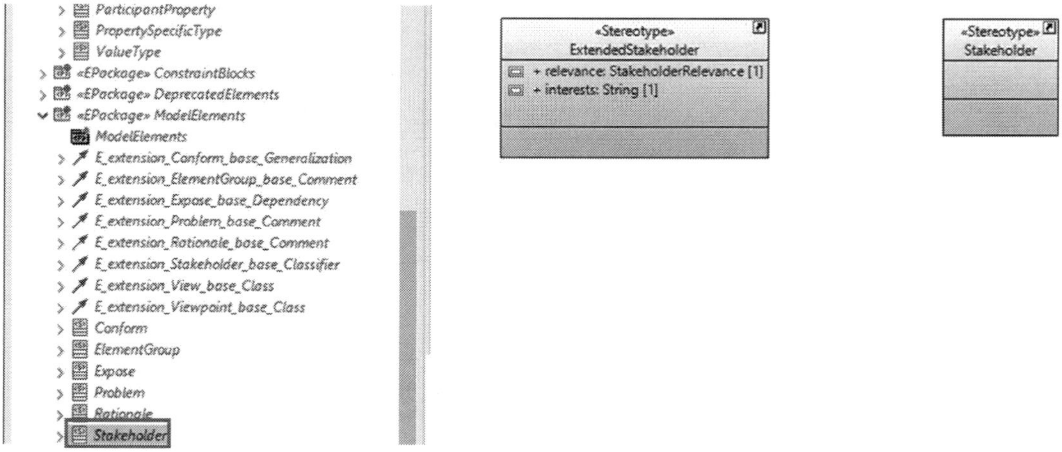

Figure 41. Add stereotypes to the profile diagram

Select the "Extension" element from the Diagram Palette and, first, click on the "ExtendedStakeholder" stereotype and then on the "Stakeholder" stereotype to model the relation between the two stereotypes.

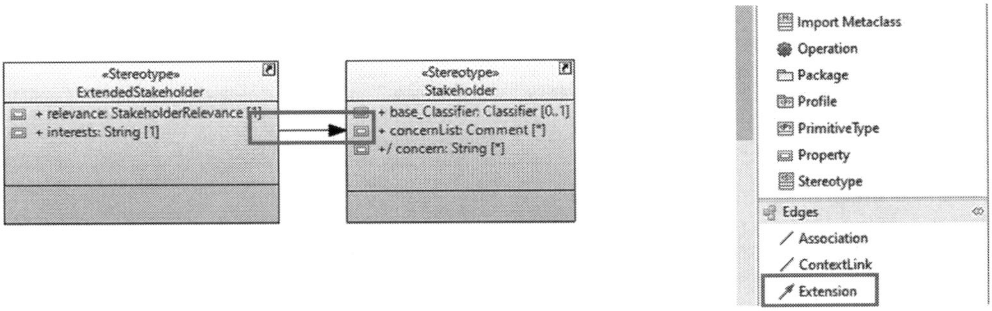

Figure 42. Create extension relation

To tell Papyrus that the "ExtendedStakeholder" stereotype can be applied to "Actor" elements, the stereotype needs to be an extension of that metaclass. Select the "Import Metaclass" element from the Diagram Palette and click on the diagram to open the "Select Metaclass" dialog.

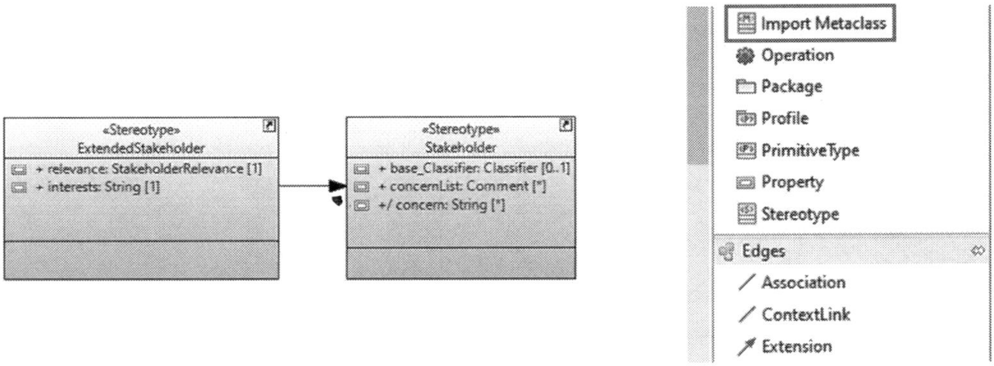

Figure 43. Open the "Select Metaclass" dialog

Select the "Actor" metaclass by moving it from the left side of the dialog to the right side using the button with the single arrow pointing to the right.

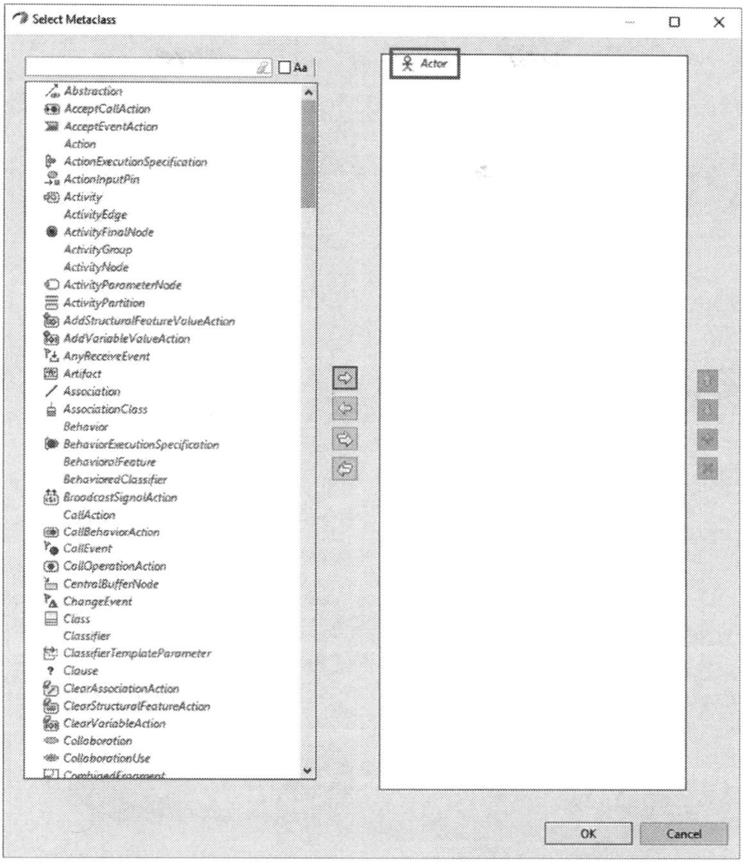

Figure 44. Select the "Actor" metaclass

The "Actor" metaclass is now shown on the profile diagram.

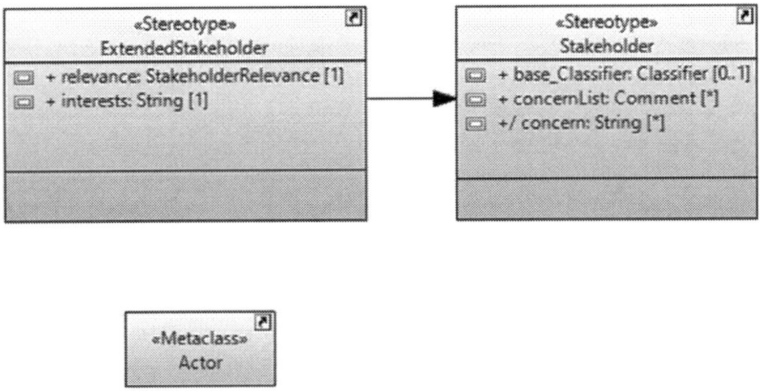

Figure 45. The "Actor" metaclass on the profile diagram

Create an extension from the "ExtendedStakeholder" stereotype to the "Actor" metaclass. To complete this part of the diagram also add the "StakeholderRelevance" enumeration to it.

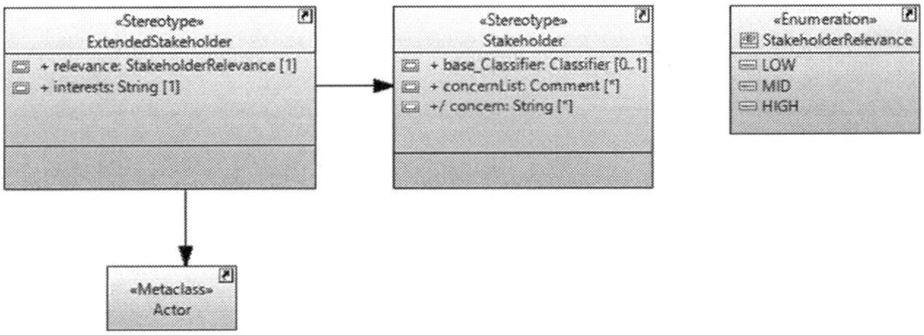

Figure 46. Relations of the "ExtendedStakeholder" Stereotype

Repeat the same process for the "StakeholderRequirement" and "ExtendedRequirement" stereotypes.

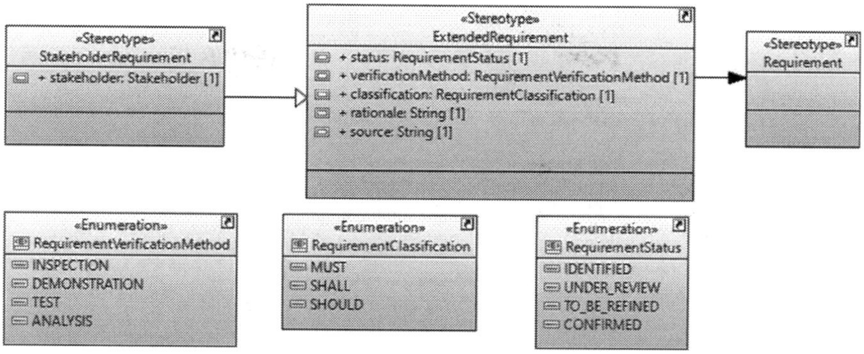

Figure 47. "StakeholderRequirement" and "ExtendedRequirement" stereotypes

Last, a stereotype called "MeasureOfEffectiveness" should be created, and the "Property" metaclass should be extended.

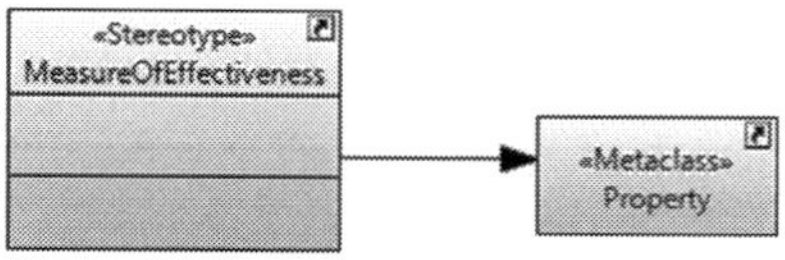

Figure 48. "MeasureOfEffectiveness" stereotype

Every time a profile project is saved, the "Profile Definition" dialog opens and asks if a dynamic definition of the profile project should be defined now. Answer the dialog with a click on the "Yes" button.

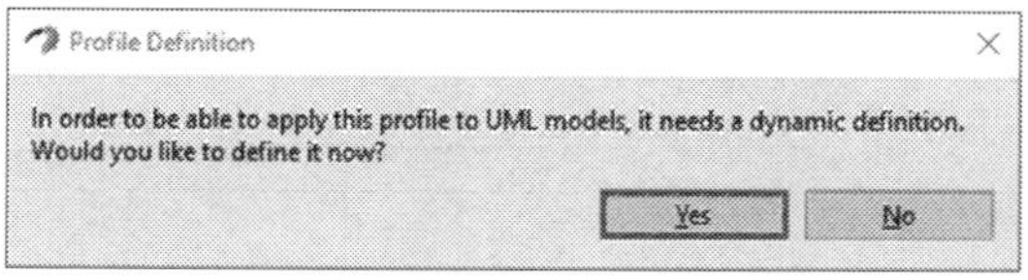

Figure 49. Dialog opened on saving a profile project

The "Profile Definition" dialog enables setting further properties of the profile project. For a development project, where the profile is used in the same workspace as the system model, no changes on this dialog are necessary.

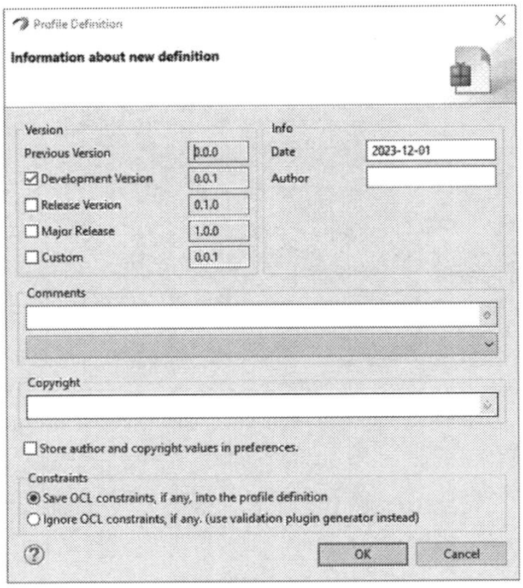

Figure 50. "Profile Definition" dialog

Note that Papyrus automatically increments the version of the profile project after each save. Incrementing the profile version number enables Papyrus to detect if a profile used in a modeling project has changed.

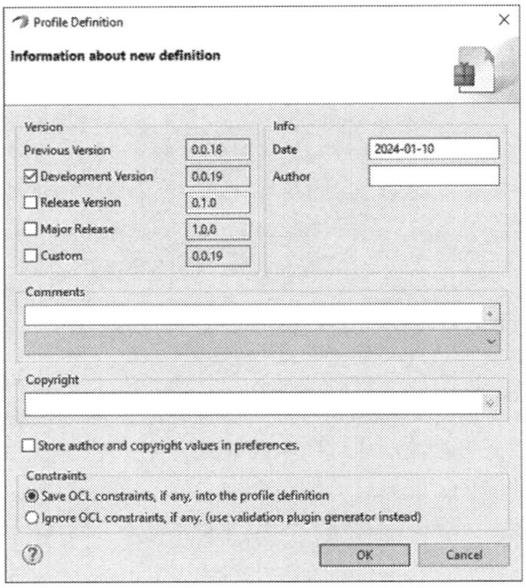

Figure 51. Profile version increments

After a Profile that is used in a system model project is changed, close the system model project and re-open it. Papyrus will then automatically detect that the profile version has changed and show a dialog that asks if the version of the profile used in the system model should be updated to the latest version.

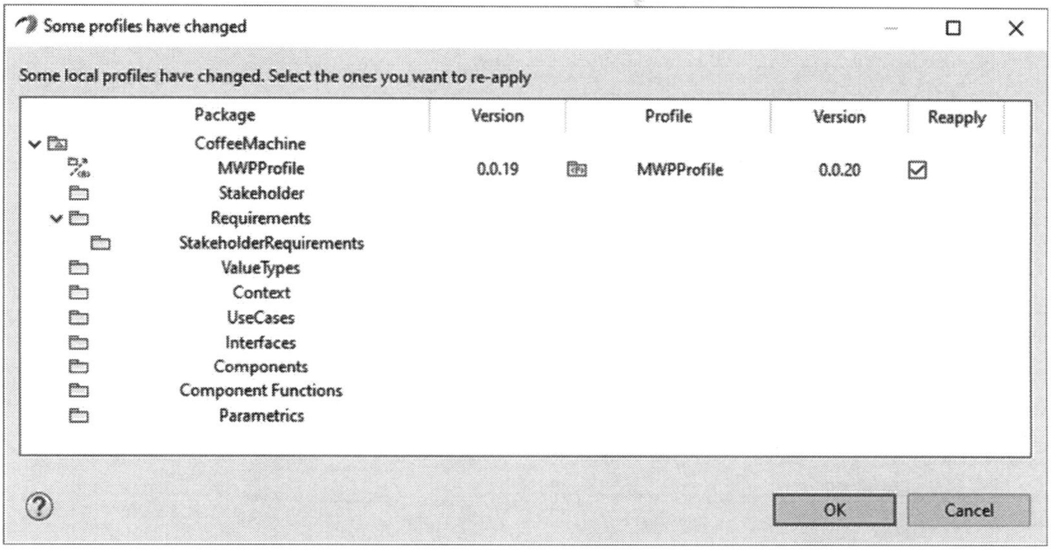

Figure 52. Notification after profile update

Acronyms and Abbreviations

AD - Activity Diagram

BDD - Block Definition Diagram

CM - Configuration Management

CAD - Computer Aided Design

CDO - Connected Data Objects

DSML - Domain-Specific Modeling Language

EMF - Eclipse Modeling Framework

GUI - Graphical User Interface

I^2C - Inter-Integrated Circuit

IBD - Internal Block Diagram

IDE - Integrated Development Environment

INCOSE - International Council on Systems Engineering

ISO - International Organization for Standardization

MBSE - Model-Based Systems Engineering

MCO - Mars Climate Orbiter

MoE - Measure of Effectiveness

MTBF - Mean Time Between Failures

MWP - MBSE with Papyrus

OCL - Object Constraint Language

OMG - Object Management Group

OOSEM - Object-Oriented Systems Engineering Method

PD - Parametric Diagram

RCP - Rich Client Platform

RFP - Request for Proposals

SD - Sequence Diagram

SDLC - System Development LifeCycle

SE DSIG - Systems Engineering Domain Special Interest Group

SOI - System of Interest

STM - State Machine Diagram

SysML - Systems Modeling Language

SYSMOD - Systems Modeling Toolbox

UC - Use Case Diagram

UI - User Interface

V&V - Verification and Validation

XML - Extensible Markup Language

References

Bertalanffy, L. von (2009) General system theory: foundations, development, applications. Rev. ed., 17. paperback print. New York, NY: Braziller.

Campion-Awwad, O., Hayton, A. and Vuaran, M. (no date) 'The National Programme for IT in the NHS'.

Chacon, S. and Straub, B. (2014) Pro Git: everything you need to know about Git. 2. ed. New York, NY: Apress/Springer (The expert's voice).

Checkland, P. (1981) Systems thinking, systems practice. Chichester [Sussex] ; New York: J. Wiley.

Delligatti, L. (2014) SysML distilled: a brief guide to the systems modeling language. Upper Saddle River, NJ: Addison-Wesley.

Eclipse Papyrus (2024) Eclipse Papyrus User Guide. Available at: https://help.eclipse.org/2022-12/index.jsp?topic=%2Forg.eclipse.papyrus.uml.doc%2Ftarget%2Fgenerated-eclipse-help%2FPapyrusStarterGuide.html&cp%3D66_0_0 (Accessed: 17 April 2024).

Estefan, J.A. and Weilkiens, T. (2023) 'MBSE Methodologies', in A.M. Madni, N. Augustine, and M. Sievers (eds) Handbook of Model-Based Systems Engineering. Cham: Springer International Publishing, pp. 47–85. Available at: https://doi.org/10.1007/978-3-030-93582-5_12.

Fosse, E. and Delp, C.L. (2013) 'Systems Engineering Interfaces: A Model Based Approach', in 2013 IEEE Aerospace Conference. IEEE, pp. 1–8. Available at: https://ieeexplore.ieee.org/abstract/document/6497322/ (Accessed: 17 April 2024).

Friedenthal, S., Moore, A. and Steiner, R. (2015) A practical guide to SysML: the systems modeling language. Third edition. Amsterdam ; Boston: Elsevier, MK, Morgan Kaufmann is an imprint of Elsevier.

Haberfellner, R. et al. (2019) Systems engineering: fundamentals and applications. Cham, Switzerland: Birkhäuser. Available at: https://doi.org/10.1007/978-3-030-13431-0.

Holt, J. and Perry, S. (2018) SysML for systems engineering: a model-based approach. 3d edition. Stevenage: The Institution of Engineering and Technology (IET professional applications of computing series, 20).

Kapurch, S.J. (2010) NASA systems engineering handbook. Diane Publishing.

Kossiakoff, A. et al. (2020) Systems engineering: principles and practice. Third edition. Hoboken, NJ: John Wiley & Sons, Inc (Wiley series in systems engineering and management). Meadows, D.H. and Wright, D. (2011) Thinking in systems: a primer. Nachdr. White River Junction, Vt: Chelsea Green Pub.

Hoffmann, H.-P. (2022) Model-Based Systems Engineering with Rational Rhapsody and Rational Harmony for Systems Engineering–Deskbook 3.1.2. Available at: https://www.ibm.com/support/pages/model-based-systems-engineering-rational-rhapsody-and-rational-harmony-systems-engineering-deskbook-312 (Accessed: 17 April 2024).

Hossain, N.U.I. et al. (2020) 'A Historical Perspective on Development of Systems Engineering Discipline: A Review and Analysis', Journal of Systems Science and Systems Engineering, 29(1), pp. 1–35. Available at: https://doi.org/10.1007/s11518-019-5440-x.

INCOSE (2019). INCOSE Definitions, The International Council on Systems Engineering (INCOSE).

INCOSE (2023) 'Guide to Writing Requirements'. INCOSE. Available at: https://www.incose.org/docs/default-source/working-groups/requirements-wg/gtwr/incose_rwg_gtwr_v4_040423_final_drafts.pdf?sfvrsn=5c877fc7_2 (Accessed: 17 April 2024).

Madni, A.M. (2018) Transdisciplinary Systems Engineering. Cham: Springer International Publishing. Available at: https://doi.org/10.1007/978-3-319-62184-5.

Meadows, D.H. (2008) Thinking in Systems: A Primer. Illustrated Edition. White River Junction, Vt: Chelsea Green Publishing Co.

Morkevicius, A., Aleksandraviciene, A. and Strolia, Z. (2023) 'System Verification and Validation Approach Using the MagicGrid Framework', INSIGHT, 26(1), pp. 51–59.

OMG (2019) 'OMG Systems Modeling Language (OMG SysML) v1.6'. Needham, MA: OMG. Available at: https://www.omg.org/spec/SysML/1.6 (Accessed: 17 April 2024).

OMG (2022) 'OMG Systems Modeling Language (OMG SysML) v1.7 beta'. Needham, MA: OMG. Available at: https://www.omg.org/spec/SysML/1.7/Beta1 (Accessed: 17 April 2024).

Roques, P. (2016) 'MBSE with the ARCADIA Method and the Capella Tool', in. 8th European Congress on Embedded Real Time Software and Systems (ERTS 2016).

Roques, P. (2017) Systems architecture modeling with the Arcadia method: a practical guide to Capella. Elsevier.

Rothenberg, J. et al. (1989) 'The nature of modeling', in Artificial Intelligence, Simulation and Modeling.

Sauser, B.J., Reilly, R.R. and Shenhar, A.J. (2009) 'Why projects fail? How contingency theory can provide new insights–A comparative analysis of NASA's Mars Climate Orbiter loss', International Journal of Project Management, 27(7), pp. 665–679.

Schlager, K.J. (1956) 'Systems engineering-key to modern development', IRE Transactions on Engineering Management, EM-3(3), pp. 64–66. Available at: https://doi.org/10.1109/IRET-EM.1956.5007383.

Stachowiak, H. (1973) Allgemeine Modelltheorie. Wien New York: Springer.

International Organization for Standardization (ISO) (2023) 'Systems and software engineering – system life cycle processes', ISO Standard ISO/IEC/IEEE 15288:2023.

Walden, D.D. and International Council on Systems Engineering (eds) (2023) INCOSE systems engineering handbook. Fifth edition. Hoboken, NJ: John Wiley & Sons Ltd.

Wasson, C.S. and Augustine, N.R. (2016) System engineering analysis, design, and development: concepts, principles, and practices. Second edition. Hoboken, N.J: Wiley (Wiley series in systems engineering and management).

Weilkiens, T. (2007) Systems engineering with SysML/UML: modeling, analysis, design. Amsterdam Boston: Morgan Kaufmann OMG Press/Elsevier.

Weilkiens, T. (2020) SYSMOD - the systems modeling toolbox: pragmatic MBSE with SysML. 3rd edition, version 4.2. Fredesdorf: MBSE4U (MBSE4U booklet series).

Weilkiens, T. et al. (2022) Model-based system architecture. Second edition. Hoboken, NJ: Wiley (Wiley series in systems engineering and management).

Index

Eclipse Papyrus 41
- Apply profile 62
- Apply stereotype 64, 74
- Automatic layout 136
- Diagram grid 78
- Export table to Excel 88
- Export to Excel 232
- Git 274
- GUI 44
- Installation 267
- Line style 136, 170, 238
- Navigation 125, 202
- Preferences 46

MBSE . *See* Model-Based Systems Engineering
Model-Based Systems Engineering 21
Modeling 17
MWP method 49
- Allocate functions to components 226
- Analyze system states 187
- Analyze use case 159, 176
- Context analysis 129
- Define initial system concept 223
- Define system goals 109
- Define system internal architecture 233
- Functional analysis 213
- Identify stakeholders 71
- Identify use cases 148
- Manage requirements 93
- Model parametrics 246
- Model SOI 57
- Modeling guidelines 265
- Overview 50

Pitfalls
- Context analysis 128
- Functional analysis 211
- Requirements management 53
- Stakeholder analysis 68
- SysML 38
- System goals definition 103
- System state analysis 185
- Use case analysis 154
- Use case identification 146

Requirements 53, 90
SysML 31
- History 32
- Misconceptions 37
- Profiles 284
- Recommended books 3

SysML diagrams 33
- Activity diagram 159, 196, 215
- Block definition diagram. 115, 129, 223, 249
- Internal block diagram 137, 233, 242, 244
- Parametric diagram 251
- Requirement diagram 98, 121, 152
- Sequence diagram 176
- State machine diagram 186, 187, 202
- Use case diagram 76, 147, 148
- Use Case diagram 76

SysML elements
- Activity 159, 164, 166, 180, 194
- Activity final node 163
- Activity parameter 164
- Activity parameter node 217
- Allocate activity partition 160
- Allocation matrix 226
- Behavior execution specification 180
- Binding connector 243, 256
- Block 129
- Blocks 56, 63
- Call behavior action 162, 215
- Change event 207
- Choice node 190

Communication path ... 149
Connector ... 235
Constraint ... 200, 248
Constraint block ... 247
Constraint property ... 246
Control flow ... 168
Decision node ... 163
Description ... 75
Enumeration ... 289
Enumeration literal ... 289
Exit point ... 203
Flow property ... 240
Full port ... 234
Generic table ... 82, 143, 213
Guard ... 172, 200
Initial node ... 163, 189
Input pin ... 216
Interaction ... 177
Interface block ... 236
Item flow ... 139
Lifeline ... 177
Merge node ... 163
Message async ... 179
Object flow ... 169, 219
Opaque expression ... 172, 173, 201, 208, 249
Output pin ... 216
Package ... 71
Parameter ... 248
Part property ... 134
Property ... 109, 132, 246
Proxy port ... 234
Quantity kind ... 106, 111
Requirement ... 93
Requirement table ... 95
Signal ... 182, 193
Signal event ... 192, 197
Stakeholder ... 72
State ... 188
State invariant ... 178
State machine ... 188
Stereotype ... 64, 283, 287
Transition ... 190
Trigger ... 191
Unit ... 107, 111
Use case ... 149
Value type ... 106, 110
SysML relations
  Allocation ... 231
  Composite association (directed) ... 132
  Extend ... 148
  Generalization ... 81, 116
  Include ... 148, 150
  Trace ... 122
Systems engineering ... 9
  Key concepts ... 13
  Origins ... 11
  Recommended books ... 3
  Recurring themes ... 14
Systems thinking ... 12
  Recommended books ... 4

Made in United States
Orlando, FL
04 November 2026

90207646R00177